Energy, Earth and Everyone

by Medard Gabel
with the World Game Laboratory

Foreword by
R. Buckminster Fuller

Afterword by
Stewart Brand

Anchor Books
Anchor Press/Doubleday
Garden City, New York
1980

4.5×10^{58} kwh = Approximate amount of total energy as mass and radiation in the Universe. (This is a small fraction of the amount of energy as gravitation in the Universe.)[1]

8.33×10^{25} kwh = Sun's daily output of energy.[2]

4.15×10^{15} kwh = Earth's daily receipt of energy.[3]

$.22 \times 10^{12}$ kwh = Humanity's daily conversion of energy.[4]

83×10^{12} kwh = Humanity's yearly (1978) conversion of energy.[4]

83×10^{12} kwh (83 trillion) kwh = 19,000 kwh per person on Earth = approximately 20 100-watt bulbs burning 24 hours per day for 1 year for each person on Earth.

4.3×10^{9} individuals = Total number of known conscious energy measuring and manipulating entities in Universe.

"The origin and destiny of the energy in the Universe cannot be completely understood in isolation from the phenomena of life and consciousness."[5]

"There is no energy shortage. There is no energy crisis. There is a crisis of ignorance."[6]

"The biggest danger in our present energy planning lies in not being imaginative enough to see the wide range of choices available."[7]

"Unprecedented problems require unprecedented solutions."

"The basic danger of the world energy situation is that it could become critical before it seems serious."[8]

"Humanity teeters on the threshold of the greatest revolution in history. If it's to pull the top down and it's bloody, all lose. It if is a design science revolution to elevate the bottom and all others as well to unprecedentedly new heights, all will live to dare spontaneously to speak and live and love the truth, strange though it often may seem."[6]

Energy, Earth and Everyone

Energy Strategies for Spaceship Earth

by Medard Gabel
with the World Game Laboratory

Library of Congress Cataloging in Publication Data

Gabel, Medard.
  Energy, Earth, and everyone.

  Bibliography.
  1. Power resources. 2. Energy policy. 3. Power (Mechanics) 4. Ecology. I. World Game Workshop, University of Pennsylvania, 1974. II. Title.
TJ163.2.G3   1980   333.7   77.92213
ISBN 0-385-14081-9

Copyright © 1975, 1980 by Medard Gabel
Foreward Copyright © 1975 by R. Buckminster Fuller
All Rights Reserved
Printed in the United States of America

Design and Production by Hal Hershey and Brenton Beck, Fifth Street Design Associates, Berkeley, with Jack Popovich, Carol Egenolf and Pamela Webster.
Illustrations by Pedro Gonzalez.
Typesetting by Polycarp Press, San Francisco.
Cover Camera Effects by Graphic Impressions, Berkeley.

Photograph of the earth on the cover and on page 12 courtesy of NASA. Photographs on pages 60, 83, 90, 126, 135, 136, 145 and 146 courtesy of U.S. Department of Energy. Cover photographs courtesy of Lick Observatory, U.S. Department of Energy, and Pacific Gas and Electric Co. Illustrations on page 105 from *Popular Science Monthly*, February 1978. © 1978 by Times Mirror Magazine, Inc. Illustration on page 180 from *Popular Science Monthly*, December 1978. © 1978 by Times Mirror Magazine, Inc. Reprinted by permission. Illustration on page 135 from "Power with Heliostat" by A. F. Hildebrandt and L. L. Vant-Hull, *Science*, Volume 197, pp. 1139–1146, Figure 1, September 16, 1977. © 1977 by the American Association for the Advancement of Science. Reprinted by permission. Illustration on page 191 from "The Amateur Scientist" by C. L. Strong. © 1974 by Scientific American Inc. All Rights Reserved. Reprinted by permission of W. H. Freeman Co. Illustration on page 119 from Wind Machines. © 1975. Used by permission of The Mitre Corporation. Illustration on page 139 from Solar Energy Utilization, Federal Power Commission, National Power Survey, Lincoln Laboratory, Massachusetts Institute of Technology. © 1973. Reprinted by permission. Illustration on page 136 from *New Low Cost Sources of Energy for the Home*, by Peter Clegg. © 1975 by Peter Clegg. Reprinted by permission of Garden Way Publishing. Illustrations on pages 160 and 195 from *The Illustrated Science and Invention Encyclopedia*, Volume 18. © 1976 by H. S. Stuttman Co., Inc. Reprinted by permission of Marshall Cavendish Corp.

The author gratefully acknowledges permission to reprint excerpts from the following:

"Transport Processes at the Interface Between Ocean and Atmosphere" by George S. Benton in *Transport Phenomena in Atmospheric and Ecological Systems* (American Society of Mechanical Engineers, 1967); *Energy For Man: Windmills to Nuclear Power* by Hans Thirring, copyright © 1958 by Indiana Press, Bloomington, reprinted by permission of the publisher; "Thermal Energy—The Neglected Energy Option" by Robert Rex in the *Bulletin of the Atomic Scientists*, October 1971, reprinted by permission of the *Bulletin of the Atomic Scientists*, copyright © 1971 by the Educational Foundation for Nuclear Science; "Hydrogen, Synthetic Fuel of the Future" by T. H. Maugh II in *Science*, Vol. 178, 24 November, 1972, copyright © 1972 by the American Association for the Advancement of Science; "Conservation in Industry" by Charles Berg and "Solar Energy by Photosynthesis" by Melvin Calvin in *Science*, Vol. 184, 19 April, 1974, copyright © 1974 by the American Association for the Advancement of Science; *Interim Report: The Growing Demand for Energy*, April 1971, by Deane Morris, reprinted by permission of The Rand Corporation; *Energy Resources: A Report to the Committee on Natural Resources*, Publication #1000-D, M. King Hubbert, National Academy of Sciences—National Research Council; "Energy Resources" by M. King Hubbert in *Resources and Man*, National Academy of Science—National Research Council; "World Energy in the Balance" by Peter Odell in *Geographical Magazine*, Vol. XLVI, No. 8, May 1974; "Hydrogen—Fuel of the Future?" by Alden P. Armagnac in *Popular Science*, January

1973, copyright © Popular Science Publishing Co., reprinted by permission; "Solar Sea Power" by Clarence Zener (supported by National Science Foundation, RANN Division, under Grant No. GI-39114) in *Physics Today,* January 1973; "An Agenda for Energy" by Hoyt C. Hottel and Jack B. Howard in *Technology Review,* ed. at the Massachusetts Institute of Technology, copyright © 1972 by the Alumni Association of the M.I.T.; "Wind Power Potential for the United States" by K. H. Bergey in *Aware,* October, 1974, reprinted by permission of the author; "Energy in the Universe" by Freeman J. Dyson, "Decision-making in the Production of Power" by Milton Katz, "Energy and Power" by Chauncey Starr, and "The Conversion of Energy" by Claude M. Summers in *Scientific American,* September 1971, copyright © 1971 by Scientific American, Inc., all rights reserved.

Author's Acknowledgments

This book would not have been possible without Buckminster Fuller. It is largely based on his philosophy, theories, and methods; inspired by his untiring enthusiasm for life; and cultivated by his generosity, warmth, and friendship. My patient friend and colleague at Earth Metabolic Design—Howard Brown—deserves my warmest thanks for his support and the fruitful discussions we have had over the years that have enriched this book.

Most of all, I would like to thank the participants of the 1974 World Game Workshop who tirelessly worked on the forerunner of this book—Bill Albin, Tell Anderson, Michael Berz, Bob Cook, Larry Crane, Bob Crews, Larry Dumoff, Paul Fagan, Ron Goodfellow, Joe Johnson, Joe Lang, John Lambie, Debbie Prindle, John Ussery, and Dane Winberg—as well as the participants of the Energy Group of the 1977 World Game Workshop—Charlie Cary, Paul Davit, Steve Mosenson, Scott Plummer, Bob Schiffner, Tony Smith, Dane Winberg, and Dorothy Newsome. In addition to his work at the World Game Workshop, Charlie Cary was invaluable in the ensuing months assisting in the research for this edition of *E3*. The energy and ideas of all these people have contributed greatly to the content of this document.

I would also like to thank the participants at previous World Game Studies Workshops (1969–1973), especially those of 1969, 1971, and 1972, who helped develop the study of global energy problems. Neva Kaiser's warm and many-faceted assistance was particularly helpful to not only this document but the Workshops as well. The lecturers from the two different Workshop's first weeks—Ed Schlossberg, Gene Youngblood, Ian McHarg, Wilfred Malenbalm, Michael Ben-Eli, Bill Perk, and Russ Kolton from 1974, and, from 1977, Russell Ackoff, Elizabeth Mann-Borgese, Edwin Schlossberg, Hazel Henderson, John Platt, Erich Jansch, J. Baldwin, John Todd, as well as Buckminster Fuller who lectured at both of these workshops—helped enormously to set the context for the work which was undertaken. In addition, Ed Schlossberg provided insightful feedback on the preliminary edition of this document. Much help and assistance was provided during the crucial period of writing and rewriting this document by Margaret Cullen and by the staff at Dr. Fuller's office—Shirley Sharkey, Timothy Wessels, Janet Bregman-Taney, Anne Mintz, and Peter Kent. I would also like to thank Kiyoshi Kuromiya and Dan Daniels who did a magnificent job in typing the manuscript; my editor, Bill Strachan, without whose extraordinary help this book might never have gotten out; Brenton Beck and Hal Hershey whose elegant graphic design has greatly improved the clarity of everything I wanted to say; and Stewart Brand who published an excerpt of the original *E3* in the Spring '75 *CoEvolution Quarterly*.

To my parents,
whose quiet dignity and integrity
has helped me throughout my life

Contents

Preface 6
Foreword *by R. Buckminster Fuller* 8

Energy for the Earth
1. Overview 14
2. Review 17
3. Energy 18
4. The Fundamentals 21
5. The Liabilities 33
6. Trends, Assumptions, and Key Indicators 38
7. Decision-Making Criteria/Economic Accounting Systems 39
8. Summary 44
9. References 46

Energy Sources
1. The Role of Design Science 50
2. Energy Sources, Old and New 52

Capital Energy Sources
3. Coal 54
4. Peat 66
5. Petroleum 67
6. Natural Gas 82
7. Nuclear 89

Income Energy Sources
Introduction: Old Myths and New Structures 102
8. Hydroelectric 104
9. Geothermal 111
10. Wind 118
11. Solar Energy 131
12. Bioconversion 146
13. Wood 152
14. The Tides 157
15. Waves 161
16. Ocean Currents 170
17. Temperature Differential 172
18. Hydrogen 176
19. Animate Energy 182
20. Cogeneration 182

Exotic Energy Sources
21. Water Salination 186
22. Nuclear Fusion 188
23. Electrostatic Energy 190
24. Deep Ocean Pressure 192
25. Phase Transformations 194
26. Seebeck Effect Power 195
27. Humid Air Power 196
28. The Purple Membrane 197
29. Energy Fluctuation 197
30. Gravity 198
31. Glacier Power 199
32. Other Pies in the Sky 199
33. Energy Conversion Techniques and Engines 201
34. References 208

Making the World Work: Global Strategies for a Regenerative Energy System
1. Introduction 216
2. Blueprints for Energy Survival 220
3. Energy Uses and Abuses/Conservation and Ephemeralization 224
4. Critical Paths for the Global Energy Development Strategy 233
5. Global Scenario 246
6. References 255

Appendix
A. Design Initiative 256
B. Glossary 258
C. Conversion Tables 262

Afterword *by Stewart Brand* 264

Charts and Illustrations

Energy for the Earth
World Population 1980 17
U.S. Energy End-Use Requirements 19
Earth System Inputs and Outputs 22
Energy Flows Through the Earth—Quantitative Input and Output 23
Present Energy System—Qualitative Inputs and Outputs 25
Distribution of Energy for Typical Industrialized Region: 1977 26
Existing Energy System—Constraints and Forcing Functions 27
World Energy Production 28
Primary Sources of Energy for Developed Industrialized Region: 1973 28
Energy Consumption in Developed Industrialized Region 29
Primary Energy Consumption in Developed Industrialized Region, Broken Down to Show Electric Production 29
Distribution of Total Energy 29
World CO_2 Emissions per Year from Fossil Fuel Combustion 22
Pollutants Released into the Air in 1970 in Millions of Tons per Year 33
Air Pollution: Sources of Pollution and Means of Control 34
Fuel Inputs and Emissions Control from 1,000 MW Conventional Electric Power Stations 35
Projected Nonenergy Consumption of Fossil Resources 36
Lifetimes of World Fossil Fuel and Nuclear Fuel Supplies 37
Energy-System Efficiency 42

Energy Sources
Design Science/Energy System Interaction 51
Scheme for Energy and Protein Production by Covered Agriculture 55
World Recoverable Coal Reserves, 1973 56
Major Chemical Products Derived from Coal 58
Coal: Potential Environmental Impacts Summary 63
Above Ground Coal Gasification 64
Below Ground Coal Gasification 64
Low-BTU Gas for Electric Power Generation 65
High-BTU Gas 65
The Geology of Oil 68
World Crude Oil Production 68
World Petroleum Consumption 1976 68
World Total Petroleum Consumption 68
Fossil Fuels in Historical Perspective 69
Crude Petroleum and Some of Its Products 70
World Proven Crude Oil Reserves, by Region 73
World Petroleum Refining Capacity 78
Secondary and Tertiary Methods of Oil Recovery 78
Oil: Potential Environmental Impacts Summary 84
World Natural Gas Consumption 84
World Natural Gas Consumption 1976: Total 84
World Natural Gas Consumption 1976: Per Capita 84
World Proven Natural Gas Reserves 85
Natural Gas: Potential Environmental Impacts Summary 87
Fuel Cycle of Light Water Reactor 91
Fuel Cycle of Fast Breeder Reactor 91
World Nuclear Production 1976 93
Nuclear Power Production 93
Nuclear Fuels: Potential Environmental Impacts Summary 100
Tidal "Lift Translator" 105
Cross Section of Low-Head Dam 105
World Hydroelectric Production 106
World Hydroelectric Production 1976 106
World Water-Power Capacity 108
Areas Suitable for Oceanic Hydroelectric-Power Generation 108
Sandia Magma Energy Power Plant Concept 112
Model of a High-Temperature Hot-Water Geothermal System 113
System for Extracting Energy from a Dry Geothermal Reservoir 113

Geothermal Energy: Potential Environmental Impacts Summary	115
Wind Machines Taxonomy	119
Two Modes of Utilizing Wind Machines	120
Wind Power Possibilities	122
The General Circulation of the Wind in Schematic Representation	124
Solar Energy Use in Perspective	132
Flat Plate Solar Heat Collectors	132
Solar Energy Harnessing Artifacts Taxonomy	133
Solar Energy Harnessing Systems Taxonomy: Passive	134
The 100-megawatt (Electric) Heliostat Power Plant Concept	135
Roof-type Solar Still	136
Simple Solar Heating System	136
A Naturally Air-Conditioned Building	138
The Power Mountain Concept	138
Photovoltaic Cell Cost 1954–1985	139
Schematic Diagram of a Solar Energy Collector	139
Two Light Shafts	141
A Sun Tracker	141
Light Output from Pipes	141
Two Types of Diffusers	141
Solar-Powered Total Energy System	142
Schematic Diagram Illustrating the SSPS Concept	145
Agricultural Waste Useful for Energy Production	147
Methods of Fuel Production by the Use of Solar Energy	147
World Fuelwood Production 1976	154
World Fuelwood Production 1976: Per Capita	154
One-Way, Single-Basin Tidal-Power Installation	159
Linked-Basin Tidal-Power Installation	159
Wheel for Raising Thames Water at London Bridge	160
Oscillating Water Column	162
Contouring Rafts	162
Wave Rectifier	163
Pliable Strips Device	163
Siphon Effect Device	164
Floating Tubes Device	164
Nodding Ducks	165
Wave Pumps	166
Floating Breakwater Device	167
Power Generation Utilizing Momentum Transport of Shoaling Waves	167
Underwater Vanes for Tapping Ocean Currents	170
Ocean Current Tapping System	170
OTEC Power Plant	173
Closed Rankine Cycle	173
World Hydrogen Production	177
Industrial Processes and Products Dependent on Hydrogen	178
Fossil Fuel and Hydrogen Flow	179
Energy Density Characteristics of Fuel	179
Combustion Characteristics	179
Boeing 747 Modified to Carry All Liquid Hydrogen Fuel	180
Subsonic Cargo Planes	180
Closed-Cycle Hydrogen Home	180
Cogeneration Plants	183
Osmotic Salination Energy Converter	187
Dialytic Battery	187
Comparison of Energy Available in Selected Salt Dome	187
Laser Fusion Process	188
Benjamin Franklin's Electrostatic Motor	191
Device to Utilize Deep Ocean Pressure	192
Nitnol Engine Schematic	194
Simple Russian Thermoelectric Device	195
Natural Draft Tower	196
Expansion-Compression Tower	196
Methods of Energy Transport and Storage	202
Relative Costs of Energy Transmission and Transportation	202
Power Output of Basic Machines 1700–2000 A.D.	203
Energy Conversion Efficiency	204
World Energy Data Chart	204

Making the World Work

Strategic Themes	220
Low Pressure Cone Above Dome	227
Energy Conversion Efficiencies	229
Unit Energy Consumption in Aluminum Reduction	230
Before and After Globalization	232
Decentralized and Centralized Energy Systems	247
Critical Paths	
Outline of Strategic Moves	234
Wind	235
Solar	236
Geothermal	237
Tidal	238
Refuse Reduction	239
Temperature Differential	240
Bioconversion	240
Nuclear Power	241
Hydroelectric	242
Hydrogen	242
New Energy Systems	243
Global Utility	243
Fossil Fuels	244
Conservation	245
Storage and Transport	245
Strategy of Switchover to Income Energy	248
Proposed Energy System	249
Hydrogen Economy Structure	249
Global Energy Utility	251

Maps

Note: The maps appearing in this book are based on R. Buckminster Fuller's Dymaxion Sky-Ocean World Map. This map projection was chosen to display world resources because of its unique properties; namely, it has a minimum amount of distortion and it is possible to see all the world's land masses and resources as one, almost interconnected, island. The base maps and their component parts and methods of projection are fully protected by international copyright convention and can not be reproduced without permission of R. Buckminster Fuller. Cartographers: R. B. Fuller and Shoji Sadao. Patent Number: 2,393,676. Copyright 1967.

World Population	30
Per Capita Energy Use	31
World Electricity Production	32
Coal Deposits	56
World Solid Fuels Production	57
Oil Deposits	71
Offshore Oil and Gas Fields	72
Petroleum Reserves	73
World Crude Petroleum Production	74
World Petroleum Refinery Capacity	75
Oil Refineries	76
Major Oil Spills	77
Oil Shale/Tar Sands	80
Natural Gas Reserves	85
World Natural Gas Production	86
Gas and Oil Pipelines	87
Nuclear Weapons and Nuclear Facilities in the US	92
Uranium Resource Locations	95
World Uranium Production	96
Nuclear Research Reactors in Operation	97
Number of Power Reactors	98
Hydroelectric Power Generating Stations	107
Currently Utilized Geothermal Sources	116
Potential Geothermal Sites	117
Wind	125
Wind in the Antarctic	127
High Voltage Electric Grid	129
Mean Annual Solar Radiation	137
Algae Production	148
Fuelwood Production	155
Forests	156
Tidal Sites	158
Annual Wave Energy in Specific Areas	168
Ocean Currents in January	171
Ocean Currents in July	171
Ocean Temperature Differentials	174
Suitable Locations for Temp. Diff. Power Plants	175
Potential Sites for Deep Ocean Pressure	193

Preface

The first edition of *Energy, Earth and Everyone* (1974) sought to demonstrate the technical feasibility of powering the entire world using only non-depletable energy sources. It presented a detailed critical-path plan for the phase-out of nuclear *and* fossil fuel use and a similarly detailed ten-year plan for the phase-in of the Earth's income energy sources. The global energy development strategy of the first edition of *E3* sought to demonstrate that with then current (1974) technology, more energy than the amount the world currently uses could be produced from our non-depletable energy sources. The amount was more than enough to insure that everyone on Earth, 100% of humanity, could be provided with a clean and regenerative supply of energy for meeting all their needs. The initial *E3* was the first document that presented such a global energy development strategy that dealt holistically with the long and short-range world-wide transition to regenerative energy sources. Since that time, many studies have appeared that deal, in whole or part, with switching our energy sources from our depleting stock of fossil and nuclear fuels to our incoming energy sources either on a local, regional, national, or international scale.[9] The world needs many more.

A lot has happened since 1974 and this updated, expanded, and revised edition of *Energy, Earth and Everyone* reflects this development. The book is organized in the same way, but every section has had extensive additions; Sections II and III are almost entirely rewritten. The major thrust of the book—outlined in the above paragraph—is the same but now is even more thoroughly documented. What seemed like pie-in-the-sky speculation to some in the early 70's turns into the hard fact of the late 70's as the "energy crisis" refuses to resolve itself by disappearing like the Nixon administration, Vietnam War, or hoola-hoop; instead, the critical shortages of our fossil fuels and the dangers of nuclear power have come into sharper focus. Who in 1970 would have thought that oil prices would quadruple in one year? that coal and gas prices would also skyrocket? that nuclear power would defy all government administration and industry prognostications of exponential growth and stop in its tracks in 1975 and 1976? that CO_2 could pose a threat to the planet? that solar energy would have the widespread impact that it has had? that there would be more than 600 manufacturers of solar energy equipment by 1977? that there would be a ten-megawatt solar-powered electric power plant being built in 1978? that there would be a three-megawatt wind turbine being built in 1979? that solar cells would drop in price by over 450%? that insurance companies and banks would begin to invest in solar energy? that people would begin to question the centralized and all-electric visions of the utilities in the developed regions of the planet? that people would begin to demand "appropriate technology" for developing regions? or that someone would seriously propose that nuclear and fossil fuel consumption be phased out? The list could go on and on.

The present edition of *Energy, Earth and Everyone* documents these radical changes in our energy environment and presents global and local strategies for dealing with the world's energy problems and prospects as well as background data for every available energy source. It attempts to treat the whole Earth, all of its energy resources, everyone on the Earth, and their total energy needs as one functional unit. The first section presents an overview of the planet's energy problems, the nature of energy, the limits on our present ways of utilizing it, and what the world's energy situation *should* look like given a different set of values than those that brought the existing energy system to its present state. The second section deals with all the known sources of energy on planet Earth, their history, locations, uses, advantages, and disadvantages. The third section presents energy development strategies and the fourth section is an appendix with glossary and conversion tables.

Both editions of *Energy, Earth and Everyone* have benefitted from the energies and intellects of those individuals who have participated in the World Game Workshops of 1974 and 1977. These workshops are sponsored by the research, planning, and educational organization, Earth Metabolic Design, Inc., of New Haven, Connecticut, and Buckminster Fuller. They are intensive, month or longer, international, multidisciplinary events where people of highly varied ages, geographical and cultural origins, and academic training gather to learn about their planet, its problems, prospects, and ways of effectively participating in the resolution of those problems. The workshops are organized around what Buckminster Fuller calls "design science"—the conscious resolution of global humanity's problems through the comprehensive and anticipatory application of the principles and findings of science. This approach involves understanding the critical interrelated nature of our problems and their global scope, the inability of present, locally-focused planning methods to deal effectively with these problems and new approaches and specific strategies for recognizing, resolving, and preventing our present and anticipated problems.

The method of analysis and synthesis employed in these studies was one of always starting with the whole and working towards the particular. Energy was dealt with in the context of all the energy in the Universe; the energy problems unique to humanity's life-support systems on Spaceship Earth were dealt with within the context of the Sun/Moon/Earth system; the specific energy problems of humanity's self-made energy production, distribution, and conversion equipment were dealt with within the context of 100% of humanity, its total history, and all of its needs.

A major objective of the World Game is to furnish the participants with the perspectives and conceptual tools that are needed for recognizing and resolving problems. It is also intended to give the participants the ability to identify the necessary steps and needed artifacts along the way to a problem's resolution that they themselves could design, develop, prototype, test, and make available to the world. In a sense, this document is the beginning of the work.

—Medard Gabel

Foreword

This book makes it incontestably clear that it is feasible to harvest enough of our daily income of extraterrestrial energy as well as of the surface eruptive steams of internal Earthian infernos all generated at an inexorable, nature-sustained rate to provide all humanity and all their generations to come with a higher standard of living and greater freedoms than ever have been experienced by any humans and to do so within ten years, while completely phasing out all further use or development of fossil fuels, atomic and fusion energies.

This means it is possible for Earthian humanity to live on its daily energy income as generated by star radiation and cosmic gravity—primarily that of the star Sun—rather than: (a) by exhaustion of the millions of years of celestial energy photosynthetically impounded by the terrestrial vegetation and deposited into Earth's crust as a cosmic savings account possibly to be used many billions of years hence to convert planet Earth into a star; or (b) by burning up the atoms of which spaceship Earth is structured; or (c) by fusion's disruption of the biosphere's delicate hydro and thermal balancing, the incisive integration of which governs the comprehensive metabolic chemistries of terrestrially regenerative ecology.

Three quarters of our planet Earth's surface is covered by water. Water constitutes about 60 percent of the physical substance of planet Earth's biological organisms. Sixty-five percent of the human body consists of water. The Earth's oceans contain 97 percent of all our planet's water. The surface of the Earth's waters are being continually vaporized into clouds to be redistributed around Earth as rain or snow. Sumtotally, Earth's waters are being continually recirculated throughout its combined ecological and geological biospheric system. Water is the "blood" of Earthian life. The average depth of its oceans is less than one four-thousandth of the Earth sphere's diameter. It is a gossamer film so thin that it is proportionately less than the depth of the blue-ink printing of the oceans on a 24-inch Earth globe. This almost ethereal film is kept from instant evaporation by the Sun only through the energy-reshunting properties of the plurality of additional concentrically enshrouding chemical and electromagnetic spherical mantles of Earth's biosphere.

The entropic energy losses to Universe occasioned by our emergency-urged fortuitous exploitation of spaceship Earth's inventory of integral atoms either by fission or fusion will probably violate the integrity of the complex cosmic design for successful maintenance of human life aboard Earth.

If you were a poor farmer and an oil gusher suddenly spurted on your land and further geological investigation showed that you could realize a million dollars a day for the rest of your life, it would be difficult for you to convince yourself that you ought to forego such monetary advantage. Human reasoning is compromised by such circumstances. Most individuals would probably rationalize, "If I am wise I can do more good for humanity with that wealth than I can as a farmer." Having persuaded themselves to accept such fortune, the oil-rich discover many complexities of tax accounting requiring a host of professional lawyers and wealth managers who recommend various foreign residences, etc. There are now millions of such supra-national millionaires. Through many such spontaneous human rationalizations the multi-eons of cosmic energy wealth deposited by Universe as a future-evolution-accommodating savings account in this little planet is being depleted in one spendthrift century.

Ownership of the as yet remaining fossil energy wealth of our planet Earth has now become concentrated in the control of a very few individuals. Arabia embraces by far the largest remaining petroleum reserves. The majority of Arabian oil land-owning monarchs are beset by the world's most powerful, quick but shortsighted profit exploiters of human needs, and the latters' terrestrial resource gratifications whose accreditation to the oil owners is the exploiters' world distribution system of tanker fleets, pipelines, storage tanks, refineries and metered retail filling stations as well as their proven know-how to take oil out of the ground and convert it into money.

The grand strategy of the world's free-enterprise cartels is determinedly set to establish comprehensive atomic energy power before their exhaustion of the world's petroleum. The atomic energy know-how which they exploit was developed by the U.S. government at a 200-billion-dollar cost to its citizen-taxpayers. A third of a century's skillful manipulation of governmental evolution has shifted proprietorship of atomic power in the noncommunist economies from the general citzenry's ownership perogative to that of supranational private enterprise corporations. The grand strategy of the free-enterprise cartels is predicated on their conviction that humanity cannot conceive of any possible alternative to their atomic energy development plans.

But the omni-humanity advantaging alternative option does exist. That is what this book is about.

Discovery and proof of the existence of an eminently feasible, cosmic energy income alternative required exhaustive research. It required full inventorying of all the known and physically proven daily income of inexorably operative cosmic energy behavior, such as the omni-energy conserving terminal turnaround intertransformability of entropic radiation and syntropic gravity, as complexedly and complementarily manifest for instance in the Sun and Moon's pulsatively rhythmic, gravitationally pulled and precessionally elevated tidal waters of Earth; or in the Sun-generated winds; or in the syntropic intertransforming of world resources by cosmic receipts whose chemical processes produce methane, hydrogen, alcohol, etc.; or by inexorably operative thermal-steam generation by natural subterranean inferno sources available around our planet.

Proof of the existence of an onmi-humanity-supporting cosmic energy income required a fully documented inventorying of the prime annual energy income sources as well as the known rates of toolability of the proven technical processes of employment of those proven natural daily energy income resources of our Earth.

The whole task needed the same kind of exhaustively thorough assessment of resources and capabilities that went into the scientifically designed and plotted "critical path" system from bare start to ultimately realized dispatching of humans to the Moon and returning them safely to their mothership Earth.

What are all the first-things-first that must be done? How long does each one take? How do they overlap without delaying interferences? At what stages of each of the first-things-first can each of the second things be inaugurated? The critical path and the countdown of its accomplishments for getting humans safely to the Moon and back aboard planet Earth was in the magnitude of two million items that had to be realistically accomplished. The Cape Kennedy "countdown" was not from 10 to 1, but from 2 million to 1.

The basis for the energy study presented herewith was accomplished by the 1974 World Game seminar led by Medard Gabel and his Earth Metabolic Design associates, all hosted by the University of Pennsylvania. They undertook the exhaustive inventorying of proven daily income energy resources, proven know-how technologies and proven rates of realization of the full complex of developments. Their work was governed by all the technological realities of sustainably energizing humanity's highest-known standards of living and accommodating not only all its further exploratory needs but also accommodating adequately all the generations of humans to come. In 1977–79 the study was completely updated and expanded.

We are confident that this report shows how all this can be accomplished within ten years. It will bear the most rigorous examination by Russian, Chinese, and Western scientists, engineers, and operating managements. Because it clearly evidences a previously unknown but workable alternative option in the provision of adequate energy for foreseeable human needs and desirable development, it invalidates the general systems assumptions underlying the grand strategies of the present world-power masters of free enterprise.

Because it is youth who must always cope with tomorrow's problems, it is of most intimate concern to all of humanity's youth and to all youth henceforth to come that the energy needs of world society be completely supported by the daily income energies of our planet. This report is of number one concern to all youth.

Humanity is now deeply but silently apprehensive about exhaustion of its petro-energy resources. Humanity is also intuitively apprehensive about burning up the irreplaceable atoms of which its spaceship Earth is comprised, but 99 percent of humanity knows of no alternatives and can only protest that the petro-atomic exploitation seems unwise.

It is quite another matter for humanity to learn that it has indeed a clearly cut, feasible, and omni-desirable alternative. With proven alternatives youth's spontaneous stratagems become holistically constructive. Negatives beget negatives. Positives beget positives. Instead of using their heads for battering rams in bloody revolution, humans can use them to carry forward the design science revolution which alone leads to the comprehensive success for all humanity for all time.

The frequently negative and ofttimes lethal experiences of humanity recurring throughout all history have brought forth an ever-evolving succession of political leaders. Those same negative experiences persuaded the political and economic leaders that there is a fundamental inadequacy of life support on our planet. Assuming fundamental inadequacy also hypothesizes that: (a) many humans are to perish short of their potential life span; and (b) vast numbers are destined to poverty and all the physical deprivations that go with it; or (c) are destined to die in trials of arms in evolution's process of selecting those fittest to survive. It is only because of the universal assumption of the lethal inadequacy of life support that Earthians have politicians.

Because seemingly only a fortunate few are eligible for survival, politicians are inherently biased. Each political ideology assumes that it has the most logical and fair method of coping with fundamental inadequacy of life-support resources.

All politics assume ultimate trial by armed might to determine which political group is to survive. Hence 200 billion dollars for arms have been jointly spent each year for the last two decades by the world's leading political powers.

Not only have the design-science explorations of the World Game seminars proven the complete adequacy of natural energy income to support all of humanity, but they have also found that all the livingry in general, the housing, clothing and all environment-controlling needs and desirables, as well as living perquisites in general favorable to humanity, can now be provided through known technology and provenly available resources in such a manner as to take care of not only all the fundamental needs, but also of an ever-increasing number of the "desirings" of humanity, and do so within ten years.

We can state informedly that we now know how to take care of all humanity at a higher standard of living than any humans have heretofore ever known and can do so within ten years. We now know there is no fundamental inadequacy of life support. There was only inadequacy of human knowledge and experience. Knowledge and experience now make it clear that humanity was potentially designed to be a success and not a failure. We now know that all "you or me" political theories are obsolete. We now know that all warring is obsolete. We now know that all weaponry is obsolete. Universe is trying to make humanity an important success aboard our spaceship Earth. We can therefore assume that Universe has important need for the successful functioning of humanity's highest capabilities which are its mental abilities.

Because of the powerfully conditioned reflexes of humanity; because of the fear-born inertias of bureaucracies; because of the lack of common knowledge that humanity has the option of sustainable physical and economic success; because evolution is moving inexorably to make humans successful despite the opposing factors, humans on Earth are now entering a period of total revolution. If it is bloody because humans resign themselves to the solution of survival problems only by physical power, humanity will disqualify itself from further continuance aboard our planet.

What is unique about humans, in contradistinction to others aboard our planet, is our human mind's access to scientific knowledge of the operating principles governing eternally regenerative Universe itself.

Infinitesimally small humans on tiny planet Earth of medium-sized star Sun of our 100-million-star Galaxy, have been able to discover cosmic principles and have been able to employ those generalized principles to develop astrophotography. Infinitesimally minute humans employing generalized principles of optics have telescopically discovered and photographed a billion other celestial galaxies in our thus far 11½-billion-light-year radius sphere of local Universe observation. These little humans on planet Earth through their gift of mind have been able to discover the unique electromagnetic frequencies of each of 92 chemical elements, and little human beings on planet Earth have been able to take inventory of the relative abundance of all the chemical elements present in the 22-billion-light-year diametered sphere of Universe thus far humanly observed, 99.9 percent of which phenomena are utterly invisible to the naked eye of humans.

Up to the 20th century, reality was everything humanity could see, smell, touch and hear. Humans knew nothing of the electromagnetic spectrum, of X-rays or the radio. Now 99.9 percent of all the evolutionary endeavors of humans are being conducted within the electromagnetic realms of Universe, directly unapprehendable by humanity's naked senses. The 99.9 percent sensorially undetectable realities have been reached and are being explored and employed by humans only through their faculty of mind and the latter's capability to understand eternal, utterly incorruptible, principles.

Yet the majority of little individual humans are as yet struggling against one another for survival. Little groups of individuals struggle for survival. Corporations and states struggle against one another for survival. These are the conditioned reflexes of humanity as we enter the great revolution.

Their physical prowess being greatly inferior to donkeys, humans are clearly present in Universe for their mind functioning. If humans are to pass their final examination for continuance on board our planet, it will be because weightless metaphysical mind and all the synergetic truths that mind discovers have come into sustainable command of brain and muscle in the resolution of humanity's continuance.

If the design science revolution is conducted by mind not for the advantage of the few but only for the advantage of all humanity, humanity will pass its final examination and thereby earn cosmic accreditation to its further functioning in the great scheme of maintaining the integrity of eternal cosmic regeneration.

R. Buckminster Fuller
Philadelphia, Pennsylvania
Spring 1979

The following section outlines, in a game-type format, the perspective used in the writing of this document. Readers are encouraged to participate in this game or planning exercise by adding any new, unasked questions of their own that they think are essential to making a decision that will affect their survival in the given situation. "Dare to be naive" is the planning exercise's motto.¹⁰

Energy for the Earth

Overview

Suspend disbelief for a moment. Imagine that you are on a spaceship. You are captain of that spaceship. You are in trouble. Something or many things have gone wrong. Your energy life-support systems are malfunctioning: You don't know what is happening, where it is happening, when it happened, how it happened, or whom it affected or affects. You don't know where you are or where you are going.

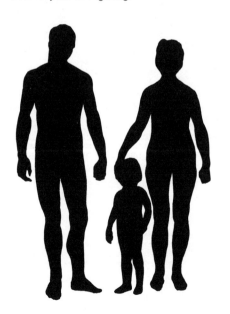

What information do you need to make a rational decision about your survival?

What is the problem (or problems) that confront us?

What area of the spaceship is affected by the problem? Is the problem local or universal?

How much time do we have to solve the problem? How long before the problem becomes critical?

What will the effects be if the problem is not dealt with?

How many people does it affect and to what degree?

What is the spaceship's life-support system? Is it working properly?

How can we organize ourselves to do what needs to be done?

Are there any other ships? rescue ships?

Who built the spaceship? Does it have a guarantee?

Where is the spaceship going?

Where has the spaceship been?

What are the spaceship's needs? What are its energy needs?

What is energy? What is power?

How much energy is there flowing through the spaceship? Where does it come from?

What is energy used for on the spaceship? How much utilizable energy is there on the spaceship?

Where is all the energy located and converted on the spaceship?

How much energy is needed by the spaceship? by humanity? by humanity to have all the life-support facilities that are needed?

What are all the sources of energy? How long will they last? What sources are presently being used?

Which sources could be used?

What are the efficiencies of using various forms of energy?

Who needs energy? How much? What kind of energy? For what purpose? For how long?

What are the tasks we wish to have performed by energy? Which are crucial?

What is the scale, type, quality, and degree of clustering of energy needs?

How is energy stored and transported?

How much of our energy problem is the result of our accounting system?

What are the critical energy needs? How do we determine what is critical?

Do we have a goal? How far do we have to go to reach it? How far to we want to go? How far are we able to go?

What are our alternatives?

What are the limits to our solutions? What is feasible?

What resources are available to solve the problem?

What is the history of the energy problem? What caused it?

Has this problem happened before?

What were the effects of past "solutions"?

How would the proposed energy strategy be implemented?

***Developing a new approach to the energy problem and its social ramifications has only become possible as new insights have shown us what questions to ask. Often the questions are so simple that we never thought of asking them before.*[11]**

What new problems will arise as a result of implementing the proposed energy strategy?

How can the energy sources be utilized in a better way? How can we do more with less?

Are the energy resources we have enough to enable us to reach our goal?

What have we forgotten?

When should we stop asking questions and act?

We are all on a spacecraft; we have known it for some time now, and we need to ask and answer such questions as those above if we are to put our energy-related problems in a perspective that will allow us to recognize, define, and solve these problems rather than merely to treat their symptoms.

There is an important distinction between the swift (66,000 mph), efficient, and regenerative mobile-home Earth and human-designed spaceships: Spaceship Earth has the automated capacity to continue on without its passengers or crew. Its physical trajectory is completely automated and is not something with which the crew or passengers would realistically wish to tamper, even given the capability (which it seems we may have).

In addition, as Margaret Mead points out, the "spaceship" we are on was not designed and built by us and accordingly we need to treat it with a great deal more respect and care than those built by humans. It is not as simple (luckily) as our machines, or else we would have fouled up its operation long ago. What we are concerned with, and what we do have the capacity to consciously participate in and positively effect, is our voyage aboard our spacecraft, for the trajectory of human evolution is not as assured as is Earth's path around the Sun. We have critical choices to make that will greatly affect our future voyaging. The vast interrelatedness and magnitude of effects which human-engendered life-support systems are now having on the Earth's ecological life-support systems are forcing us to become aware of our actions and their consequences before we take them. We are moving from making choices by default to making decisions by design. In effect, we are choosing survival and evolution or oblivion.

Metabolism

To survive, we need food. For a society to survive, it needs energy. Food is the energy for one's personal life-support mechanism; energy is the food for society's life-support mechanisms. Human needs seem at present to have no upper limit, but they do have a lower limit—the minimum food and warmth necessary to maintain life. Human society also has minimum lower limits of energy that it needs to remain viable. The quantity and quality of energy needed is a function of how developed and extensive the technology is for the particular area under question.

Only man is able to utilize energy in excess of that necessary to satisfy his nutritional needs.[12]

"A modern industrial society can be viewed as a complex machine for degrading high-quality energy into waste heat while extracting the energy needed for creating an enormous catalogue of goods and services."[13] The human body can likewise be viewed as a complex machine which breaks down high-quality energy as food (catabolism) into heat while extracting the energy needed for growth, repair, and maintenance of the body (anabolism). *Metabolism* is the process that describes the entire cycle.

Income Energy

Society's 95% reliance on "capital" marketed energy sources (non-renewable fossil and nuclear fuels—coal, gas, oil, uranium—in contradistinction to "income" energy sources, which are constantly renewing, like sunlight, wind, and tides) is analogous to an individual relying solely on the limited food found in his or her storage cellar, without gathering or planting any more food. Sooner or later he would run out of food and starve, unless he invested his "free" time in exploring ways of obtaining new sources of food when the present cellar was empty. He could look for other cellars, or he could begin to develop alternative sources of food.

Society has been looking for, and finding, other cellars for a long while now. We are just beginning to look for feasible means of harnessing and harvesting alternative sources of energy. The nondepletable income energy sources become more and more attractive the larger the problem becomes—larger in the extent of global concern with energy shortages and inefficiencies ("pollution"), and larger in time, i.e., where are our children and our children's children going to get their energy?

Like the individual in the example, society seems headed toward the need to harness and harvest nondepletable energy supplies. As prehistoric man turned to the constantly replenishing source of agriculture to supply his energy/food needs, modern society seems destined to turn to sources of energy that are nondepletable. Some of the problems associated with harnessing these energy sources for society are paralleled in prehistoric man's attempts to master agriculture. Just as crops would fail from drought, flood, or poor soil, so harnessing solar or wind power is presently curtailed by their lack of sustained intensity and duration.

There is no one answer to the world's energy problems, e.g., the breeder reactor, solar power, etc. Nature teaches us that such myopic views are dangerous, that the more diversity there is within a system, the more adaptive and hence stable the system is. As nonadaptivity leads to extinction, adaptive stability is an obvious virtue of global energy system design. The more diverse our sources of energy are, the more dependable will be the flow of energy. If one source should temporarily (or permanently) "run dry," the rest of the system will continue to supply our energy needs.

Review

An "energy crisis" exists on Earth because the energy or power demands for the basic necessities of life—food, shelter, clothing, health care, transportation, communication, education, recreation, and logistical support—are increasing both in extent of use and in intensity of use. This is partly because world population is increasing, and, more importantly, because living standards and expectations are increasing, and humanity's intuitive value structure seems to dictate the humane objective of providing the people of the world with a quality of life as high as man's ingenuity can develop. The world's population has increased from 1.6 billion in 1900 to 4.3 billion in 1980; simultaneously, man has gone from less than 1% who would be classified as "haves," or economically successful, to over 40% "haves." The primary reason for this exponential growth of the "haves" is the huge increase in energy consumption over this same period (from 11 to 83 trillion killowatt hours total consumption).[14] Using a small portion of that energy, the "haves" developed artifacts that tended to decrease the death rate of both the "haves" and "have-nots," allowing the population to increase exponentially and put further demands on the energy supply.

World Population
1980
4.3 Billion

1900
1.6 Billion

Given these basic driving forces, the world's energy needs have increased to their present levels and are trending to increase much further.

The world's total energy resources are for, and can meet, the entire energy needs of 100% of humanity aboard Earth. We are all on the same planet, we use the same energy sources with the same technology and know-how, and we are members of one amazingly complex species. Viewing the problems and making decisions from this perspective does not make them more complex, but in a real sense simplifies them enormously. It simplifies our energy problems by putting them in their true functional relation with the Earth's total life-sustaining biosphere, and all of humanity's shared experiences as one species. Such a perspective leads to solutions that are considerate of all the world, its people, and its delicate ecology, rather than 7% or 40% of humanity and the dangerous piecemeal "local focus hocus-pocus" approach to our energy problems brought about by most present perspectives of our energy problems.

There is as yet no patron for Earth other than our mutual love and respect for one another. We are the clients of our own concern.

Energy

The Go of Things

"Energy," as James Clerk Maxwell aptly put it, "is the go of things." What this statement lacks in mathematical precision it makes up in intuitive familiarity. Everything we know of "goes"; without energy, what is wouldn't be. Energy is the input into every system which allows that system to maintain its particular pattern integrity, and, since Einstein and "$E = mc^2$," energy is now understood to also be the physical system itself.

Stated in another way, we can say that all systems—physical, biological, or social—depend on energy. The type of structure that a system has (or evolves) depends upon the energy available to it. Energy and structure are intimately related. "Energy is used for *control*—to get more energy—for *conversion*—to produce goods and/or services—and for *storage*—to preserve energy for future use. *Control* or capture implies the possession of, or ability to use, an energy source: a plant captures and controls sunlight; an oil company, petroleum. *Conversion* changes energy from one form to another by processing materials through an appropriate technology: sunlight is converted to carbohydrates in photosynthesis; coal is converted into useful work in a steam engine through combustion. *Storage* is the preservation of energy in a specific form: sunlight is stored in fossil fuels; a farmer stores food energy in his granary.[15]

"In general, most of a system's controlled energy is devoted to the operating costs of maintaining the system's structure, i.e., providing the materials (goods) and functions (services) for survival and well-being. Some controlled energy must be converted in the process of capturing more energy. What is left over after these two operations is stored or surplus energy. The energy distribution between control, conversion, and storage determines the dynamic state of the system, i.e., growth, stability, or decline."[15]

Energy is perceived as kinetic energy when it is affecting matter, and as potential energy in its capacity to affect matter. It can be defined in terms of what it does or the way it is used. Science's progressive understanding of what energy does and technology's progressive understanding of what it can make energy do traces the evolution of humanity's understanding of, and enlarging participation in Nature. We have come from energy as understood by the Greeks to Newtonian mechanics, to thermodynamics, quantitative chemistry, and electromagnetism; to relativity, quantum mechanics, and cellular metabolism; from food to fire to steam to electricity, gasoline, and rocket fuel; to nuclear reactors and photosynthesis.

Science's description of energy is an operational one; it is in terms of what energy does or can do rather than what it is. It is stated conveniently in terms of a capacity, "the capacity for doing work." Work is done when a body is moved by a force. This capacity for doing work can be transformed into other capacities. Certain kinds of energy transformation are known as power generation.

---

100 joules is the potential energy of a 22 pound weight dropped from waist-high (1 m) or the kinetic energy just before touching the foot.[16]

Power is the rate at which work is performed or energy flows; power is energy per second.

Energy is measured in ergs, joules, calories, watts, BTU's, horsepower, foot-pounds, metric tons of coal, or barrels of petroleum. The erg is a very small unit of work measurement. It is the work that is done when a body of 1.02 milligrams (.0000359 ounces) of weight is lifted vertically 1 centimeter against the gravitational force at sea level. Ten million ergs equal 1 joule; 1 joule per second is 1 watt. The practical unit of power measurement is the kilowatt, or 1,000 watts.

Energy is valuable when energy is valvable; that is, the easier it is to harness a particular energy source the more value it has. For example, gravity is found everywhere, and in great quantity, but has less value in present-day economic accounting than a lump of coal because it cannot be as readily used. Energy thus has a qualitative as well as quantitative dimension. Entropy is a measure of this aspect of energy. It measures the relative disorder of energy and energy systems. The more organized and concentrated a given energy source the more work it can be made to perform. The more disorganized and diffuse an energy source the less work a given unit can perform. "High quality" refers to concentrated heat sources such as the thousand-degree

---

A warm lake contains far more energy than a small battery, but it is difficult to power a pocket calculator with a warm lake.[17]

---

temperatures associated with coal combustion or the even higher temperatures of fissioning uranium; "low quality" would refer to the hundreds of degree temperatures obtainable from solar energy. Luckily, more than 40% (see charts) of our energy needs can be met with lower quality heat sources. Using higher quality heat sources to do lower quality heating jobs, such as having the extremely high temperatures of a nuclear reactor boil water to produce steam to generate electricity to heat water in your house is, as Amory Lovins points out, like "using a chainsaw to cut butter" or "frying an egg with a forest fire."

Energy, as the First Law of Thermodynamics states, can neither be created nor destroyed. When fuel is "consumed" energy is not. There is still the same amount of energy after combustion as before, but the amount of available, concentrated, or organized energy has decreased. As the fuel is consumed, its energy content is used to perform work. After being used, the fuel is "gone" (because it cannot be used again in the same way) but the energy still exists. Qualitatively, entropy increases when fuel is consumed, but quantitatively, there is still the same amount of energy. It is just that 50 to 60% went up the smokestack as waste heat, 30 to 40% went into the water to make steam, some more went into light, and finally, if the fuel being consumed was in a thermal electric power plant, some ended up as electricity.

"Primary energy" refers to the energy content of fuels before they are processed and converted ("consumed"). Primary fuels are crude oil, coal at the mine, natural gas at the well head, and so on. "Delivered energy" is the processed energy that is delivered to the consumer in a useable form such as electricity,

U.S. Energy End-Use Requirements[11]

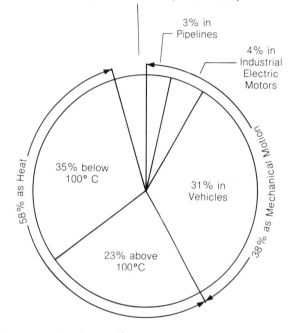

4% as Lighting, Electronics, Telecommunications, Electrometallurgy, Electrochemistry, Electric Motors, etc.; end uses that *require* electricity.

Temperature Requirements[18]

200° C could directly supply more than 40% of the total U.S. requirements.

150° C could directly supply more than 30% of the total U.S. requirements.

100° C could directly supply more than 20% of the total U.S. requirements.

gasoline, oil, and heat. The total consumption of primary energy does not tell you how much energy is actually delivered, nor does the amount of delivered energy tell you how much work is actually performed because that is dependent upon the efficiency of the end use energy converters—lightbulb, car, furnace, etc.

It's not how much you use; it's how you use what you have.

In addition, and most importantly, there is nothing in any of the figures—primary, delivered, or work performed—that tells you whether the work performed was at all necessary for increased social well-being. Did that lightbulb need to be on? Is that car necessary? Is that 4,000-pound car necessary?

"Energy is a means, not an end. It's worth derives entirely from its capacity to perform work. No one wants a kilowatt-hour; the object is to light a room. No one wants a gallon of gasoline; the object is to travel from one place to another. If our objectives can be met using a half or even a quarter as much energy as we now use, no benefit is lost."[17] Energy is consciously used by humanity only in its capacity to procure materials, food, shelter, health and medical care, transportation, communication, education, recreation, logistics, energy production and distribution, and to alter the ecological context in the form of by-products of these processes, i.e., pollution.

External Metabolics

Our discussion of "energy" is limited to that energy used in the above-mentioned areas as an input—that which gives the area its viability—but will not include that energy that makes up the area itself, as mass. There is also a distinction between animate or physiological energy, which flows through biological systems (generated in plants, animals, bacteria, molds, fungi, etc.), and inanimate or physical energy (derived from wind, water, wood, peat, coal, oil, gas, tidal motion, heat from the earth, radioactive elements, etc.), which flows through mechanical systems. This distinction can also be made through the metaphor of internal and external metabolics. As internal metabolics is the life-support system of the individual, external metabolics is the collective life-support system of humanity. "Food" fuels internal metabolics, and "energy" fuels external metabolics. The main reason for making this distinction is that food and fuel are two separate problems in today's world and, although clearly inter-related and partially overlapping, the food-need problem is distinct from the energy problem (see *Ho-ping: Food for Everyone*).

The definition of energy as used in this document is functional; that is, it describes the designed operation or role which the energy process plays in the overall scheme of things. It is a definition of the part that energy plays within the life-support systems of humanity.

Energy is that which is used to power the production, distribution, maintenance, utilization, and recycling of the materials and information used in the global energy, materials, food, shelter, health, transportation, communication, education, recreation, and logistical systems.

For example, this includes:
- The Sun;
- Electric utilities;
- Electric power plants, hydroelectric dams, transmission lines, transformers;
- Oil companies;
- Oil fields, wells, offshore wells, tankers, refineries, gas tanks, barrels, pipelines, combustion by-products;
- Coal fields, strip mines, power plants, blast furnaces, combustion by-products;
- Uranium mines, nuclear power plants, nuclear waste;
- Waste heat, sulfur dioxide, particulate matter, carbon monoxide, carbon dioxide, nitrogen oxides from combustion;
- Gas engines, jet engines, steam engines, electric motors.

Energy is used for:
- Locating, mining, processing, distributing, and recycling raw materials;
- Cultivating, planting, growing, harvesting, processing, distributing, storing, preparing, and utilizing food;
- Producing fertilizers and powering tractors and harvesters;
- Producing clothing and packaging;
- Excavation, construction, maintenance (heating, cooling, lighting), and destruction of shelter facilities;
- Powering autos, buses, trucks, trains, airplanes, and pipelines, and for building highways, railroads, airports, seaports, and pipelines;
- Powering radios, TVs, printing presses, record and tape players, telephones, computers;
- Powering hospitals, X-ray machines, examination and laboratory equipment, pacemakers, sanitation facilities, pest-control equipment, ambulances, fire trucks, funeral facilities, Red Cross and civil defense facilities;
- Powering schools, educational aids, laboratories, libraries;
- Powering recreation vehicles (snowmobiles, dunebuggies, boats, campers, go-carts);
- Powering museums, stadiums, gymnasiums, sports equipment, exercisers, amusement parks, zoos, parades, ski lifts, games;
- Powering the manufacture of all the general and specific tools for all of the above;
- And as a by-product input into the ecological context.

The Fundamentals

The "energy crisis" has little to do with gasoline or fuel shortages, or with rising prices of gas or electricity. Indeed, when viewed from a larger perspective, these so-called problems can be part of the solution to our comprehensive energy problems.

The real problem we face is how to get enough energy for 100% of humanity's life support, with equitable distribution, little or no negative environmental impact, and on a continually sustainable basis with little or no coerced human labor inputs. The present global energy system is characterized and hindered by the inverse of this preferred state; there is not enough energy for 100% of humanity's life support and there is inequitable distribution, high negative environmental impact, low life expectancy of energy sources, and a high number of coerced human labor inputs. The diagrams and charts used in Part 1 illustrate this system from the broad overview to the particular. Each aggregate level is a more specific definition of the present state.

"The Earth may be regarded as a material system whose gain or loss of matter over the period of our interest is negligible. Into and out of this system, however, there occurs a continuous flux of energy in consequence of which the material constituents of the outer part of the Earth undergo continuous or intermittent circulation. The material constituents of the Earth comprise the familiar chemical elements. These, with the exception of a small number of radioactive elements, may be regarded as being nontransmutable and constant in amount in processes occurring naturally on the Earth.

"The energy inputs into the Earth's surface environment are principally from three sources: 1) the energy derived from the sun by means of solar radiation, 2) the energy derived from the mechanical, kinetic, and potential energy of the earth-sun-moon system which is manifested principally in the oceanic tides and tidal currents, and 3) the energy derived from the interior of the Earth itself in the form of outward heat conduction, and heat convected to the surface by volcanos and hot springs. Secondary sources of energy of much smaller magnitude than those cited are the energy received by radiation from the stars, the planets, and the moon, and the energy released from the interior of the Earth in the process of erecting and eroding mountain ranges."[19]

Energy Exchange

"The atmosphere and the oceans are coupled heat engines, driven by energy received from the sun. In lower latitudes, the sun provides more energy than is re-radiated back to space; the reverse is true in the polar regions. Despite this, the average temperature of the earth neither increases rapidly at the equator nor decreases at the poles. This is possible only because the fluid portions of the globe—the oceans and the atmosphere—develop circulations which carry energy in various forms from the tropics to the polar regions.

"Over most of the earth, the atmosphere is the primary carrier for the poleward transport of energy. This does not mean, however, that the oceans do not play a fundamental role in the energy cycle. The large-scale atmospheric circulations which carry energy poleward are driven primarily by heat received from below. Since more than 70 percent of the earth's surface is covered with water, this means that the energy to drive the atmospheric circulation comes primarily from the oceans.

"The reason for this lies in the physical characteristics of the atmosphere itself. In the absence of clouds, the atmosphere is highly transparent to incoming solar radiation, a high proportion of which penetrates to the surface of the earth where it is absorbed and converted to heat. To maintain an energy balance, the earth re-radiates energy back toward outer space. But since the wavelength of emitted radiation tends

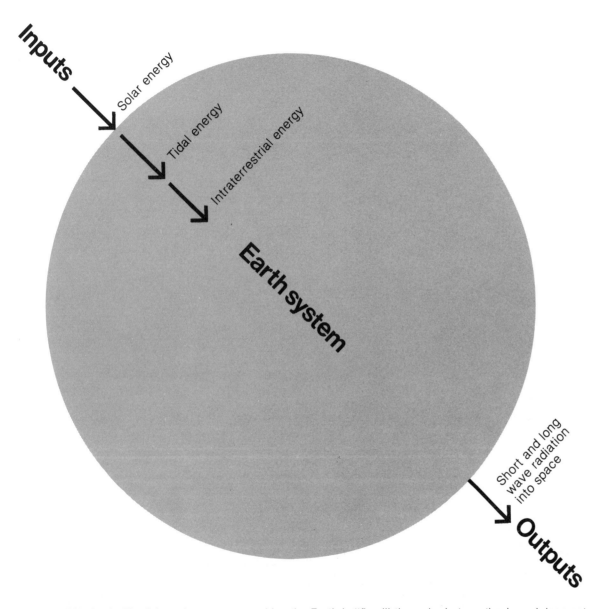

About one-fiftieth of 1% of the solar energy reaching the Earth is "fixed" through photosynthesis and does not come out right away. Because of our fossil fuel consumption though, there is more output than input.

to be inversely proportional to the temperature of the radiating surface, this outgoing radiation occurs in the infrared. Most of this longer wavelength, terrestrial radiation is absorbed by gases in the atmosphere, especially water vapor and carbon dioxide. As a result of this "greenhouse effect," the main cycle of energy is from the sun to the surface of the earth; from the earth to the atmosphere; and from the atmosphere back to space. Thus, there is a continual flow of energy from the oceans to the atmosphere in the form of infrared radiation. The flux of radiation from the air-sea interface is the major source of energy which drives the atmospheric winds.[20]

"Although the energy cycle is dominated by the radiation process, other processes are also important. In particular, substantial amounts of energy are transferred from the surface of the earth in two other ways: by the evaporation and subsequent condensation of water vapor, and by the direct exchange of energy in the form of sensible heat. Both of these forms of energy transfer are important over the oceans; both are significant factors in air-sea coupling.

"The evaporation of water is an exchange of mass. At a typical oceanic temperature, the evaporation of one cubic centimeter of water requires approximately 600 calories of energy. This results in a cooling of the ocean surface, and a transfer of energy to the atmosphere in the form of latent heat. This latent heat moves with the atmospheric circulations and is finally released when the water vapor condenses to form clouds or precipitation. The net result is a flow of energy at all latitudes from the ocean to the upper atmosphere.

"This movement of water vapor also plays a role in the latitudinal energy balance of the

Energy Flows Through the Earth—Quantitative Input and Output.[19]

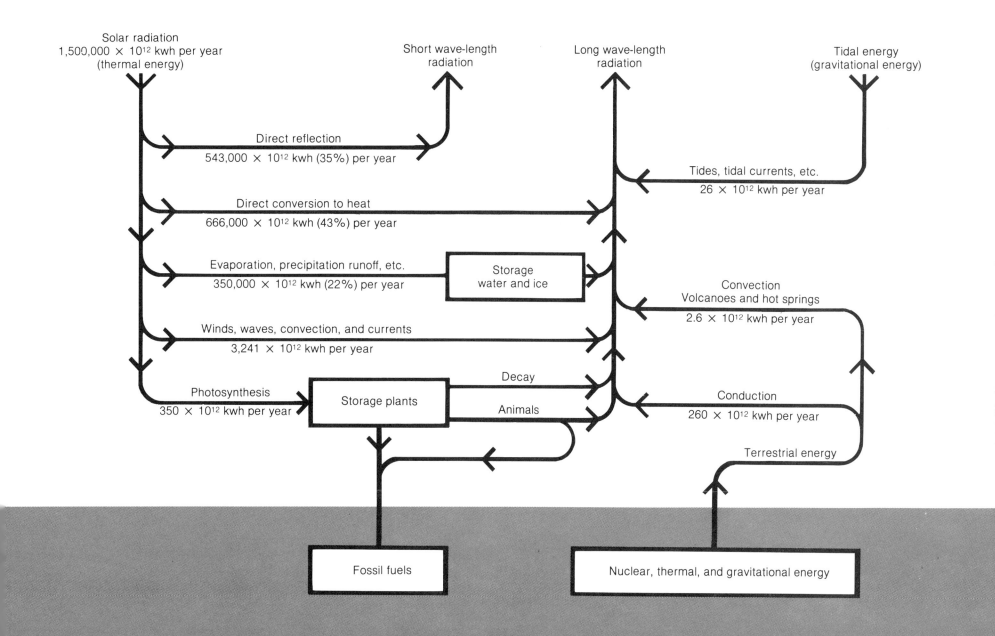

combined ocean-atmospheric system. Energy is expended in the subtropics, where evaporation exceeds precipitation. This energy is released at higher altitudes, and to a lesser extent near the equator, where precipitation exceeds evaporation. The poleward transport of energy in temperate latitudes helps to remove the excess energy supplied to the tropics by radiation.

"The final way in which important amounts of energy are exchanged between the oceans and the atmosphere is by direct heating and cooling. When warm air moves over colder water, energy is transferred from the atmosphere to the water; when cold air moves over warmer water, the reverse occurs. This is a typical boundary layer phenomenon.

"The direct exchange of energy in the form of heat is usually much smaller than the exchange due either to radiation or to the evaporation of water. On a global scale, the flow of heat is from the oceans to the atmosphere, but the total energy transferred is probably not larger than 10 to 15 percent of the exchange due to evaporation. Nevertheless, the direct exchange of energy can be important, especially in regions and for seasons of the year for which a large temperature contrast between the air and the underlying water is observed. One such example occurs over the Gulf Stream during the winter, when cold air moving out from the North American continent frequently sweeps across the warm offshore current. In this case, the transfer of energy to the air is many times the global average, and plays an important role in modifying the structure of both the Gulf Stream and the atmosphere.

"To summarize, we may note that the energy budget of the globe is strongly dependent on the exchanges of energy that take place across the air-sea interface. At all latitudes, the atmosphere is heated by the oceans. This occurs in three ways. In order of importance these are: exchange of energy by long-wave, terrestrial radiation; the flux of energy due to evaporation of water at the ocean's surface and re-condensation within the atmosphere; and the direct transfer of heat due to temperature differences between the ocean and the atmosphere above. These exchanges of energy drive the atmospheric winds, which in turn carry energy from the tropics towards the poles. The exchanges of energy also strongly affect the thermal stratification and physical composition of the oceans, and therefore play a fundamental role in the physics of large-scale oceanic currents.

"The above discussion has not included the exchange of mechanical energy. This exchange occurs as a result of atmospheric winds and pressures, which exert forces on the moving water surfaces. Despite the importance of this energy flux in generating and maintaining ocean currents and waves, it is many times smaller than the exchange of energy due to radiation or the evaporation of water.

"The energy used to generate and maintain ocean currents and waves flows along an indirect path from sun to ocean, from ocean to atmosphere, and from atmosphere back to ocean. In this process, the air-sea interface is crossed three times."[20]

Present Energy System—Qualitative Inputs and Outputs

Inputs needed for the energy transformations of the energy system:

1. Know-how
2. Human labor
3. Naturally occurring energy transformations
   (a) direct solar
   (b) indirect solar
       (1) wind
       (2) waves
       (3) fossil fuels (coal, petroleum, natural gas, peat)
       (4) ocean currents
       (5) temperature differentials
   (c) Gravity; tidal
   (d) Earth heat
   (e) Atomic fission

The energy system uses energy to:

1. Produce and process food for humanity
2. Change form of energy, i.e., oil to electricity
3. Extract and/or process raw materials for:
   (a) fuel/power for itself (energy system) to extract more fuel and power
   (b) fuel for humans
   (c) fuel/power for manufacturing subsystems
4. Manufacture tools and equipment for its own and/or human use
   (a) environmental controls
   (b) communication
   (c) education
   (d) health/medical care
   (e) recreation
5. Transport humans
6. Transport materials

Outputs from the energy transformations of the energy system:

1. Increased know-how
2. Spent human labor (tired people)
3. Returns to natural energy processes and flows:
   (a) heat
   (b) molecular transformations (i.e., coal to heat, smoke, and ash)
   (c) atomic transformations and waste by-products
   (d) spatial relocations
4. Products and services for human life support

Distribution of Energy for Typical Industrialized Region: 1977

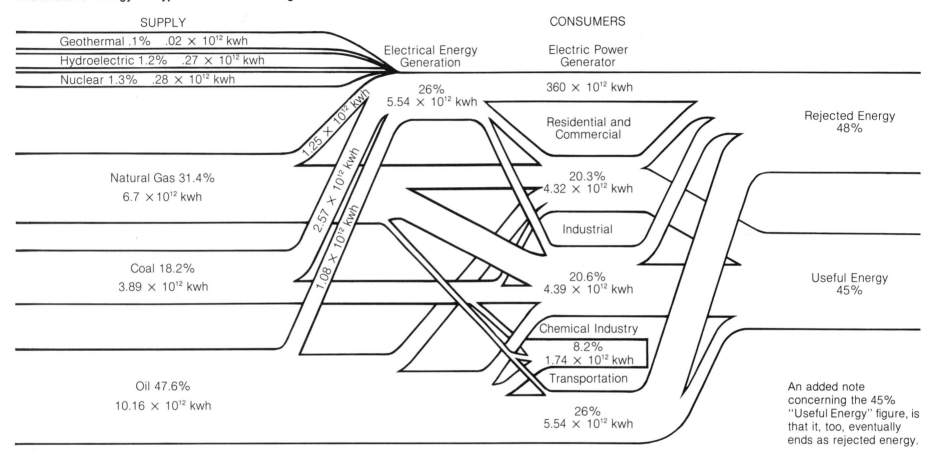

A subtle but fundamental distortion is found on charts like the one above. Since energy is measured in a time frame, energy use is not to be considered as static. The "single frame" of charts and graphs distorts the true nature of the process of transformation which is a fundamental characteristic of all energy use. A single frame will not tell you a caterpillar will become a butterfly, and a single frame of that butterfly will not tell you that the butterfly can fly. "Input/Output" are machine-type models or formulations. The sun, from which almost all of our energy comes, shines continually; it is not a machine, it has no off state. In formulating energy problems as well as proposing alternatives it helps to view ourselves and our energy-using artifacts as part of a process or "field" rather than as part of a machine.

Existing Energy System—Constraints and Forcing Functions

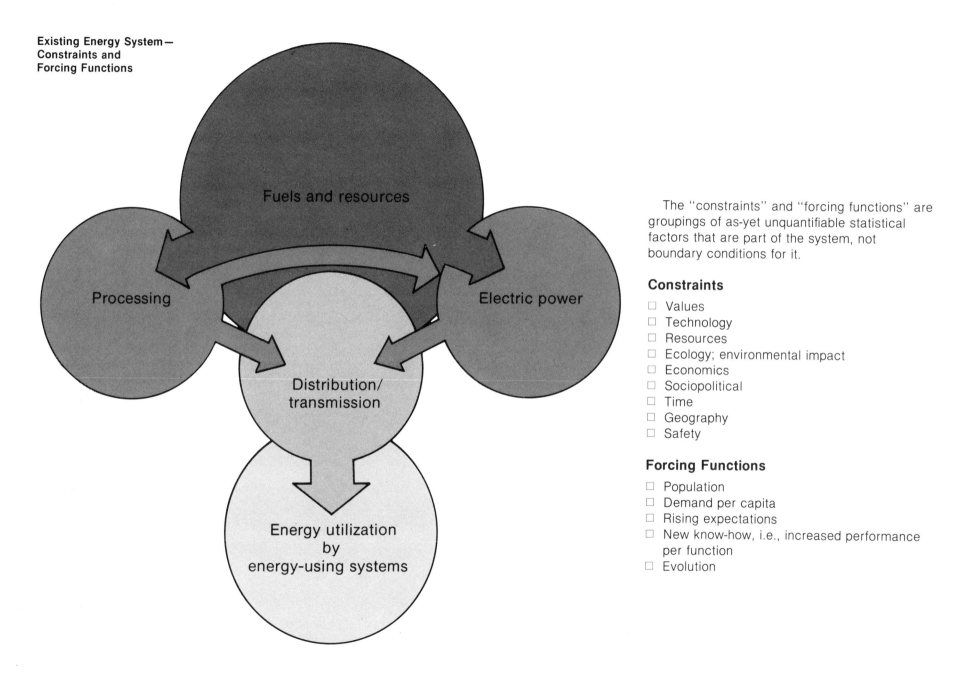

The "constraints" and "forcing functions" are groupings of as-yet unquantifiable statistical factors that are part of the system, not boundary conditions for it.

Constraints

- Values
- Technology
- Resources
- Ecology; environmental impact
- Economics
- Sociopolitical
- Time
- Geography
- Safety

Forcing Functions

- Population
- Demand per capita
- Rising expectations
- New know-how, i.e., increased performance per function
- Evolution

World Energy Production[14]

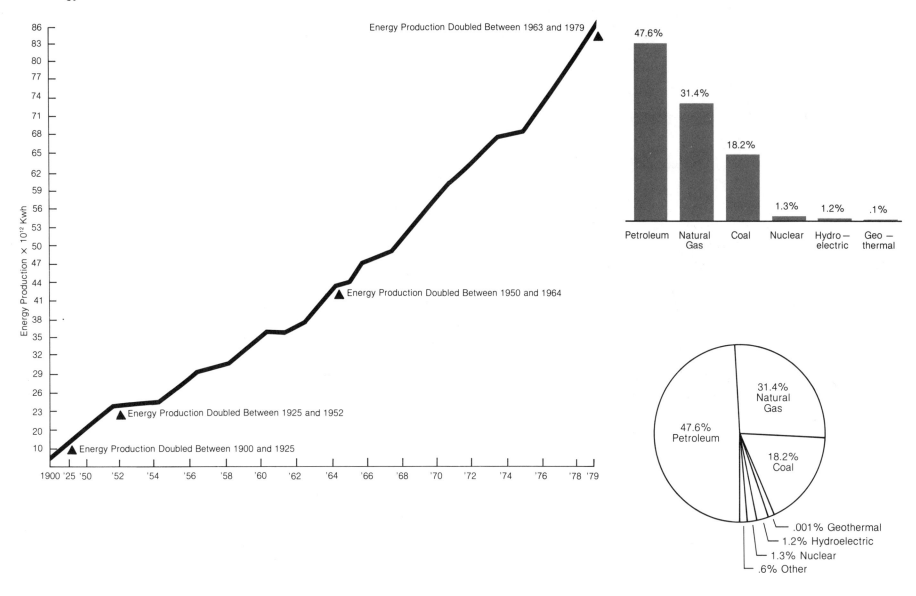

Energy consumption in Developed Industrialized Region[21]

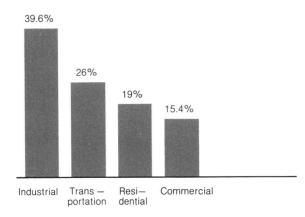

Primary Energy Consumption in Developed Industrialized Region, Broken Down to Show Electric Production[21]

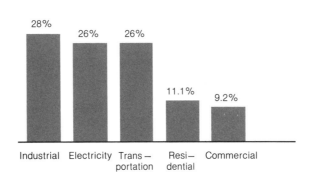

Distribution of Total Energy[21]

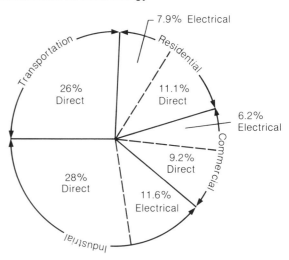

A-D percentages are of the entire energy consumption system.

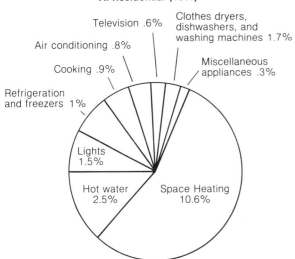

A. Residential (18%)

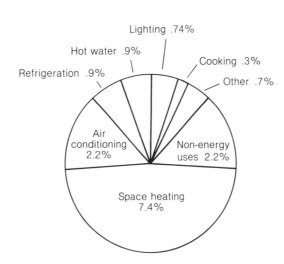

B. Commercial (15.4%)

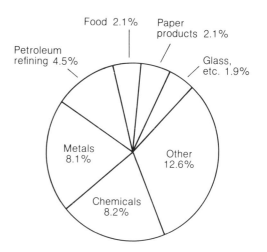

C. Industrial (39.6%)

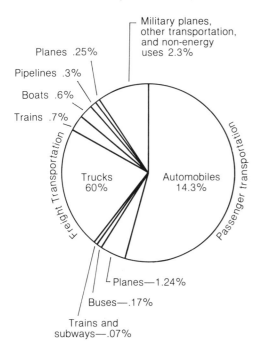

D. Transportation (26%)

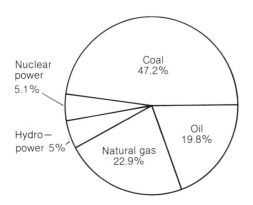

E. Electric Power Generation.[21] Distribution by amount of energy consumed in generation of electricity.

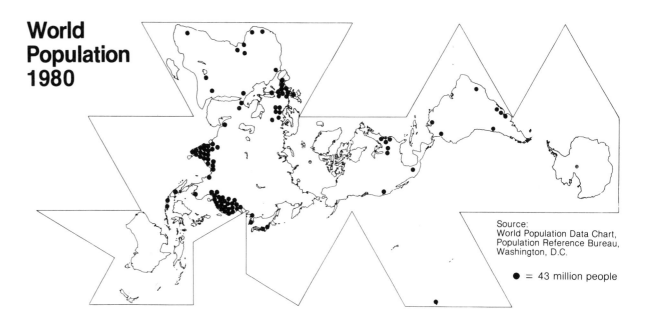

World Population 1980

Source:
World Population Data Chart,
Population Reference Bureau,
Washington, D.C.

● = 43 million people

Energy consumption figures broken down by geographical distribution can be misleading. For instance, due to the vast quantities of food, material goods, and services that are exported from North America for use by less energy-intensive regions, not all the energy utilized in North America is for North Americans. Any indictment of North American overconsumption should be tempered (but not eliminated) by this fact. (An interesting point in regard to energy consumption is that the highest per capita energy consumption figure is not that of North America, but of Antarctica. The figures are not even close: the per capita figure for the Antarctic researcher is the equivalent of 250 100-watt light bulbs, over twice the figure for North America.

Per Capita Energy Use

Source:
World Energy Supplies, 1972–1976, U.N., New York, 1977.

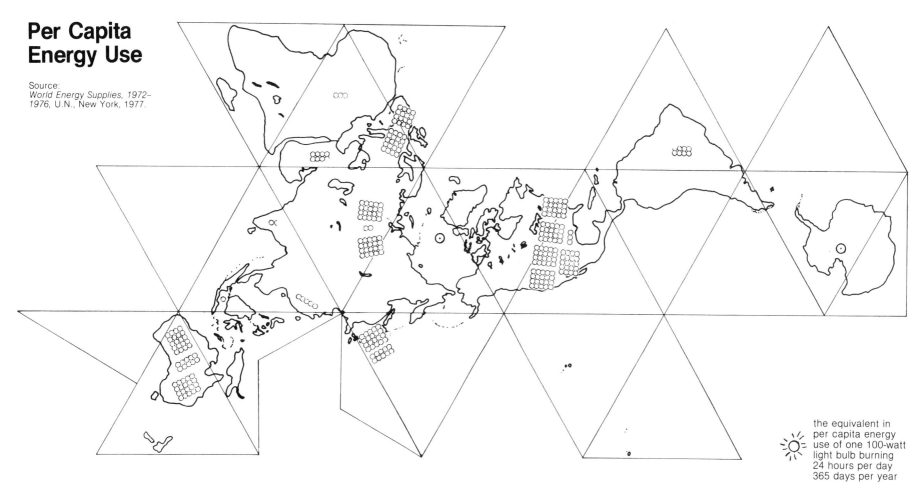

☀ the equivalent in per capita energy use of one 100-watt light bulb burning 24 hours per day 365 days per year

The amount of energy used by an individual's internal metabolics is roughly equivalent to that used by a 100-watt light bulb—this amounts to 800 kwh per year. The total amount of energy used by humanity in 1979 to power its externalized life-support system is about 83 trillion kwh (283×10^{15} BTU) annually. Divided evenly, this is about 19,000 kwh per person on Earth each year, which is roughly twenty 100-watt light bulbs burning 24 hours per day 365 days per year. The twenty light bulbs shown on the right, then, represent the average for each human being's annual share of all the energy used by all human-made energy-using devices.

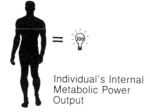

Individual's Internal Metabolic Power Output

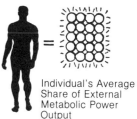

Individual's Average Share of External Metabolic Power Output

World Electricity Production

Source:
World Energy Supplies 1972–1976 (United Nations, 1978).

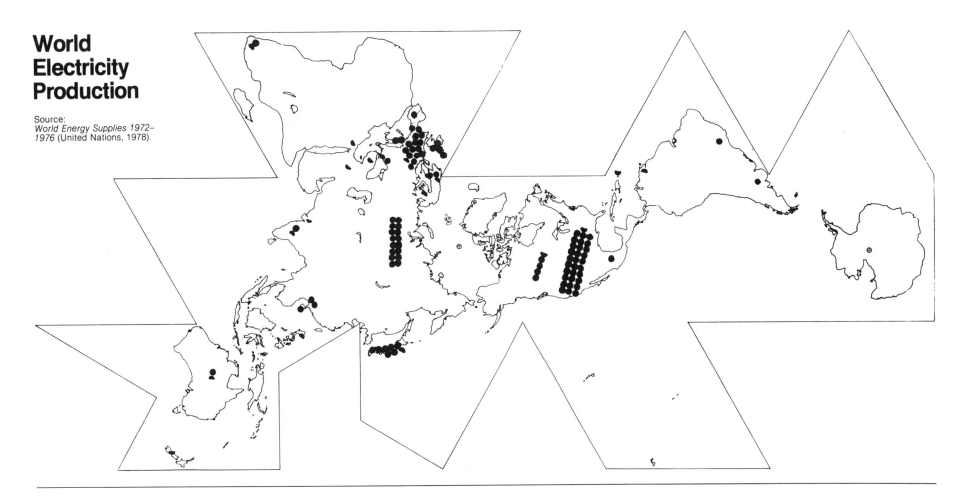

● = 1% of World Electricity Production
● = 69.2 × 10⁷ million kilowatt hours

The previous pages illustrate how the Earth and its life within the biosphere receives and utilizes energy, its flows, the uses of energy by humanity, the inputs needed to harness energy use, where the energy for humanity's life-support processes comes from, how it is allocated, and some of the constraints on these allocations. It is descriptive of the present state. The following pages help document the disadvantages of an energy system organized, as is presently the case on Earth, around the burning of the fossil fuels coal, petroleum, and natural gas, and a look at the dangers present in switching to nuclear power.

The Liabilities

From a whole system's perspective, it is hard to imagine anything more abysmally wasteful and ignorant than the total destruction of a hundred-million-year-old complex hydrocarbon. Considering how long it took to produce, burning a complex hydrocarbon is analogous to dissolving a pearl in vinegar. Anyone astounded at the corpulent decadence of Shakespeare's Cleopatra dropping a string of pearls into a glass of wine so that Caesar will have a more expensive drink ought to examine more closely how he or she gets his electrical energy or travels about in a car.

In addition to the waste inherent in burning fossil fuels when they could be used instead for recyclable raw materials, their combustion has considerable negative impact on the environment. The non-CO_2 chemical emissions from fossil-fuel combustion came close to 150 million tons in 1970 for the United States alone.[22] (See chart.) These emissions are composed of alchemydes, carbon monoxide,

---

Knowing that a ton of SO_2 or a curie of plutonium is emitted into the environment is a long way from having a measure of the harm to human health or ecosystem that may result.[25]

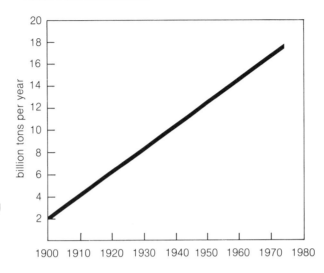

World CO_2 Emissions per Year from Fossil Fuel Combustion[8]

particulate matter (such as soot and smoke), hydrocarbons, nitrogen oxides, and sulfur dioxide. CO_2 emissions are particularly troublesome because they pose the greatest threat to global climate. Every ton of fossil fuel that is burned releases about two to three tons of CO_2 into the atmosphere.[23] So far, humanity's combustion of coal, oil, and gas has raised the CO_2 content of the atmosphere by 13%; 5% in the last five years.[24] About 125 billion tons of carbon have been introduced into the atmosphere in the last 110 years.[26] If fossil fuels continue to be exploited at increasing rates, as current plans project, then global average temperatures could rise as much as 6°C (11°F) with even higher rises in higher latitudes, resulting in vast effects on polar caps, the oceans, agriculture, plant photosynthesis, and life in the very populous arid and semi-arid regions of the planet. Existing technology for emission control of hydrocarbon pollutants is expensive. Some programs for the control of industrial air pollution have been legislatively postponed to cut power costs during fossil fuel shortages.

Thermal pollution of streams, lakes, the ocean, and atmosphere is caused by the discharge of waste heat from industry and power plants. Use of naturally occurring income energy sources, such as winds and tides, would not add to the Earth's native heat budget.

The danger of oil spills in the shipping and offshore mining of fossil fuels and the resulting slaughter of marine life and wildfowl, as well as contamination of the beaches, is a recurrent negative environmental impact, appalling in its effect on wildlife and perhaps on the life of the seas themselves.

Liquid natural gas (LNG) involves high risks to safety in being transported, and may cause vapor clouds, fire, or flameless explosions.

Pollutants Released into the Air in 1970 in Millions of Tons per Year[22]

Source	Carbon monoxide (CO)	Particulates	Sulfur oxides (SO_X)	Unburned hydro-carbons (HC)	Nitrogen oxides (NO_X)
Transportation	111.0	0.7	1.0	19.5	11.7
Fuel combustion in static sources, power generation	.8	6.8	26.5	.6	10.0
Industrial processes	11.4	13.1	6.0	5.5	.2
Solid-waste disposal	7.2	1.4	.1	2.0	.4
Miscellaneous	16.8	3.4	.3	7.1	.4
1970 totals	147.2	25.4	33.9	34.7	22.7
Totals in 1940	85	27	22	19	7

Air Pollution: Sources of Pollution and Means of Control[27]

Pollutant	Produced by:	Important Sources	Control Methods
SMOKE	Combustion of any fuel with high-ash content	Coal and oil-fired power plant using high-ash fuel Open fires burning coal	Use of low-ash fuel Electrostatic precipitators on stacks Use of smokeless solid fuels (manufactured)
	Incomplete combustion of hydrocarbon fuel	Diesel engines Refuse incineration	Careful maintenance and adjustment After burners to complete combustion Filters and precipitators on stacks
DUST	Handling of solid fuel	Coal mining Coal processing	Water spraying Precipitation and/or filtration
HYDROGEN SULPHIDE AND OXIDES OF SULPHUR SO_x	Burning of any fuel containing sulphur SO_x	Coal and oil-fired power plant using sulphur-containing fuel	Use of low-sulphur fuel Desulphurizing fuel before use Dispersion by use of high stacks Catalytic oxidation or reduction Absorption in dolomite or limestone Wet scrubbing of stack gases
		Domestic and industrial heaters and boilers Petroleum refineries	Use of low-sulphur or desulphured fuel Desulphurization prior to combustion or cracking Chemical absorption Catalytic oxidation or reduction
		Refuse incineration	Chemical absorption or wet scrubbing of flue gases
HYDROGEN SULPHIDE AND MERCAPTANS	Release of natural sources H_2S and SO_x	Geothermal energy plants	Control of and absorption of sulphur oxides from exit gases
	Processing of sulphur-containing hydrocarbons	Petroleum processing	Wet scrubbers Catalytic oxidation or reduction
OXIDES OF NITROGEN NO_x	Burning of any fuel in air at high temperature	Internal combustion engines	Lowering of flame temperatures Restricting of oxygen supply by enriching fuel/air mixture After burning of exhaust gases in combustion chamber Catalytic re-conversion to N_2 and O_2
		Fossil fuel power plants Petroleum processing	Use of two-stage combustion Exhaust gas recycling Lowering of flame temperatures in crackers
CARBON MONOXIDE CO	Burning of carbonaceous fuel with restricted oxygen supply	Internal combustion engines	Maintenance of correct air/fuel ratio Design to ensure complete combustion Catalytic oxidation of exhaust gases
		Fossil fuel burners	Design to ensure complete combustion Operate at correct air/fuel ratio
ORGANIC COMPOUNDS HYDROCARBONS	Evaporation/losses of hydrocarbon fuels	Evaporative losses from fuel tanks	Use of unvented tanks with activated carbon vapour adsorption
		Crankcase 'blowby' in piston engines Petroleum processing	Vent crankcase to inlet manifold Careful maintenance of seals and valves Absorption/adsorption of off gases Cryogenic removal from exit streams
ORGANIC COMPOUNDS HYDROCARBONS	Incomplete combustion of hydrocarbon fuels	Internal combustion engines	Combustion chamber design to ensure complete combustion Use of weak fuel/air mixture Catalytic converters to oxidize exhaust gases
ORGANIC COMPOUNDS ALDEHYDES		Internal combustion engines	Maintenance of correct air/fuel ratio Catalytic oxidation of exhaust gases
BENZO PYRENE		Internal combustion engines	Use of gasolines low in aromatic content
LEAD (in particulate form)	Burning of fuels containing lead-based anti-knock additives	Internal combustion (spark ignition) engines	Filtration of exhaust gases Use of low lead fuel (with high aromatic content Use of lower compression ratios
		Refuse incineration	Efficient precipitation of fly/ash
HYDROGEN CHLORIDE	Burning of Hydrocarbon fuel containing chlorine	Refuse incineration (source of chlorine mainly PVC)	Wet scrubbing of exit gases

Pollutant	Produced by:	Important Sources	Control Methods
RADIOACTIVE NUCLIDES Kr-85 Kr-87 Kr-88 Xe-133 Xe-135 Xe-138 H-3	Fission products of radioactive decay	Leaking fuel rods in nuclear reactors	Rapid detection and removal of leaking fuel rods (gas cooled reactors) Adsorption of emissions on activated carbon to delay release beyond several half-lives of decay products Detention of gaseous emission products in gas storage (pressurized water reactors)
Kr-85 H-3		Fuel reprocessing	Cryogenic removal Adsorption in activated carbon
Ar-41	Product of neutron activation of argon in air	Nuclear reactors using air cooling in high-neutron areas	Cryogenic removal Adsorption in activated carbon
Ra-226 Ra-228 U-238 Th-230 Th-232 K-40	Burning of fossil fuels	Fly-ash from fossil-fuelled power plants	Efficient precipitation of fly-ash Dilution of effluent by use of tall stacks
HEAT	Use of any non-ambient energy source	Fossil-fuelled and nuclear power stations Geothermal power plants	Dispersion of power plant reduces local effects of thermal pollution Reduction in waste heat output in energy conversions Effective mixing of waste heat streams reduces ecological effects at a local level

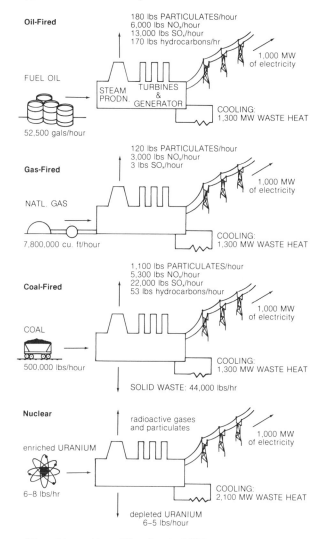

Fuel inputs and emissions from 1,000 MW conventional electric power stations

Oil-Fired
FUEL OIL
52,500 gals/hour
180 lbs PARTICULATES/hour
6,000 lbs NO$_x$/hour
13,000 lbs SO$_x$/hour
170 lbs hydrocarbons/hr
STEAM PRODN.
TURBINES & GENERATOR
1,000 MW of electricity
COOLING: 1,300 MW WASTE HEAT

Gas-Fired
NATL. GAS
7,800,000 cu. ft/hour
120 lbs PARTICULATES/hour
3,000 lbs NO$_x$/hour
3 lbs SO$_x$/hour
1,000 MW of electricity
COOLING: 1,300 MW WASTE HEAT

Coal-Fired
COAL
500,000 lbs/hour
1,100 lbs PARTICULATES/hour
5,300 lbs NO$_x$/hour
22,000 lbs SO$_x$/hour
53 lbs hydrocarbons/hour
1,000 MW of electricity
COOLING: 1,300 MW WASTE HEAT
SOLID WASTE: 44,000 lbs/hr

Nuclear
enriched URANIUM
6-8 lbs/hr
radioactive gases and particulates
1,000 MW of electricity
COOLING: 2,100 MW WASTE HEAT
depleted URANIUM
6-5 lbs/hour

SO$_x$: sulphur oxides   NO$_x$: nitrogen oxides

Source: "The Energy Crisis," AAAS Symposium, Philadelphia, December 1971

Strip mining of coal or oil shale damages the environment, produces soil-waste problems and acid drainage, and leaves visibly ugly and agriculturally unproductive terrain. To date, restoration of such land is at best only partial and is very expensive. Deep mining of coal still engenders a dangerous and unhealthy occupational environment for human beings.

Oil-shale retorting to produce synthetic liquid fuel has been much publicized by the big oil companies as a viable strategy. This process produces a high salt concentration in its waste water, which would have a negative impact on water bodies into which it would be released. Saltwater drainage would render the soil it touches sterile and unfit for agriculture. The oil shale would be stripmined, if exploited, *increasing* the volume of the original rock material by 50% in the retorting process, and places would have to be found to dump the overburden and mine tailings. An average retorting plant would generate over 40 acres of spent shale, spread 1 inch deep, *per day*. Revegetating and controlling leaching of these spent shale deposit sites would be difficult to achieve.

The depleting stock of fossil resources is needed as raw material for manufactured industrial products. These nonfuel uses for our fossil resources include: (from oil) waxes, paving, naphthas, medicines, lubricants; (from coal) fibers, insecticides, margarine, tar, dyes, drugs, nylon, perlon, glue, cosmetics; (from both oil and coal) fertilizers, and plastics.

It seems clear that if other sources of fuel are available that meet the criteria of availability, low environmental impact, safety, and efficiency, then coal and oil should be conserved for use for needed industrial raw materials.

Nuclear power has unprecedented dangers associated with its use, whether limited or widespread. Humanity has never before faced such a challenge to its foresight, design competency, or humility in admitting colossal mistakes. The evidence so far indicates we have yet to deal adequately with these challenges. Reactor safety, spent-fuel transport, long-term (centuries or more) storage of high-level radioactive wastes, recycling of nuclear facilities (e.g., decommissioning technologically obsolete nuclear power plants), public health implications of a plutonium energy economy, social well-being issues of a centralized power structure, diversion of nuclear materials to weapons, and sabotage are all deadly serious problems without safe or even adequate solutions. The 861 "abnormal occurrences" at operating nuclear plants in 1973 alone—about half of which had direct or potential safety significance,[28] the fact that there is no long-term storage plan for radioactive wastes, the fact that nuclear reactors and fuel reprocessing plants are not designed to be decommissioned, the fact that plutonium is one of the most toxic substances known (20,000 times more toxic than cobra venom), and the fact that a single individual with 20-30 pounds of plutonium could build himself or herself a crude atom bomb from studying unclassified publications—all attest to our utter failure to maintain necessary and reliable levels of quality control, foresight, design competency, and even ethical considerations.

Besides considerations of the relatively local objectionable biospheric and human conditions

Projected Nonenergy Consumption of Fossil Resources[29]

Year	Bituminous coal and lignite		Petroleum		Natural gas		Total		
	Non-energy use (b)	% of total coal	Non-energy use (b)	% of total petroleum	Non-energy use (b)	% of total natural gas	Non-energy use (b)	% of total fossil fuels	% of total energy (c)
1973(a)	0.14	1.1	3.40	10.5	0.69	2.9	4.30	6.0	5.7
1975	0.15	1.1	3.80	10.8	0.70	2.8	4.65	6.3	5.8
1980	0.20	1.2	4.46	10.6	0.75	2.8	5.41	6.3	5.6
1985	0.33	1.5	4.85	9.6	0.80	2.8	6.35	6.3	5.4
2000	1.40	4.5	8.44	11.8	0.90	2.6	10.74	7.8	5.6

(a) Extrapolated from 1971 data and projected 1975 data
(b) 10^{15} BTU Equivalent
(c) Total energy used by U.S. all sources

which the burning of fossil and nuclear fuels presents, there is also another view—a longer, or cosmic, view. Present-day economic accounting deems these fuels to be "free" for the taking. "Cosmic accounting" holds them to be worth considerably more. If humanity wanted to build its own fossil fuels, it would have to spend more than $1 million per gallon of petroleum to duplicate the enormous pressures over the eons of time that nature "spent" to build up the fossil fuels. From this perspective, each commuter is in the ludicrous position of spending a few million dollars every day to make about $50. The Earth's fossil fuels are like a battery in a car; they are the "self-starter" of our external metabolic system. The battery cannot drive the car cross-country or even cross-town, but it can start the main engine. It is time for humanity on spaceship Earth to switch from its storage batteries to its nondepletable main engines.

The short-range or narrow view does not lead to long-term survival. The long-range view is mandatory for a system seeking to evolve. Cosmic accounting is the long view; our energy transactions must be based on this viewpoint if they are to be regenerative.

Lifetimes of World Fossil Fuel and Nuclear Fuel Supplies (at Current Mineable Grade and Rates of Consumption)[30]

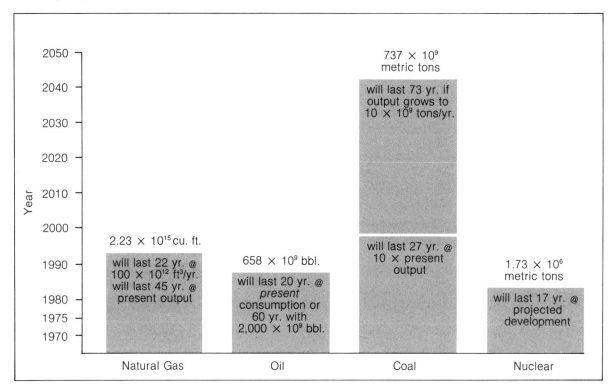

Trends, Assumptions, and Key Indicators

Trends assumed most likely to continue through 1985 and in high probability to 2000:

- Overall population will continue to increase.
- Rate of population growth will continue to decrease in developed regions and more and more developed regions will arrive at net zero population growth.
- Per capita energy demand and use in developed regions will not continue to rise, but reach a plateau before the year 2000.
- Production of goods and services will continue to increase, with a growing emphasis on quality.
- Energy conversion inventions will continue to proliferate.
- Fossil fuel reserves will continue to diminish and costs continue to rise.
- Lag between invention and implementation of energy technology will continue to decrease.
- Technology will continue to do more with less; energy input per unit of output will continue to decline.
- State-of-the-art energy conversion efficiency will be used.
- Existing energy converters will be replaced by more efficient energy converters as they become obsolete and/or worn out.
- Large power plants will continue to exist.
- Income energy sources will be increasingly used.
- Discovery of new types and sources of energy and power conversion will continue to increase.
- Impact of energy system on the ecological context will continue.
- World mobility and use of energy for transportation will continue.
- World cooperation and globalization of institutions in regard to energy will increase.
- Coordination and standardization of tools and contact (end use) products regarding energy will increase.
- We will learn more and more about energy needs and uses.

Other Assumptions

- Total energy and per capita energy consumption cannot continue to increase forever; energy use will reach a steady state in this century or shortly thereafter.
- Viable alternatives must be technologically feasible before they are economically feasible.

When planning in and for complex systems the desired effects of one's plan need to be explicitly stated, and stated in a time frame, i.e., what should happen when. These planned-and hoped-for results must be constantly compared with actual developments. Actual developments are monitored through indicators much in the same way the medical doctor monitors the indicators of pulse, blood pressure, and reflexes to measure the health of the individual patient. "Key" indicators are correlated with a set of other variables, thus giving an accurate assessment of the functioning of the entire system. The indicators need to be monitored as a set, not alone. As in internal metabolics, pulse might be normal, but the patient might have cancer.

Key Indicators

Key indicators for the health of collective humanity from an energy-use perspective are:

- Total population and rate of growth.
- Energy use per capita per quality of life.
- Fossil-fuel consumption and remaining supply.
- Energy conversion efficiencies and their ratio of change.
- Impact of energy system on ecological context, e.g., thermal burden, CO_2 content of atmosphere.
- Amount of energy sources feeding energy system.
- Amount of human labor involved in energy system.
- Number of humans involved in the decision-making of the energy system.
- Amount of integration of total energy system.
- Life expectancy, level of education, leisure time.

Decision-Making Criteria/Economic Accounting Systems

"The decision-makers in the production of power are many and diverse. Each has its own view of the objectives to be pursued, its own formulation of the issues, its own choice of the criteria for decision and the priorities among them, its own selection and analysis of pertinent facts and its own art of applying the criteria to the issues on the basis of the facts in pursuit of the objectives.

"Who are the decision-makers? In the first instance they are the enterprises—private companies, local government agencies, and such national instrumentalities as the Tennessee Valley Authority—that build and manage the power-generating facilities. The initiating enterprises draw others into the decision-making process, notably their sources of capital and commercial finance (investment and commercial banks for private companies, budget offices and appropriations committees for government agencies) and the state and national regulatory agencies that issue certificates or permits required by law or administer legal measures for safety, health or environmental protection. Often the courts, state and federal, are drawn into the decisional process. Their authority can be invoked through statutory procedures to review the orders of regulatory agencies or through independent proceedings initiated by complainants. Still other decision-makers participate in a more general and diffuse but nonetheless critical way: Congress, legislative committees, and the President; the legislatures and governors of the states; scientific, engineering, legal and other academic or professional organizations or groups, and the general public, which operationally tends to mean active citizen groups and powerful individuals. At times these officials and groups are directly articulated into the decision-making process, but even when they are not, they influence the process by modifying the societal medium in which it takes place."[31] In centrally planned economies the decision-making is done not by the commercial enterprises but by the state and bureaucracy.

Decision-making criteria are guidelines for reaching a goal or preferred state, as well as a) values that formulate a preferred state made more explicit; b) success criteria; c) general performance specifications for the designer; and d) the framework for an accounting system. More often than not most people's and institution's decision-making criteria are not stated or even known. Nevertheless, even if unconscious, they exist. There are a number of advantages to stating them explicitly, not the least of which is that they are then subject to review by all those who will be affected by decisions resulting from their employment. Another advantage in stating decision-making criteria explicitly is that it helps to clarify the values of those doing the stating.

Every person and institution has their own values, goals, and procedures—be they conscious or unconscious—and hence decision-making criteria. Many or most of the values are commonly shared with other peoples and institutions but there are often unique orientations, emphases, and values that distinguish one individual or group from another. In addition, there are some decision-making criteria (and values) imposed upon everyone and every institution by the social structure the individual or institution is within. Some of the most powerful of these social-structure forces are found in the prevailing economic accounting system.

If an individual or institution is to survive in a specific economic structure then the rules of that structure that determine survival apply, no matter what the values of the individual or group may be. An accounting system is basically a value system; one counts what one values. One values what experience has taught favors survival. Because humanity learns more and more as time goes on, what has survival value increases in scope. However, the present-day economic accounting system has not similarly expanded its horizons but is "frozen" into an economic system by non-adaptive laws, rules, regulations and customs. The present-day accounting system was formulated, evolved,

---

Present-day economics counts money; and "Money devalues what it cannot measure."[32]

> *Only economists still put the cart before the horse by claiming that the growing turmoil of mankind can be eliminated if prices are right. The truth is that only if our values are right will prices also be so.*[33]

and codified before the value of the environment was understood, before the value of our limited fossil fuels and resources were known, before the value of flexibility and adaptability were known, and before we knew much about the larger energy systems of which the Earth is a part. The expansion of what humanity has learned to value and the subsequent conflict and forced incorporation of these values into the existing economic accounting system is one of the causes of our economic upheavals.

The following decision making criteria were used in formulating the energy development strategies found in *Energy, Earth and Everyone*. As such they form the value basis for a different accounting system than the present-day economic system, which could be aptly characterized as being a "Consumption Accounting System": who consumes what, where, and how much and how often. A "Regenerative Accounting System" would more accurately reflect the values in the decision-making criteria listed in the following pages.

Energy Criteria

Maximum value placed on doing the most with the least amount of energy.

Maximum use of reusable materials and packaging.

Maximum diversification and interdependence of energy sources.

Maximum availability and distribution of usable power.

Maximum concentration of energy-intensive activites.

Maximum interlinkage of energy-intensive activities.

Maximum use of low-impact decentralized energy-harnessing artifacts.

Maximum energy conversion and transport efficiency use.

Minimum use of energy-intensive materials.

Minimum use of nonreusable materials and packaging.

Minimum energy use in construction, maintenance, and recycling.

Minimum dependence on one source of energy.

Minimum heat discharge into environment.

Ecological Context Criteria

Maximum value placed on virgin areas of globe.

Minimum topographical, geological, hydrological, physiographical, limnological, meteorological, soil, vegetation, and wildlife disturbance.

Minimum use of land, water, water space, air, and air space.

Minimum input of solid, liquid, gaseous, and heat waste into ecological context.

User Criteria

Maximum value placed on 100% of humanity as energy user: sufficiency—enough energy for everyone; accessibility—enough distribution to everyone.

Maximum quality control of energy artifact, system, or service.

Maximum reliability of energy artifact, system or service; maximum use of back-up systems.

Maximum durability of energy artifact, system, or service.

Maximum ease, simplicity, and clarity of use of energy artifact, system, or service.

Maximum stability and consistency of output, artifact, system, or service.

Maximum cultural, esthetic, and individual human option diversity.

Maximum value placed on user's time and energy.

Maximum comprehensive responsibility and responsiveness to the needs of energy user by energy supplier.

Safety Criteria

Maximum value placed on human life.

Maximum safety in construction, operation, maintenance, and recycling.

Maximum designed-in safety for emergencies and breakdowns.

Maximum safety for future generations.

Adaptability Criteria

Maximum value placed on adaptive stability.

Maximum responsiveness to short-term energy demand changes.

Maximum expandability/contractibility (responsiveness to long-term energy demand changes).

Maximum reserves of emergency supplies and facilities.

Maximum flexibility and adaptability to new technology, needs, and know-how.

Efficiency Criteria

Maximum value placed on doing more with less.

Maximum use of minimum number of energy artifacts, systems, and services.

Maximum energy output per invested man-hours, materials, and energy input.

Maximum ease, simplicity, and clarity of repair, replacement, and recycling in minimal time.

Maximum use of modularity of construction where applicable.

Organizational Criteria

Maximum compatibility between parts of the energy system, as in the interfaces between public and private utilities.

Maximum compatibility or standardization of all similar parts of the energy system.

Maximum compatibility between different energy systems, as autonomous energy units designed for use by single families, schools, health units, etc.

Maximum decentralization of information and decision flow.

Maximum use of feedback.

Maximum indexing and cataloging of energy systems, parts, services, and outputs.

Maximum knowledge about energy system interactions with all other systems, especially the ecological context.

Maximum centralization of coordination functions, maximum decentralization of decision-making functions.

A regenerative accounting system that would incorporate the above criteria would need to have at least the following seven interconnecting and overlapping aspects: 1) Life Cycle Costing, 2) Ephemeralized Replacement, 3) Conservation, 4) Comprehensive Resource Accounting, 5) Ecological Accounting, 6) Production Accounting, and 7) Sociocultural Accounting. What is meant by these terms follows:

Life Cycle Costing takes into account the cost of the energy, materials, and time needed to produce an artifact, the energy needed to run and maintain it for its entire life time (including increasing inefficiencies and repair as it wears out), the energy needed to dismantle and recycle it (if possible) and the energy needed to recover, control, and/or recycle wastes. Also included would be credit for reused or scrapped materials when the artifact is recycled.

Ephemeralized Replacement refers to "doing more with less." In this context it refers to the process of redesigning artifacts and processes to perform a given function or series of functions, or a broader scope of functions, with greater efficiency by using less materials and energy. As an artifact reaches the end of its useful life or becomes glaringly obsolete, it is replaced by an "improved" (one that does more with less) model (if such improvements have been developed). Ephemeralization may be achieved by the design or redesign through invention, miniaturization, consolidation, or rearrangement of an artifact, process, or its use. In regenerative accounting this evolution is to be encouraged rather than resisted. Examples of ephemeralized replacement include the 1/4 ton communications satellite that outperforms the 175,000 ton transoceanic transmission cable; the one-pound pocket calculator that outperforms the 1950's three-ton computer; the five-ton geodesic dome that outperforms the twenty-five-ton house, and the contemporary electric turbine that produces a kilowatt-hour of electricity with less than a pound of coal while the turbine of 1900 took about seven pounds of coal to produce one kwh.

As it is used here, *Conservation* is primarily a term that describes the retro-fitting of already existing artifacts to make them more efficient. In contra-distinction, ephemeralization is the design of new artifacts or processes. Because we learn more every time we use our know-how to organize resources into technology to meet human needs, we are able to continually (so far) "do it better" the next time. Hence, there will always be a standing backlog of already existing artifacts that might be amendable to modification that will make them more efficient. "Conservation," which means "to preserve from waste," is the process of bringing older artifacts up to the efficiency performance of the latest artifacts or processes. It may take several forms. For example, insulation of existing buildings is an add-on to an existing artifact that increases its energy-use efficiency; car/van pooling is a rearrangement of the use-patterns of the existing commuter transportation network that would also save energy. The elimination of commuting to work through a major redesign of the city so that people would walk or bicycle to work would be an example of ephemeralized replacement. Ephemeralization primarily involves structural alterations, while conservation deals with content.

Comprehensive Resource Accounting refers to the accounting of both energy and materials and their flows and cycles. It would take account of how much energy and materials the final product or service uses in terms of its entire development; e.g., how much energy and materials does it take to produce a nuclear power plant or skyscraper or windmill?; how much energy does it take to mine and refine the metals and other materials involved?; how much energy does it take to actually build the artifact in question?

On the finite globe it is possible to go only so far before one begins running in circles.

The entire system's efficiency would be measured, not just each part in isolation. Energy system efficiency is a measure of the amount of actually useful work performed for an entire system rather than one facet of it. For example, automobile engine mechanical efficiency (considering losses of useful work done on fan, water pumps, and generator) averages 71%. However, the energy system efficiency of crude petroleum to actual useful work in the transportation sector, taking into account losses from oil extraction, refining and transportation, the thermal efficiency of the engine, and the above-mentioned mechanical losses, plus the "rolling" losses of the transmission, rear axle, and tire deformation, is only 5%. "The net result is that 95% or more of the energy contained in the original amount of crude petroleum is used not to move the car along the road, but to extract and refine the crude oil; to

Energy-System Efficiency[35]

	Efficiency of each step, in %	Efficiency, including all preceding steps, in %
Production of crude petroleum	96	96
Refining of petroleum	87	83–5
Transportation of petroleum	97	81
Thermal efficiency of engine	29	23–5
Mechanical efficiency of engine	71	16–7
Rolling efficiency	30	5

carry the gasoline to the filling station; to heat the water in the radiator and the gas in the exhaust; to operate automobile auxiliaries; and to overcome friction in gears and tires."[35] It should also be borne in mind that rolling efficiency depends, among other things, on the condition of the road and even on the weather. In addition, engine efficiency is by no means a given fixed property but is a function of its operating condition such as load and engine "tune-up." Until very recently motorists have not realized any increase in miles per gallon; indeed until 1975, average miles-per-gallon went down because the running of ever-more powerful engines under lighter partial loads and the hauling of more tons of automobile at higher average speeds had more than overshadowed any increase in engine (or other) efficiency. "For example, a car designed for a maximum speed of 100 miles per hour must have 3.5 times as much installed horsepower per ton as a car designed for a maximum speed of 50 miles per hour."[35] Measuring energy system efficiency leads to higher end-use efficiency because it points out the need for the appropriate matching of energy quality with energy use. Measuring whole system efficiency points out the efficiency losses along the way to the end-use.

Another reason for measuring whole system efficiency has to do with a seeming paradox which states that the behavior of whole systems exhibits behaviors not contained in any of the system's parts taken separately. For example, as a whole system, you can run or write a poem but your legs or hands by themselves can do neither. In terms of energy systems, this has enormous relevance because any one facet or sub-system of a whole energy system could have a very high efficiency, such as an electric water heater, but the whole system efficiency, from coal mine to power plant to electric grid to water heater to water to shower, is much lower. Correspondingly, one part of a whole energy system could have a very low efficiency, such as a 5% efficient solar collector, but counting the entire system's efficiency and thereby including the free fuel cost, makes this low efficiency either irrelevant, or, minimally, not a valid reason for rejecting such a system as "inefficient."

Ecological Accounting refers to the incorporation of costs (or damages) incurred by nature from human economic activities into the accounting system. Biological (and obviously human) activities and systems can be measured by their complexity. The more complex and integrated a whole system is, the more adaptive and stable it is. The more adaptive and stable a

---

***. . . the most revolutionary consciousness is to be found among the most ruthlessly exploited classes: animals, trees, water, air, grasses.*[36]**

system is, the higher are its survival capabilities. When a human economic activity loses some of its adaptability it also loses some of its economic survivability. Bankruptcy and no-adaptability are almost synonymous in an economic sense. Presently, when nature loses some of its adaptability no one pays; contrary to a popular saying, nature *has* been providing free lunches for quite a while. We cannot continue to take out loans from nature and then default. Either nature will go bankrupt and the whole economic system will crash catastrophically, or nature will send around her bill collectors—usually in the guise of ecological disasters—and we will pay the price (and debt service) of the costs we have incurred against nature.

Ecological accounting seeks to incorporate nature into the economic accounting system. When an activity decreases the complexity of nature (in other words, when something is "bought" from nature), the human activity doing the buying should pay. Ecological accounting refers to more than just the closing of the industrial valves that cause environmental pollution. Recycling waste is clearly an important part of any ecological accounting but of much wider significance is the idea of paying nature not just for damages incurred through pollution but for services and resources provided. For example, Buckminster Fuller points out that "Scientific calculation shows that the amount of time and energy invested by nature to produce one gallon of petroleum 'safety deposited' in subterranean oil wells,

---

Any attempt to solve a problem at the expense of the planet will create a larger problem later.

---

when calculated in foot-pounds of work and chemical time converted to kilowatt hours at the present commercial rates at which electricity is sold, amounts to approximately *one* million dollars per gallon of petroleum as cosmically developed prior to its discovery and exploitation by humans. When humans discovered petroleum, they wrongly assumed that it was absolutely free and belonged to the finder. Humans take into account only the cost of pumping, processing, and distributing the oil. Anyone should be able to sell a million dollars for fifty cents!" Prices should reflect the replacement value of a fuel (or environmental service) to society and the total energy it took (nature and humanity) to produce that fuel or service. If society had to provide the services that the environment provides it would be faced with the largest bill in the history of life on this planet. An immediate and strict implementation of the above would be pretty much impossible (as pointed out earlier, getting to work in a gasoline-powered automobile would be quite expensive) but the concept provides a frame of reference for the redesign of our present accounting system.

Production Accounting involves the comprehensive accounting of resources in the industrial process. The industrial system can be viewed as a complex biological process (an external metabolic system) wherein the wastes of one organism or process becomes the input of another. Industrial processes are viewed from the whole industrial system perspective rather than in isolation as discrete and complete-unto-themselves parts. Cogeneration, where electrical power is produced with steam that was formerly a wasted by-product of the industrial process and where the waste output

---

No human activity is "economically" viable at the time of its inception.

---

from one process, for instance sulfur removed from coal, becomes the raw material input for another industrial process, is an example of the results of more comprehensive production accounting.

Sociocultural Accounting refers to the accounting of the value of a particular culture or society and the damages that could result from a particular development strategy. Putting a highly-centralized, technocratic-elite-run, nuclear-weapon-producing, nuclear power plant in a developing country would have enormous sociocultural impacts. Power would be concentrated in fewer hands, the type of energy made available would be inappropriate to the needs and culture of the area, and the relative social well-being of the society as a whole would be diminished. Sociocultural accounting would attempt to deal with the more ethereal values associated with human well-being. It would deal with such questions as: Should a particular resource using system be built? Will this increase human and cultural well-being? Will this increase humanities collective wealth? or forward survival potential? or degrees of freedom?

A regenerative accounting system should embody the above concepts to operationalize the values in the preceding decision-making criteria. In essence, what is needed is an accounting system that is based on energy, materials, and satisfying human needs and realizing human potentials rather than on dollars and profit maximization.

Summary

Preferred States

The most important consideration in the successful treatment of any unhealthy patient is to have a good idea of what a healthy patient is. For a medical doctor to treat a patient for pneumonia, acne, or a broken arm without knowing what a human being "looks" like without pneumonia, acne, or a broken arm is ludicrous. For a nation-state, regional, or local system to devise an energy plan to treat its energy disorders without having explicit ideas of how it should function, or of what a healthy system is, is even more ludicrous. Not only do the nation-state and most other organizational systems of today have no clear idea of where they are going, or want to go, or can go, but they are largely unconscious of themselves as part of the larger, global system. In an interdependent system, such thinking can be fatal.

One useful technique for determining how a healthy societal life-support system should function, or could function, is to determine its "preferred state." The preferred state is where you want to go, a goal; and as such helps clarify the path to its attainment. A preferred state is actually a set of performance specifications and is designed by starting from scratch, that is, by designing the ideal system without regard to the existing system and its inherent political and economic constraints. The only constraints on the design of a preferred state are technological feasibility and ecological compatibility. After a preferred state of functioning has been defined, it is relatively easy to identify the things in the present energy system that could be changed or need to be developed or phased-out to bring about the preferred state. The opposite approach, that of starting with the present-day set of problems, limits possibilities and prejudices the problem-solver to deal with the difficulties of the system instead of its possibilities. Dealing with the preferred state and working back is dealing with the behavior of the whole system; dealing strictly with the problems of the present system (as most energy "plans" do) is dealing with parts.

The following fourteen points outline the qualities of one such "preferred state" for our global energy system. The items are all interrelated and partially overlapping. Together, they describe a holistic perspective of how our energy systems should be functioning and, as such, help clarify what could be done and what needs to be done to resolve our present energy difficulties. The "ideal" aspects of each point are not meant as concrete goals but rather as ephemeral directions which the systems involved could move towards, but, perhaps, never actually attain. It should also be borne in mind that a preferred state is an evolving definition of success and that what is today can and will in all likelihood be improved upon tomorrow. As a preferred state is approached, new vistas of what is desirable and possible will come into view.

A preferred state of functioning to the present global energy system is one in which there is:

1. *Regenerative availability* of enough energy for 100% of humanity's life support; ideally, there should be enough energy so that any system on the local, regional, or global scale can operate so as to service the needs of the people in the respective areas. Also, it is not satisfactory just to have "enough" energy; we need an energy supply that will be available for our grandchildren's children as well as ourselves. Ideally, our energy supply would be non-depletable and regenerative.

2. *Equitable distribution of energy;* again, it is not satisfactory to have just enough energy if that energy is not available to everyone, the conditions for its availability are unacceptable, impose an unjust economic strain, or it is used as a political weapon. Ideally, anyone anywhere would have equal access to the energy needed for their life support.

3. *High efficiency of energy use;* there should be a progressive increase in the amount of energy output or functions performed from a continuingly decreasing amount of energy input. Ideally, this "more-with-lessing" would continue so that what appears like "everything" would be accomplished with next to "nothing."

4. *Low or no negative environmental impact;* there should be the least possible impact on the air, water, and land quality by our energy system; ideally, our energy use would have a positive, beneficial effect upon the environment, like certain farming techniques in Japan and Europe have upon the lands that they are practiced upon.

5. *Complete safety* to energy producers, users, and future generations; energy systems should be resistant to accident and sabotage; ideally, human energy use would be as safe as nature's use of energy.

6. *Low or no use of depletable (fossil and nuclear) energy resources*, i.e., high use of non-depletable energy sources; ideally, all our energy use would be from our income, with our capital stocks safely deposited for future contingencies.

7. *High diversity, adaptability, resiliency, and flexibility* of energy sources and systems; energy systems should be able to absorb external perturbations without failing; and, if they do fail, they should fail "gracefully" rather than "catastrophically"; as well as be adaptable to local conditions, resources, and control; ideally, our energy system and energy use would increase the options available to the individual and society, as well as the energy systems themselves.

8. *Integrated, decentralized, preferably two-way energy systems with a high level of autonomy.* Ideally, all our energy systems would be interconnected yet able to be self-sufficient.

9. *Standardized outputs* of energy and energy using artifacts; ideally, any energy using artifact made anywhere would be able to be used anywhere.

10. *Low or no use of coerced human physical input;* ideally, no one would have to do anything he or she did not wholeheartedly want to do.

11. *Minimal maintenance, maximum simplicity;* ideally, every part of our energy system would be maintenance-free and simple enough to operate so that virtually anyone could operate it.

12. *Maximum availability of accurate, up-to-date, clear energy information;* ideally, all the information concerning our energy systems on all levels of operation would be available to everyone, anywhere, anytime.

13. *Maximum amount of recycling of products and by-products;* ideally, all parts and products of our energy systems would be designed from their beginnings to be totally recyclable.

14. *Maximum amount of participatory decision-making;* ideally, everyone who wanted to be would be involved in every major decision.

Using the above "preferred state" as a guide for present day decisions for resolving the planet's energy problems leads to very different strategies than those currently being employed by nearly all the decision-makers of local, regional, national, and international governmental bodies and organizations of the world.

Problem State

The main impediments to evolution in the present energy state have brought about a situation in which there is:

☐ Not enough energy available for 100% of humanity's life support, e.g., forced fuel rationing, materials shortages, forced shorter workweeks, "blackouts," "brownouts," etc., in developed countries, and little or no industrial energy available to construct and develop tools for life support in developing countries.

☐ Inequitable distribution of energy consumption; for example, the United States, with 7% of the world's population, consumes 32% of the world's energy.

☐ Low efficiency of energy conversion, such as appliances that waste electricity, cars that consume too much fuel, materials that require a lot of energy used in place of low-energy-costing materials,, uninsulated structures, etc. Present-day energy converters average 4-5% overall thermal and mechanical efficiency.[35] For every 100 barrels of oil produced by nature, 95 go down the drain. An overall efficiency of at least 12-20% is feasible with present-day design and engineering know-how.

☐ High negative environmental impact of energy inputs and outputs of the external metabolic system; e.g., resource depletion, waste, pollution of air, water, and land by unwanted chemicals, heat, artifacts, and noise; and disruption of ecological cycles through strip mining, pipelines, etc. In short, an environment whose capacity to provide what we are demanding of it, and to absorb what we are injecting into it, is rapidly becoming insufficient.

☐ High use of short supply energy resources, such as fossil and nuclear fuels.

☐ Low diversity/redundancy of energy sources and systems; e.g., most of our "eggs" are in one basket: oil.

☐ High use of coerced human physical labor input.

☐ Centralized and one-way energy systems, i.e., energy flows from monopolistic utilities and corporations to individual consumers, without the inverse option.

☐ Energy supply used as a weapon.

☐ No emphasis on simplicity; high degree of maintenance.

☐ Low guaranteed safety to energy producers, users, and future generations.

☐ Little standardization of energy outputs and energy-using artifacts on a world-wide basis.

☐ Low level of recycling of energy by-products and energy-using artifacts.

☐ Low amount of public decision-making.

References

1. Assuming $E = MC^2$; that the Sun is an average star; that there are a billion stars in our galaxy; and that there are a billion such galaxies.

2. Based on a mean value for the solar constant of 1,395 kw/square meter and $2,488 \times 10^{21}$ square meters of the Sun; figure from Hubbert, M. K., "The Energy Resources of the Earth," *Scientific American*, Sept. 1971.

3. Based on $173,000 \times 10^{12}$ watts of solar radiation × 24 hours; figure from Hubbert, M. K., "The Energy Resources of the Earth," *Scientific American*, Sept. 1971.

4. Based on U.N. figures for total commercial energy consumption of $8,555 \times 10^6$ mt coal equivalent in 1975; from *World Energy Supplies 1972-1976*, Series J, No. 21, U.N. New York, 1978, and other data sources for non-commercial energy. See World Energy Data Chart at end of Section II for details.

5. Dyson, F. J., "Energy in the Universe," *Scientific American*, Sept. 1971, p. 5l.

6. Fuller, R. B.

7. Peden, W., "The Renewable Energy Handbook," Energy Probe, University of Toronto, Canada.

8. Wilson, C. L., *Energy: Global Prospects 1985-2000*, Report of the Workshop on Alternative Energy Strategies, M.I.T., New York, McGraw-Hill, 1977.

9. See *Transition: A Report to the Oregon Energy Council*, prepared by the Office of Energy Research and Planning, Office of the Governor, State of Oregon, 1975; Bockris, J. O. M., *Energy: The Solar-Hydrogen Alternative*, Halsted Press, John Wiley & Sons, 1975; Sorenson, Bernt, "Energy and Resources," *Science*, July 25, 1975, Vol. 189; Lovins, A. B., *Soft Energy Paths: Towards a Durable Peace*; Hayes, D., *Rays of Hope: The Transition to a Post-Petroleum World*; Weingart, J. M., "The Helios Strategy: A Heretical View of the Role of Solar Energy in the Future of a Small Planet," Mitchell Prize Winner, 1977; Craig, P., et at., *Distributed Energy Systems in California's Future: Interim Report*.

10. One way of viewing this book is as the attempt to answer the questions generated by this "game." Readers are encouraged to send their participation in the game/questions to the author at The World Game, 3500 Market St., Philadelphia PA 19104, for possible inclusion in any subsequent editions of this book (and for the author's own education).

11. Lovins, A. B., *Soft Energy Paths: Towards a Durable Peace*, Ballinger, Boston MA, 1977.

12. Steinhart, C. E., Steinhart, J. S., *Energy: Sources, Use, and Role in Human Affairs*, Duxbury Press, North Scituate MA, 1974.

13. Summers, C. M., "The Conversion of Energy," *Scientific American*, Sept. 1971, p. 149.

14. *World Energy Supplies 1972-1976*, Series J, No. 21, U.N., New York, 1978.

15. Ryan, C. J., "The Choices in the Next Energy and Social Revolution," Mitchell Prize Winner, 1977, Woodlands Conference.

16. Eidson, W., Department of Physics, Drexel University, Philadelphia PA, personal communication.

17. Hayes, D., *Rays of Hope: The Transition to a Post-Petroleum World*, W. W. Norton & Co., New York, 1977.

18. Duffield, C., *Geothermal Technoecosystems and Water Cycles in Arid Lands*, Arid Lands Resource Information Paper No. 8, University of Arizona, Tucson AZ, 1976.

19. Hubbert, M. K., "Energy Resources: A Report to the Committee on Natural Resources," National Academy of Sciences—National Research Council, Publication 1000-D, Washington DC, 1962. Reprinted 193: National Technical Information Service, U.S. Department of Commerce, Publ. PB-222401, Springfield VA 22151.

20. American Society of Mechanical Engineering, *Transport Phenomenon in Atmospheric and Ecological Systems*, A.S.M.E., New York, 1967, pp. 21-87.

21. Romer, R. H., *Energy: An Introduction to Physics*, W. H. Freeman, San Francisco CA, 1976.

22. Clark, W., *Energy for Survival: The Alternative to Extinction*, Doubleday, New York, 1974, p. 103.

23. "Coal and the Coming Superinterglacial," *Science News*, 6-4-77, Vol. 112, p. 68.

24. "Coal and Climate: A Yellow Light on CO_2," *Science News*, 7-30-77, Vol. 112, p. 68.

25. Craig, P., et al., eds., *Distributed Energy Systems in California's Future: Interim Report*, Vol. I, p. 217. U.S. Dept. of Energy document HCP/P7405-03, May 1978.

26. Sullivan, W., "Scientists Fear Heavy Use of Coal May Bring Adverse Shift in Climate," *New York Times*, 7-25-77.

27. *McGraw-Hill Encyclopedia of Energy*, Lapedes, D. N., editor-in-chief, McGraw-Hill, New York, 1978.

28. *A Time to Choose: Energy Policy of the Ford Foundation*, Ballinger, Cambridge MA, pp. 204-18.

29. *Terrastar*, Final Report, NASA Grant No. NGT-01-003-004, Auburn University, School of Engineering, Auburn AL, Sept. 1973

30. Total quantities from Wilson, C., *Energy: Global Prospects 1985-2000*, Report of the Workshop on Alternative Energy Strategies, M.I.T., New York, McGraw-Hill, 1977.

31. Katz, M., "Decision Making in the Production of Power," *Scientific American*, Sept. 197l, p. 149.

32. Illich, I., *History of Needs*, Pantheon Books, New York, 1978.

33. Georgescu-Roegen, N., *Energy and Economic Myths*, Pergamon Press, New York, 1976, p. xix

34. Brown, H., Carpenter, J., and Gabel, M., "Working Document on Decision-Making Criteria in Design Science," Earth Metabolic Design, Box 2016, Yale Station, New Haven CT 06520, 1974.

35. Thirring, H., *Energy for Man: Windmills to Nuclear Power*, Rep. of 1958 ed., Greenwood Press, Westport CT, 1968.

36. Snyder, G., "Revolution in the Revolution in the Revolution," *Regarding Wave*, New Directions, New York, 1970.

General References

Committee on Science and Technology, *Energy Facts II*, U.S. Government Printing Office, Washington D.C., 1975.

Cook, E., *Man, Energy, Society*, W. H. Freeman & Co., San Francisco CA, 1976.

Daniels, F., *The Direct Use of the Sun's Energy*, Yale University Press, New Haven CT, 1964.

Miller, G. T., *Living in the Environment: Concepts, Problems, Alternatives*, Wadsworth, 1975.

National Academy of Sciences, *Resources and Man*, W. H. Freeman & Co., San Francisco CA, 1969.

Penner, S. S., Icerman, L., *Energy Vol. I, II & III*, Addison-Wesley Publishing Co., Reading MA, 1974.

Portola Institute, *Energy Primer*, Portola Institute, Menlo Park CA, 1974.

Scientific American, "Energy and Power" issue, Sept., 1971.

Steinhart, C. E., Steinhart, J. S., *Energy Sources, Use, and Role in Human Affairs*, Duxbury Press, North Scituate MA, 1974.

Thirring, H., *Energy for Man: From Windmills to Nuclear Power*, Reprint of 1958 ed, Greenwood Press, Westport CT, 1968.

ENERGY SOURCES

The Role of Design Science

Resources organized by human intellect into technology help meet humanity's needs for energy, food, shelter, health care, transportation, communication, education, recreation, and more resources. From every employment of resources into technology, human intellect learns more about technology, resources, and human needs.

Each technological process has an associated efficiency that measures its performance. Each efficiency and thus each performance can be improved each time new technology is used because of the increase in know-how. Because of entropy, efficiency will never be 100%, but waste can continually be reduced.

Design science deals with the inefficiencies of the specific processes and the entire process of technology by applying increased know-how to further the performance of technology (see chart). By this means technology is continually able to get more from each unit of invested kilowatt-hour of energy or ton of material.

The physical resources of the Earth are finite and limited. Humanity's needs are ever-increasing and evolving, e.g., from food and basic necessities of life to educational and recreational needs. Like humanity's needs, design science is an open system. A designer or decision-maker with energy problems chooses the preferable energy sources, location and extraction technologies, conversion techniques and engines, storage techniques, and modes of transport. Each facet of the total energy system has unique characteristics or performance efficiencies associated with it. It is the task of the design scientist not only to maximize the existing technological efficiencies within the characteristics of the existing or presently known energy sources, but also to redesign and invent new technology for existing and potential energy sources. (In a sense, a new technology for harnessing an energy source is a new energy source. For example, an orbiting solar power station taps an energy source previously unavailable, thereby "creating" a new source. New information and new integrations of existing information lead to new sources of energy.)

The design scientist seeking to solve energy problems designs, tests, and reduces to ongoing industrial practice artifacts that are progressively more efficient extractors, producers, converters, distributors, and storers of energy, thereby increasing the overall life-support capabilities of humanity.

Briefly, design science is a methodology for recognizing, defining, and resolving global problems. It is an integrative and holistic discipline rather than a specialization. It works from a goal or preferred state back to the present or problem state. That is, its first focus is on what the world should be like, based on explicitly stated values, and not on what may be current symptoms of any one problem. Design science seeks to design viable alternatives to present-day life support systems that increase the relative freedoms of humanity, decrease the constraints, and utilize fewer resources while performing more and more. It also seeks to invent the means of realizing the design.[1]

Design Science/Energy System Interaction[2]

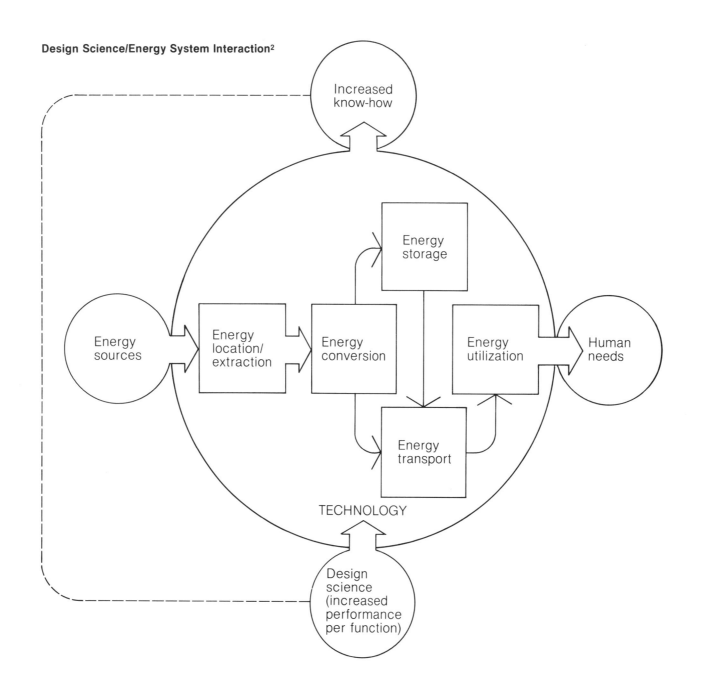

Energy Sources, Old and New

New sources of energy need to be harnessed, not only to replace the dwindling fossil fuel supplies but also to provide energy that will have less impact on the environment.

All current and potential sources and systems of energy production can be loosely classified under two headings: capital energy sources and income energy sources. Capital energy sources are finite in quantity and cannot be replaced once used. Income energy sources are regenerative and are limited by rate of use rather than supply.

Under capital energy sources are listed:
- ☐ Coal
- ☐ Petroleum, oil shale, and tar sands
- ☐ Natural gas
- ☐ Nuclear fission
- ☐ Nuclear fusion

Income energy sources include:
- ☐ Falling water (hydroelectric)
- ☐ Geothermal
- ☐ Solar energy (terrestrial and extraterrestrial)
- ☐ Bioconversion
- ☐ Wind
- ☐ The ocean tides, waves, and currents
- ☐ Hydrogen
- ☐ Temperature differential
- ☐ Animate

This section describes the above sources of energy and others whose uses, advantages, and disadvantages have been identified. The World Energy Data Chart at the end lists these same energy sources, their current consumption or use per year, the total remaining supply, their life expectancy at projected-use rates, ultimate potential available per year, potential realizable or harnessable in ten years with present-day know-how, amount realizable in twenty-five years, amount realizable in fifty years, current largest-size and smallest-size power plant, current planning and construction time of power plant, current life-span of power plant, and costs.

All this information was gathered with these questions in mind: What information is needed to determine if it is possible for 100% of humanity to have all of its energy needs serviced? What information will be needed to make rational decisions and to formulate strategies for the realization of this option?

Strategies for global development follow in Section III.

Capital Energy Sources

3 Coal

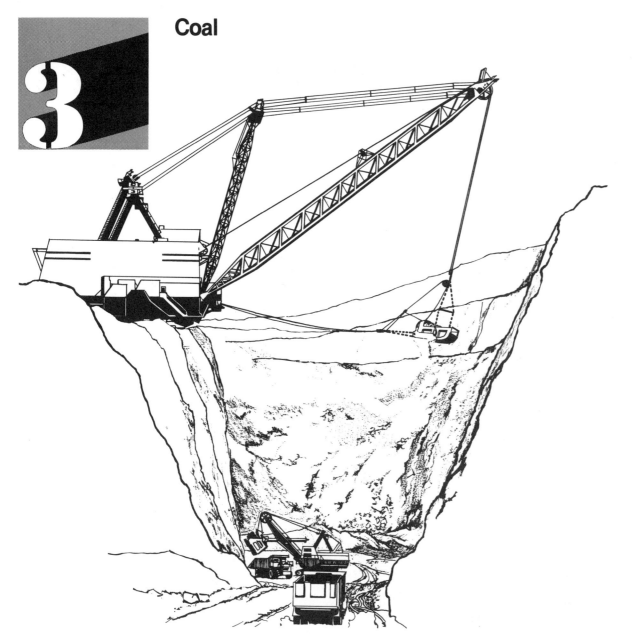

Sleeping Diamonds

Coal is the most abundant of the fossil fuels; it was formed from plants that grew on Earth millions of years ago. The five to ten feet of plant material needed for each foot of coal took 9,000 to 12,000 years to accumulate.[1] Millions of years more, at tremendous pressures, were needed to transform the original plant matter into the concentrated hydrocarbon form we recognize as lignite, bituminous, and finally anthracite coal. The geological processes that formed coal also produced methane gas. Each ton of underground coal contains about 200 cubic feet of methane.[2] There is an estimated 1,500 trillion cubic feet of methane trapped in and around the world's coal reserves (about 15% of the total world gas supply). The U.S. has an estimated 800 trillion cubic feet. This is more than three times the present proved gas reserves in the U.S.[3] Because of methane's extreme hazard to miners, due to the explosions and fire caused by its ignition, nearly 250,000 cubic feet of methane (enough to heat about 800,000 homes) are vented to the atmosphere each *day* from coal mines in the U.S. Recent work has led to practical ways of recovering at least one third of this coalbed methane.

Another way of utilizing coal is by adding it to

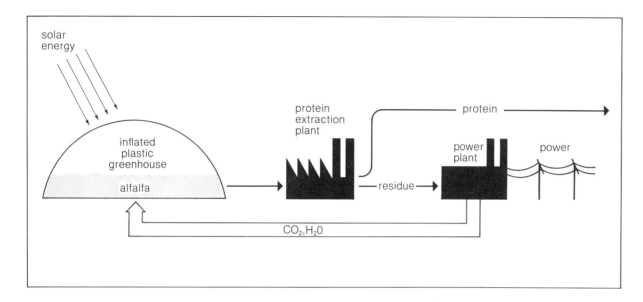

Scheme for energy and protein production by covered agriculture. Alfalfa, grown under transparent cover year-round with CO_2 enrichment, would be harvested and processed to remove some protein as a valuable product. The residue would be used as animal fodder or, in the version shown here, as fuel for power plants. Combustion CO_2 and H_2O from this and fossil fuels would be returned to the greenhouses.[4]

oil. The resulting mixture would increase the liquid amount of fuel. For example, treating the U.S. production of 10 million barrels per day of crude oil with one million tons of coal will increase the yield of oil products to 13 million barrels per day.

In most places where coal is or has been mined, there are huge piles of coal waste that litter the landscape. In northeastern Pennsylvania alone, there are some 900 million cubic yards of hard coal debris that is equal to the energy of 250 million tons of coal. The technology now exists to cleanly burn this waste, thereby simultaneously producing power and reducing some of the health and safety problems caused by these eyesores.

One way of dealing with the CO_2 problem that results from coal combustion is to link the coal combustion with controlled-environment agriculture. (See diagram.) Plants would use the CO_2 to enhance their growth.

Uses

Coal is combusted to produce heat for:

1. Electric power generation
2. Industrial processes
3. Transportation
4. Space heating

Coal is used as a raw material to produce:

5. Coke for steel production
6. Coal tar
7. Light oil
8. Ammonia
9. Coal gas

(All by heating in the absence of air.)

These products are in turn used as a raw material to produce:

10. Plastics
11. Synthetic fibers such as nylon
12. Dyes
13. Drugs
14. Detergents
15. Motor Fuels
16. Solvents
17. Explosives
18. Refrigerants
19. Fertilizers
20. Insecticides
21. Margarine
22. Glue
23. Cosmetics.

(See Chart on page 58.)

Coal Deposits

SOURCE:
Oxford Economic Atlas, 1972
Business Week (Feb 16, 1976), p. 38

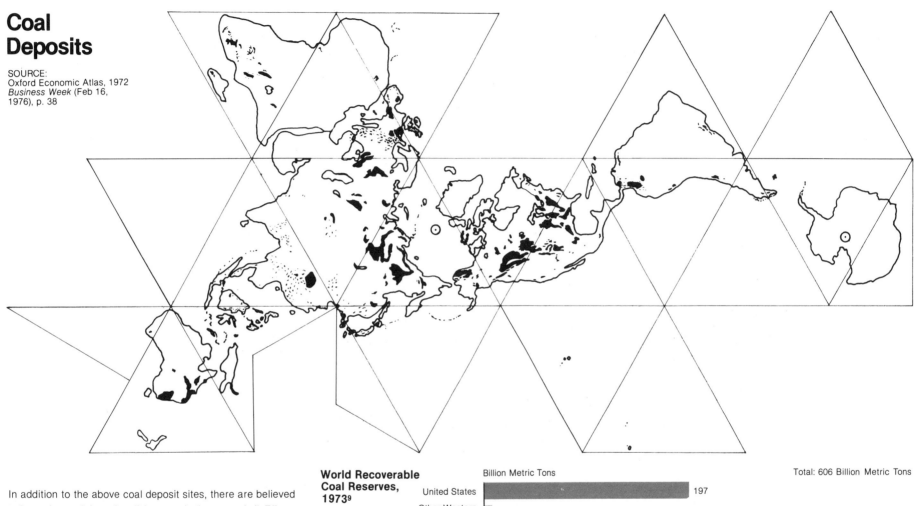

In addition to the above coal deposit sites, there are believed to be major coal deposits offshore, as in the case of oil. Efforts to mine offshore coal require modification of existing systems rather than the invention of new systems.[10] (Undersea coal mining in northeast England has been going on for over ninety years; to date, over 375 million tons of coal have been mined from this location.)

World Recoverable Coal Reserves, 1973[9]

Total: 606 Billion Metric Tons

Region	Billion Metric Tons
United States	197
Other Western Hemisphere Countries	8
Western Europe	65
Africa	15
Middle East	0
Far East & Oceania	42
Sino-Soviet Bloc	278

World Solid Fuels Production*

Source:
World Energy Supplies 1972-1976 (United Nations, 1978).

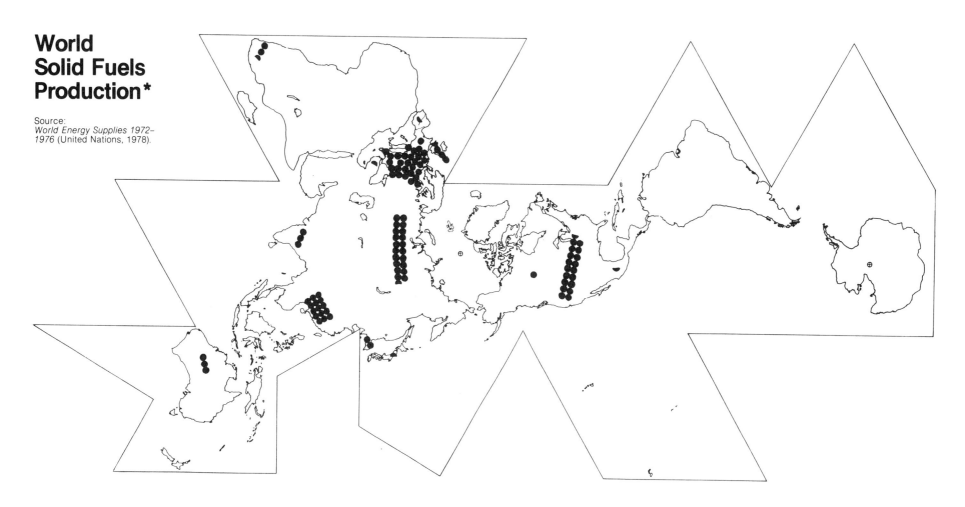

● = 1% of World Solid Fuels Production
● = 34.3 × 10^6 metric tons

*includes coal, lignite, and brown coal

Major Chemical Products Derived from Coal

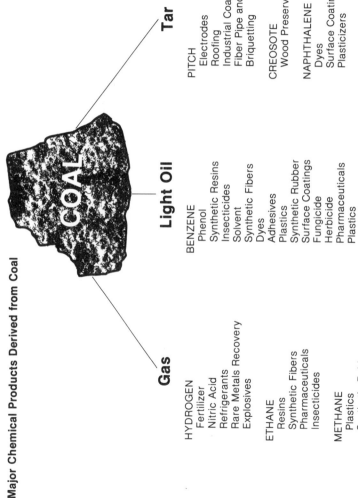

Gas

HYDROGEN
  Fertilizer
  Nitric Acid
  Refrigerants
  Rare Metals Recovery
  Explosives

ETHANE
  Resins
  Synthetic Fibers
  Pharmaceuticals
  Insecticides

METHANE
  Plastics
  Synthetic Rubber
  Alcohols
  Liquid Fuels
  Protective Coatings
  Refrigerant
  Solvents
  Anesthetic

ETHYLENE
  Plastics
  Anesthetics
  Solvents
  Anti-Freeze

HYDROGEN SULFIDE
  Sulfuric Acid
  Refrigerants
  Petroleum Refining
  Synthetic Fibers and Films
  Fertilizer

HYDROGEN CYANIDE
  Insecticides
  Synthetic Resins

AMMONIA
  Fertilizer
  Neutralizing Agent

Light Oil

BENZENE
  Phenol
  Synthetic Resins
  Insecticides
  Solvent
  Synthetic Fibers
  Dyes
  Adhesives
  Plastics
  Synthetic Rubber
  Surface Coatings
  Fungicide
  Herbicide
  Pharmaceuticals
  Plastics
  Plasticizers
  Detergents
  Explosives

TOLUENE
  Solvent
  Explosives
  Surface Coatings
  Plastics

XYLENE
  Solvent
  Synthetic Fibers
  Plasticizers

DICYCLOPENTADIENE
  Insecticides

CRUDE NAPHTHA
  Asphalt Tile
  Varnishes

Tar

PITCH
  Electrodes
  Roofing
  Industrial Coatings
  Fiber Pipe and Conduit
  Briquetting

CREOSOTE
  Wood Preservative

NAPHTHALENE
  Dyes
  Surface Coatings
  Plasticizers

PHENOLS—CRESOLS
—CRESYLIC ACIDS
  Gasoline Additives
  Plasticizers
  Hydraulic Fluids
  Resins
  Varnishes
  Adhesives

PYRIDINE BASES
  Fabric Waterproofing
  Solvent
  Pharmaceuticals

Flow of Materials Through Coke Plant[5]

In the process of making coke, coal is distilled in the absence of air at temperature of 1,900° to 2,000° Fahrenheit in the coke ovens, slot-shaped chambers lined with refractory brick. These chambers are heated by combustion gases which pass through flues between the individual ovens.

The volatile materials from the coal pass out of the coke ovens through vertical standpipes into a collecting main. Here most of the tar condenses from the vapors. Further cooling removes additional tar and ammonia liquor.

Tar can be used directly as a fuel or can be further refined by fractional distillation to obtain products such as creosote, pitch, naphthalene, phenol, cresylic acids and pyridine bases.

The ammonia liquor is treated to liberate ammonia which is combined with coke-oven gas and treated with sulfuric acid to form ammonium sulfate. The ammonium-sulfate liquor is processed for recovery of pyridine bases.

The cleaned coke-oven gas is cooled and treated with absorbing oil which removes the light oil. The light oil is separated by steam distillation and either refined by washing with sulfuric acid or hydro refined prior to fractional distillation. Benzene, toluene, xylene and crude solvent naphthas are the products of this distillation.

The coke-oven gas remaining, still rich in hydrogen, methane, ethane, and ethylene, is either used in the steel mill as a fuel or separated into its components.

History

Bronze Age (2000 B.C.)	Wales: Coal burned in funeral pyres.
100 B.C.	China: Use of coal as heat source.
100 A.D.	Italy: Romans use coal as heat source.
832	England: Monasteries use coal as a heat source.
1000	U.S.A.: Hopi Indians mine coal and use it to heat homes and fire pottery.
1100	England: Coal discovered on northeast coast; known as sea coles.
1180	England: Coal used in iron works.
1200	Europe: Commercial coal mining begins.
1239	Europe: Coal used by smiths and brewers.
1300	England: First coal used for general home heating.
1325	England: First export of coal from Newcastle to France.
1550	England: Large-scale mining.
1670	England: Clayton reports generation of luminous gas when coal is heated in chemical retort.
1679	U.S.A.: Coal is discovered in Illinois by French explorers.
1709	England: Coke used by Derby.
1710	England: Steam engine used for mine pumping.
1735	England: Derby blast furnace.
1745	U.S.A.: Coal mining begins in Richmond, Virginia.
1792	Scotland: Murdoch illuminates his home with coal gas.
1808	England: Clegg invents coal gas manufacture.
1832	U.S.A.: First strip mine.
1850	Coal replaces wood as major energy source.
1856	U.S.A.: Coal tar industry begins.
1862	France: First systematic study of the action of solvents on coal.
1890	England: First large-scale offshore coal mine.
1900's	Development of railroads made bulk transport of coal possible.
1910	Germany: Fisher-Tropsch process invented for large-scale synthetic fuels production.
1914	London, England: Pipeline used to pump coal slurry and water into London.
1920's	Germany: Internal combustion engine that ran on powdered coal developed by Powlikowsky.[6]
1920's	Germany: Production of liquids from coal initiated.
1920's	U.S.A.: Large power generation via pulverized fuel firing begins.
1922	Germany: Winkler patents fluidized combustion of coal.
1920's–1930's	Germany: Pott-Broche, Berquis-Pier, Lurgi processes invented to produce gas and oil from coal.
1930	U.S.A.: Nylon invented.
1930's	Natural gas begins to cut into the coal gas business.
1946–48	U.S.A., Forges, Alabama: Field studies of underground coal gasification.
1950's	U.S.S.R.: Two full-scale electric generating plants fueled by coal gas produced underground (still operating).
1951	U.S.A.: Coal conversion to oil plants in Texas and Missouri.
1952	Finland: First commercial Kopper-Totzek coal gasification plant.
1955	South Africa: Largest (5 million tons per year) coal-based synthetic fuels plant in world.
1957	Ohio, U.S.A.: 108-mile pipeline for transporting coal slurry.
1970	U.S.A.: 273-mile pipeline for transporting coal slurry from Arizona to Nevada.
1976	U.S.A.: E.R.D.A. funded development of coal/oil slurry fired electric power plant.
1976	U.S.A.: E.R.D.A. funded development of "methanol from coal" plant and coal liquification plant.
1977	U.S.A., West Virginia: First fluidized bed boiler (30 MW) in U.S.
1984	South Africa: Two additional coal-based synthetic fuels plants produce more than 100,000 barrels of hydrocarbons per day.

World Coal Production*

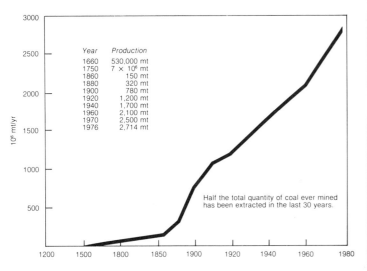

Year	Production
1660	530,000 mt
1750	7 × 10⁶ mt
1860	150 mt
1880	320 mt
1900	780 mt
1920	1,200 mt
1940	1,700 mt
1960	2,100 mt
1970	2,500 mt
1976	2,714 mt

Half the total quantity of coal ever mined has been extracted in the last 30 years.

*Source: *UN World Energy Supplies 1971–1975,* and Dorf, R.C., *Energy, Resources & Policy.* Addison-Wesley.

Four Types of Bituminous Coal Mines[7]

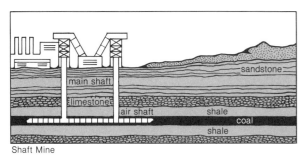
Shaft Mine

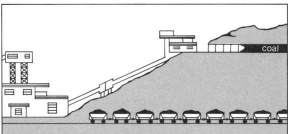
Drift Mine

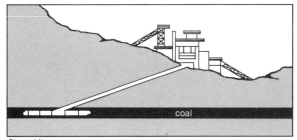
Slope Mine

Surface Mine

In addition to these methods of mining coal, there is one other: "in situ" coal production. In this method, a solvent is injected into the coal seam and the resulting liquid is pumped out to a processing plant where it is filtered, desulphured, and solidified.

World Total Coal Consumption:[8] **2,696.0 × 10⁶ metric tons**

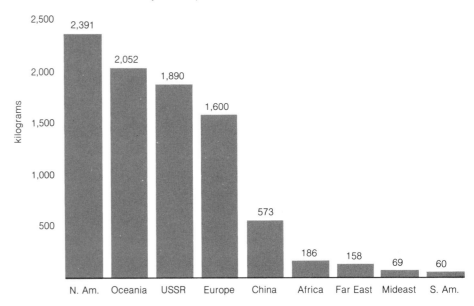

World Consumption of Coal 1976: 671 kilograms per capita[8]

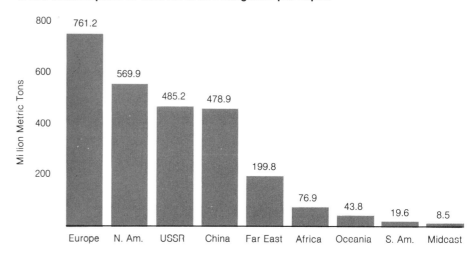

Advantages

1. Coal is in relatively plentiful supply.
2. Coal's stored energy is easily relased.
3. Coal is easily stored, transported, and controlled in large volumes.
4. Coal reserves are more widely distributed around the world than oil or gas.

Disadvantages

1. Once combusted, coal is gone forever.
2. Coal combustion results in:

   a) *Air Pollution.* About 380,000 tons/yr. for a 1,000 MW power plant (load factor .75)[11] (Each 1,000 MW coal-fired electric power plant consumes 3 million tons of coal per year, or 100 million tons over its 30-year lifetime.)

Aldehydes	.01 kg/t
CO	1.1 kg/t
Hydrocarbons	.4 kg/t
NO_x	44.1 kg/t
Sulfur	83.8 kg/t
Dust	374.8 kg/t
Radioactivity	traces

   Coal combustion is the largest single source of sulfur oxide pollution.[12] More than 50×10^6 tons of sulfur was discharged by the combustion of fossil fuels in 1972.[13] One and six tenths billion tons of carbon dioxide are released into the atmosphere each year in the U.S.; this could result in a 0.001°C warming of the environment.[14] Coal combustion also releases large amounts of waste heat (56–58% of the energy consumed), toxic metals (more mercury is released in coal combustion than in any other human activity), and carcinogenic organic coumpounds.[15] There are methods and devices for limiting or controlling pollution during combustion, such as fluidized bed combustion, where finely powdered coal is burned in a fire chamber suspended on air, as well as eliminating pollution from stacks after combustion with devices such as electrostatic precipitators and "scrubbers." Scrubbers are expensive ($75–$125 per installed kilowatt), consume as much as 5% of the power output of a given generating plant, are extremely complicated, and produce large quantities of wet sludge.[16] Nevertheless, scrubbers are better and cheaper than no pollution control.

   b) *Water Pollution.* About 7,000–41,000 tons per 1,000 MW power plant/yr. (load factor .75), depending on whether coal is deep mined or strip mined.[11]

   c) *Land Use.* About 120 (deep mined coal)–140 km^2 (strip mined) for all operations with 4 km^2 for power plant (1,000 MW)[11]. Surface (strip) mining of coal can seriously damage the land by producing soil waste problems, acid drainage, unproductive land, and visibly ugly terrain. Strip mines directly disturb at least 153,000 acres (62,000 hectares) per year in just the U.S. Soil erosion from mining reportedly pollutes 12,000 miles (19,300 km) of streams.[17] Strip mining does not have to leave a scarred landscape in its wake. It is possible to leave strip-mined land in better condition than it was before mining, but this needs the necessary committment of resources to do it. In Germany, the Rheinbraun Coal Company has been returning lands to their original or better condition for quite some time. Reclaimed lands have been covered with 53 varieties of trees (2.5 million are planted each year), new fields, parks, wildlife preserves, and about 40 lakes[18].

   d) *Solid Waste.* About 600,000 (deep mine)–3,250,000 tons (strip-mine) for all operations of 1,000 MW power plant. Coal combustion by-products from a 1,000 MW power plant produce, each year, 71,000 tons of boiler ash, 281,000 tons of flyash, and 212,000 tons of sulfur waste[19].

3. Coal is dangerous to human health. Mining of coal in underground mines has resulted in more than 100,000 men losing their lives in coal mines in just this century in the U.S.; more than a million were permanently disabled in mine accidents, even more contracted black lung disease.[19] For each two trillion kwh of electricity generated by coal combustion in the U.S., an estimated near-zero to 6,000 *deaths* result due to air pollution; near-zero to 1,250 deaths from occupational accidents and disease; 10,000 to one million cases of lower-respiratory-tract disease in children; 60,000 to six million cases of chronic respiratory disease; 600,000 to 60 million person-days of aggravated heart-lung symptoms in the elderly; and 100,000 to 10 million asthma attacks.[14]

4. Present-day technology for coal combustion has a low overall efficiency; more than 50% of the energy stored in coal goes up the stack as waste heat. The efficiency from an underground mine to the user of electricity is 18%; from surface mine to user of electricity, it is 25%. (Recovery of coal from the ground varies widely—25–70% in deep mines and 85–95% in surface mining[10].)

Coal: Potential Environmental Impacts Summary[20]

	Water	Air	Land	Solid Waste	Noise	Aesthetics
Production	Acid mine drainage	Mine fires	Strip mining damage	Waste from underground mining	Fairly high—machines used to rip overburden; detonations	Very serious: excavated areas waste, destruction of flora, etc.
	Leaching of waste piles	Waste pile fires	Total destruction of flora and fauna	Large quantities of waste from strip mining; reusable for fill in		
	Erosion and silting of streams	Dust	Diversion of watercourses, access roads, etc.			
		Exhaust gases from machines used				
		Atmospheric oxidation of material removed				
Upgrading	Preparation plant effluent streams	Particulates from fine coal drying		Waste from coal cleaning		
	Leaching of waste piles	Nitrogen oxides				
		Waste bank fires				
Transportation/ Utilization	Thermal pollution	Aldehydes, Carbon dioxide, Hydrocarbons, Sulfur Oxides, Nitrogen Oxides, Particulates, Radioactivity — Power plants		Disposal of fly ash and slag	Not considered serious: 30 db at 100 m from power plant	Problems due to size of installations, height of stacks and cooling towers, ash waste
		Particulates, Hydrogen Sulfide, Carbon Monoxide, Hydrocarbons — Coke ovens				

Thermal efficiency in electric power plants is 40–42% or 290–300 grams of fuel per kwh produced[11].

5. Combusting coal at 1,000° C to heat water or air to less than 100° C, as it often is, is inefficient.
6. Widespread surface mining in previously unmined areas could have undesirable social and economic impacts, especially in a sparsely populated and agriculturally-oriented society of small towns and isolated ranches.[10]
7. Coal pollution could have serious adverse effects on agriculture: sulfur dioxide stunts plant growth, reduces seed output, and weakens a plant's natural ability to resist pests, disease, and drought. Additionally, "acid rain," caused by sulfur dioxide combining with rain and forming sulfuric acid leaches nutrients from the soil, stunts plant growth (including forests), and reduces aquatic life in lakes and ponds.[19]

Coal Gasification/ Coal Liquification

Coal can be gasified or liquified and used in place of natural gas or petroleum. There are numerous methods for gasifying/liquifying coal. Long before natural gas became available in large volume at low cost, "coal gas" was manufactured in great quantities in the U.S. and elsewhere. More than 150 companies around the world manufactured coal gasification equipment in the 1920's.[1] The "gaslight" era was fueled by coal gas. In the 1920's and 30's Germany developed a number of processes that were eventually exploited during World War II to produce about 80,000 barrels of oil per day.[2] Since 1975 South Africa has used almost the same method to produce gasoline, motor fuels, pipeline gas, ammonia, and other products.[3] It is now in the process of building a much larger plant that will supply almost one-third of its present consumption of motor fuel. Besides South Africa, coal is gasified today in China, India, Korea, Morocco, Spain, Scotland, Turkey, Yugoslavia, and the United States.

Another method receiving attention today is underground or in-situ coal gasification. In this technique, coal is combusted underground, preferably in places too deep, thin, dangerous, or uneconomical to mine, by igniting the coal and then pumping air down a well to keep the fire going and convert the coal into carbon monoxide and methane.

Advantages

1. Underground coal mining recovers on the average only a little more than 50% of the coal; underground gasification on the average could recover a larger percentage of the heating value of the coal.[4]

Above Ground Coal Gasification

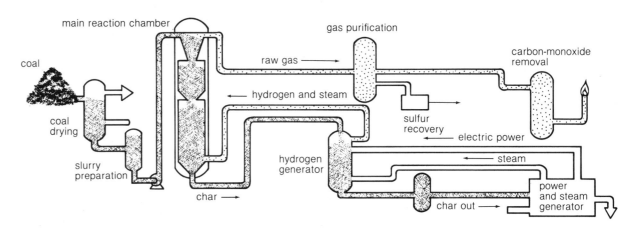

Below Ground Coal Gasification

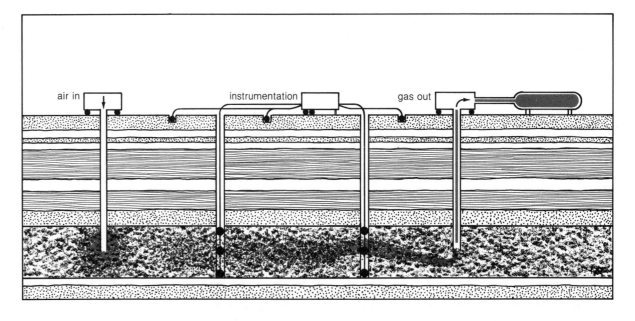

2. Gasification and subsequent burning for home heating is more efficient than burning coal to produce electricity that in turn heats a house via electrical resistance.
3. Gasification would produce less pollution than burning the coal directly; it would also simplify air-pollution control by concentrating coal use in fewer locations.
4. Gas could use the existing natural gas pipelines (see map, Gas section) which presently run at only 60% capacity.
5. Coal gasification/liquification would allow consumers to use existing oil and gas-fired equipment.

Disadvantages

1. Coal gasification is not very energy efficient: 30–40% of the energy in coal is lost in converting it into gas or oil.[5]
2. Coal gasification/liquification uses large amounts of water.
3. Coal gasification plants are very capital intensive; to produce as much gas as was consumed in the U.S. in 1975 would take 200 commercial-size plants, each costing about a billion dollars. In its mammoth capital requirements, coal gasification is comparable to nuclear power.
4. Underground coal gasification yields only low BTU gas.
5. Gas from coal is expensive: wholesale costs are around $4.00 per million BTU,[6] compared with imported gas at $3.15 per million BTU and 1977 regulated prices for domestic gas sold in interstate market of $1.42 per million BTU's.
6. Coal gasification/liquification is presently thought to entail a high probability of significant occupational exposure to carcinogens.

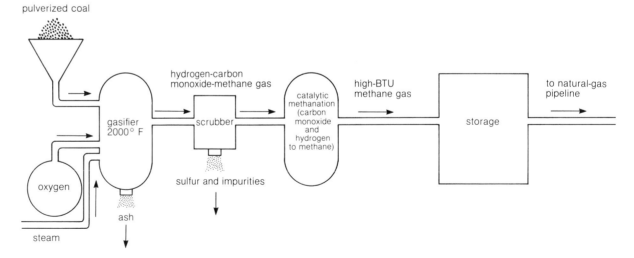

Low-BTU Gas for Electric Power Generation[7]

High-BTU Gas[7]

Peat

Peat is a type of fossil fuel which accounts for about 1.1% of the world's fossil fuel resources.[1] Referred to as "geologically young coal" it is used as an energy source in the Soviet Union, Ireland, and Finland, with Sweden, Greece, the U.S., and Canada having plans for using it in the near future. It is also a good soil conditioner and low grade fertilizer because of its 2% nitrogen content and porosity.

In 1975, the Soviet Union burned most of the estimated 70 million tons of peat that it consumed in 77 electric power plants. More than 3,500 MW of power and heat are produced by peat-fueled plants in the Soviet Union.[2] Ireland got almost a third of its total energy supply from peat, and Finland has produced about 20 new peat-fueled plants since 1972. In the U.S., where peat reserves measure about 150 billion tons, the Minnesota Gas Company wants to build a plant to convert peat to methane and a company in North Carolina is seriously looking into building a 600 megawatt power plant fueled with peat. The North Carolina firm has already purchased peat harvesting equipment from Finland and the Soviet Union.[3]

A case can be made for peat being a renewable natural resource like timber, because it continuously renews itself, but, unlike timber, peat's renewal time is measured in thousands of years. Nevertheless, each year more than 300 million tons of dry peat are added to the world's peat bogs.[4] The 15 million tons added each year to Minnesota's peat supply is enough to provide 75% of the heating value of all the natural gas consumed in Minnesota.[1]

The world's peat reserves are an estimated 340 billion tons; total energy content is about 790×10^{12} kwh (2700×10^{15} BTU).

Uses

1. Peat can be dryed and then combusted as a fuel.
2. Peat is used as a soil conditioner ("peat moss") to improve the physical and chemical properties of soils.
3. Peat bogs are used for agriculture to grow wild rice, vegetables, berries, forage grasses, and some types of timber.
4. Peat is used to produce activated carbon in West Germany and Poland.
5. Peat is used in Scotland to distill whiskey.
6. Peat can be used for waste treatment.
7. Peat can be used as a feedstock for chemical plants.
8. Peat can be used as a base for biomass "energy farms" that would grow grasses, reeds, sedges, and cattails that could then be burned for a renewable energy source.[1]
9. Peat can be used as a fuel for gasification in the same processes that turn coal into gas.

Advantages

1. Peat is easy to "mine" or "harvest" since it is located on or very near the surface.
2. Peat has a low (or no) sulfur content.
3. Peat's heating value is high; it is superior to wood, about equal to lignite, and about half of high-grade coal.
4. Peat gasification is more efficient than coal gasification; four times as much light hydrocarbon gases are produced from peat than from coal. In addition, peat gasification is less expensive than coal gasification.[5]

Disadvantages

1. The 70–95% water content of peat has to be reduced substantially before it is combusted.
2. Peat is generally located in remote areas, distant from its primary users.
3. Peat is bulky and costly to transport.
4. Mining peat causes environmental problems such as:
   a) Drainage, which leads to flooding of other lands.
   b) Water quality change that could affect fish, vegetation, and wildlife.
   c) Changes in local and regional water tables.
   d) Reclaiming mined land.
   e) Land subsistence.

5 Petroleum

Liquid Gold

Petroleum, like, coal, is a fossil fuel. It is a product of millions of years of pressure that have transformed organic materials into combustible hydrocarbons. We combust petroleum to release its stored chemical energy; it is presently the principal source of energy in the world. It is used as a fuel for space heating, transportation, and electric power generation. It is also used as a raw material in the manufacturing of many products such as fertilizer, plastics, naphtha waxes, medicine, lubricants, coke, asphalt, etc. At one time there were an estimated 2,000 billion barrels of oil on Earth. (*Proven* reserves are much smaller—660 billion barrels.) To date (1980) humanity has consumed close to 500 billion barrels.

The transition to a world with dwindling oil output is an imminent reality. The question is not whether we make the transition or not. We will make it. The only question is whether it will be a smooth one, the result of careful planning and preparation, or chaotic, the result of a succession of worsening economic and political crises. Few, if any national leaders have any visions of what their societies will look like in a post-petroleum world.[1]

The Geology of Oil[3]

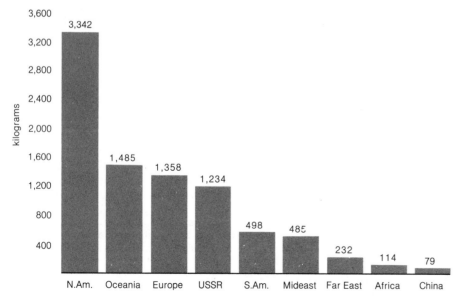

If all the oil the U.S. uses in 34 days were put into oil drums and stacked, the drums would reach the moon.

World Crude Oil Production[5]

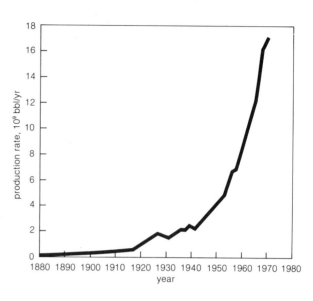

World Petroleum Consumption 1976: 603 kg per capita[4]

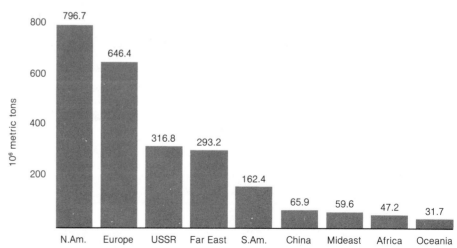

World Total Petroleum Consumption: 2422.0 × 10⁶ metric tons

68

History

6,000 B.C.	Mesopotamia: Asphalt used as fuel; pitch used for medicinal purposes and as bonding material.
3,000	Mesopotamia: Flares of natural gas lit in temples.
1,000	China: Wells drilled down to depths of 3,000 feet for producing natural gas, which was transported in bamboo pipelines for lighting and space heating.
500	Greece: Herodotus describes how natural asphalt and oil are used.
300	Mesopotamia: Fires of fluid naphtha used for military purposes.
400 A.D.	China: Natural gas used to evaporate brine from salt wells.
500	North America: Hand-dug oil wells drilled by Indians.
1,000	Mexico and Peru: Petroleum and asphalt used as fuel.
1,000	Burma: Wells drilled for petroleum production.[2]
*1300	Baku (now USSR)
1556	Germany: Term "petroleum" first used by Agricola; means literally "rock oil" from Greek & Latin *petra* (rock, stone) and *oleum* (oil).
*1692	Peru
*1750	Galicia (now Poland)
1852	Lukasiewcz invents kerosene lamp, refines crude oil.
*1859	Pennsylvania. Drake drills first oil well.
1860	U.S.: First major cross country pipeline 110 miles (177 km) in Northwestern Pennsylvania.
1865	U.S.: First railroad tank car.
1860's	U.S.: Clipper ship with cargo of oil goes around Cape Horn to California.
*1876	California
*1887	Texas
1889	U.S.: First rotary drill patent.
*1893	Sumatra
1900	Invention of internal combustion engine brings about wide scale use of oil.
*1901	Mexico
1905	Oil-burning steamer.
1907	U.S.: More than 47,000 miles of pipeline constructed.
*1908	Iran
1908	U.S.: Gasoline-powered Model-T Ford.
*1909	Trinidad
1910	Oil begins to replace coal as major energy source.
1913	Cracking process developed; gasoline yield doubles.
*1913	Venezuela; British Borneo
*1927	Iraq
*1932	Bahrein
*1938	Austria; Saudi Arabia; Kuwait
1938	U.S.: Offshore drilling begun in Gulf of Mexico.
*1940	Qatar
1940's	U.S.: World War II forces construction of two large pipelines from Texas to New Jersey.
*1967	Alberta, Canada: tar sands oil production begins.
1970	Supertankers—200,000 tons and bigger—began operating.
1973	Arab oil embargo.
1975	Mexico: large oil deposits discovered.
*1976	North Sea oil production.
1977	Alaska pipeline completed.

*Petroleum production begins

Fossils Fuels in Historical Perspective[5]

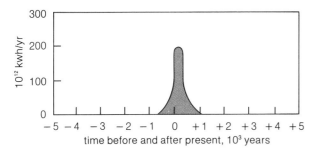

The epoch of fossil-fuel exploitation as it appears on a time scale of human history ranging from 5000 years ago to 5000 years in the future.

Trend is not destiny.
—R. Dubos

Crude Petroleum and Some of Its Products[6]

Hydrocarbon gases
- Liquified gases — Metal cutting gas, illumination gas
- Petroleum ether
- Polymers — Antiknock fuels, lubricating oils
- Alchohols, esters, ketones — Solvents
  - Aldehydes
  - Acetic acid — Resins
  - Synthetic rubber — Esters
- Acetylene
  - Acetylene black — Batteries
- Gas black
- Fuel gas — Rubber tires, inks, paints
- Light naphthas

Light distillates
- Naphthas
  - Light naphthas — Gas machine gasoline
    - Pentane, hexane
    - Aviation gasoline
    - Motor gasoline — Rubber solvent
    - Commercial solvents — Fatty oil solvent (extraction)
    - Blending naphtha — Lacquer diluents
    - Varnish-makers and painters naphtha
    - Dyers and cleaners naphtha
    - Turpentine substitutes
  - Intermediate naphthas
  - Heavy naphthas
- Refined oils
  - Refined kerosine — Stove fuel, lamp fuel, tractor fuel
  - Signal oil — Railroad signal oil, lighthouse oil
  - Mineral seal oil — Coach and ship illuminants, gas absorption oils
  - Water gas carburetion oils
  - Metallurgical fuels
  - Cracking stock for gasoline manufacture
  - Household heating fuels
  - Light industrial fuels
  - Diesel fuel oils

Intermediate distillates
- Gas oil
- Absorber oil — Gasoline recovery oil, benzol recovery oil

Heavy distillates
- Technical oils
  - White oils
    - Technical — Tree spray oils
      - Bakers machinery oil, fruit packers oil
      - Candymakers oil
      - Egg packers oil
      - Slab oil
    - Medicinal — Internal lubricants, salves, creams, ointments
  - Saturating oils — Wood oils, leather oils, twine oils
  - Emulsifying oils — Cutting oils, textile oils, paper oils, leather oils
  - Electrical oils — Switch oils, transformer oils, metal recovery oils
  - Flotation oils
- Paraffin wax
  - Candymakers and chewing gum wax
  - Candle wax, laundry wax, sealing wax, etchers wax
  - Saturating wax, insulation wax — Match wax, cardboard wax, paper wax
  - Medicinal wax
  - Canning wax
  - Paraflow
  - Fatty acids — Grease, soap, lubricant
  - Fatty alchohols and sulfates — Rubber compounding, detergents, wetting agents
- Lubricating oils
  - Light spindle oils — Steam cylinder oils
  - Transformer oils — Valve oils
  - Household lubricating oils — Turbine oils
  - Compressor oils — Dust-laying oils
  - Ice machine oils — Tempering oils
  - Meter oils — Transmission oils
  - Journal oils — Railroad oils
  - Motor oils — Printing ink oils
  - Diesel oils — Black oils
  - Engine oils — Lubricating greases

Residues
- Petroleum grease — Petrolatum
  - Medicinal — Salves, creams, and ointments
    - Petroleum jelly
  - Technical — Rust-preventing compounds
    - Lubricants
    - Cable-coating compound
- Residual fuel oil — Wood preservative oils
  - Boiler fuel — Gas manufacture oils
  - Roofing material — Metallurgical oils
- Still wax — Liquid asphalts — Roofing saturants, road oils, emulsion bases
  - Binders
  - Fluxes
- Asphalts
  - Steam-reduced asphalts — Briqueting asphalts
    - Paving asphalts
    - Shingle saturants
    - Paint bases
    - Flooring saturants
    - Roof coatings
  - Oxidized asphalts — Waterproofing asphalts
    - Rubber substitutes
    - Insulating asphalts

Refinery sludges
- Coke — Carbon electrode coke, carbon brush coke, fuel coke
- Acid coke — Fuel
- Sulfonic acid — Saponification agents
  - Demulsifying agents
- Heavy fuel oils — Emulsifiers
- Sulfuric acid — Refinery fuel
  - Fertilizers

Oil Deposits

Source:
Oxford Economic Atlas, 1972

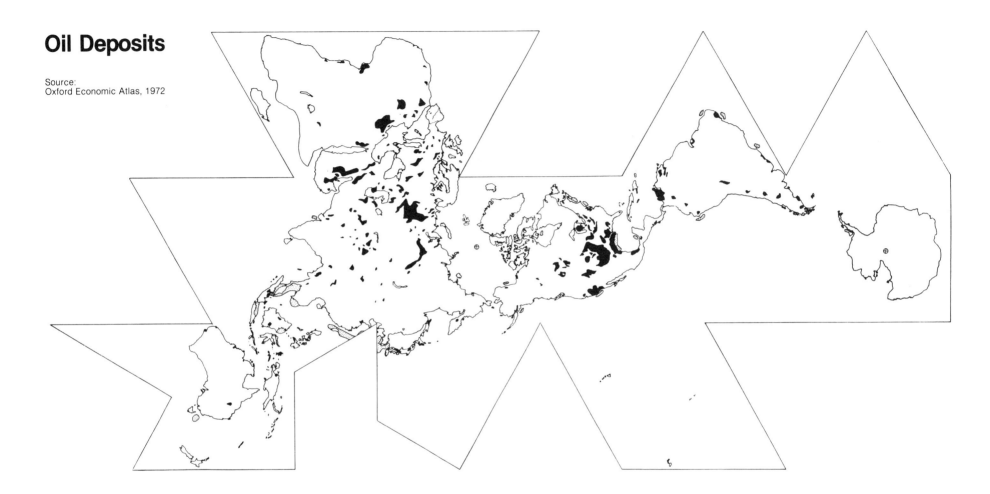

■ under land
▨ under water

Off-Shore Oil and Gas Fields

Source:
The World Energy Book, Crabbe & McBride. Nichols Publishing Co., 1978.

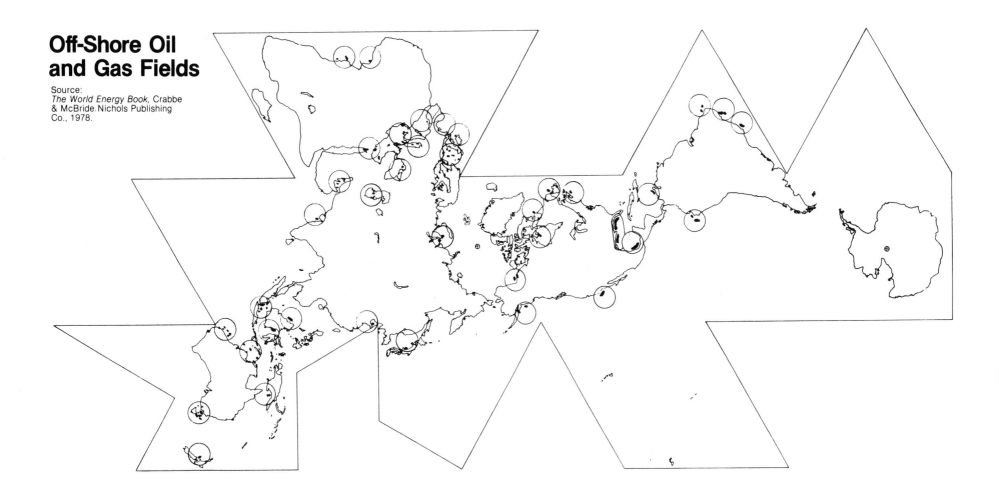

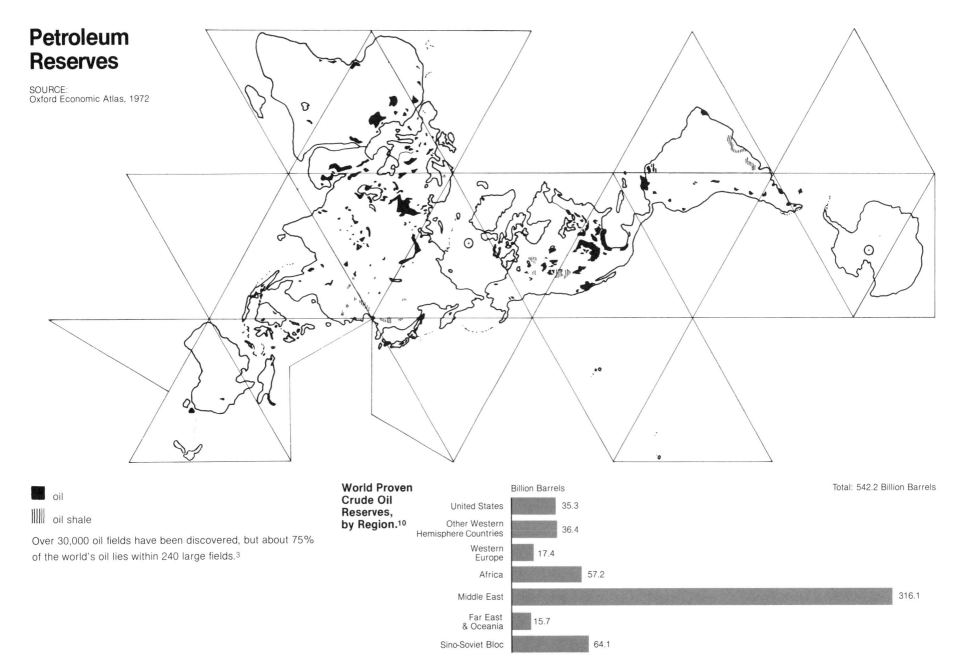

World Crude Petroleum Production*

Source:
World Energy Supplies 1972–1976 (United Nations, 1978).

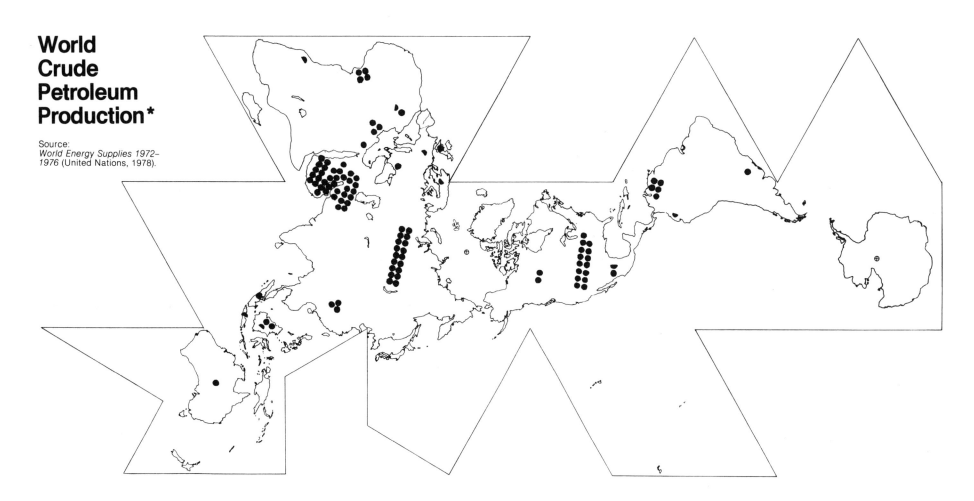

● = 1% of World Crude Petroleum Production
● = 30.5 × 10⁶ metric tons

*includes Natural Gas Liquids

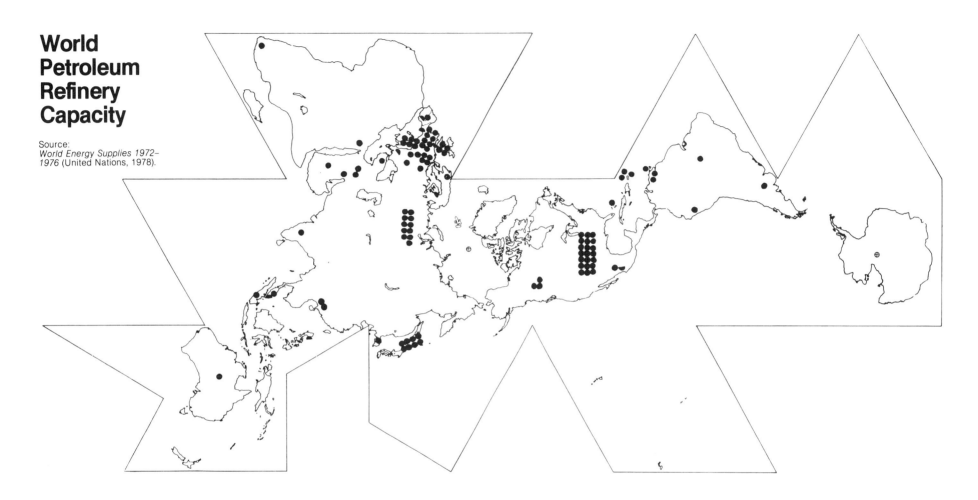

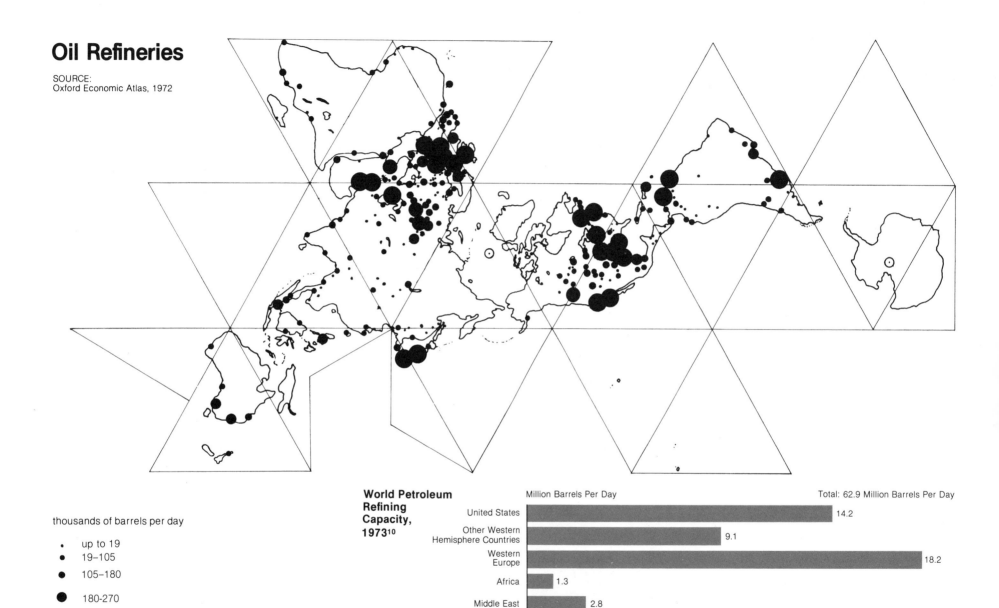

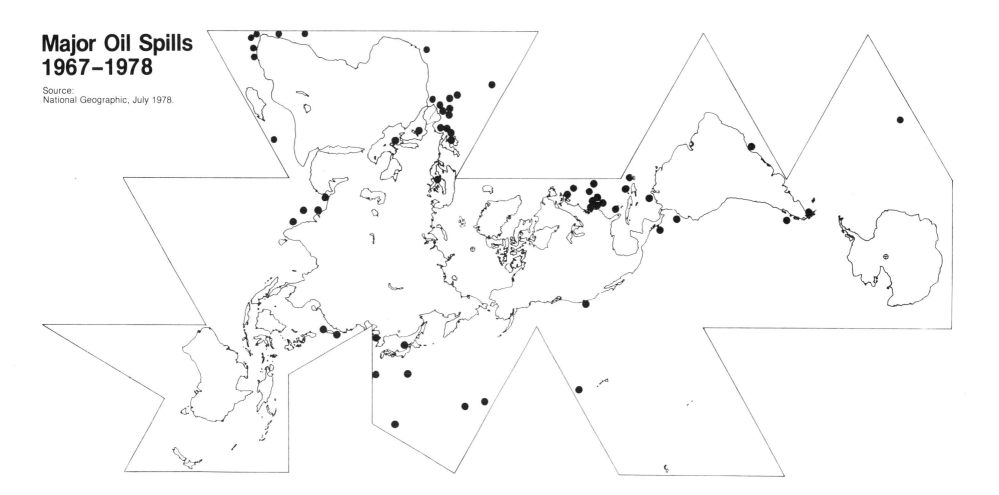

Because of its finite supply and the increasing difficulty of extracting what remains, each succeeding barrel of oil is more expensive. "...unlike many manufactured goods, which decrease in price as more are produced..." non-renewable resources increase in price.[9]

Secondary and Tertiary Methods of Oil Recovery[7]

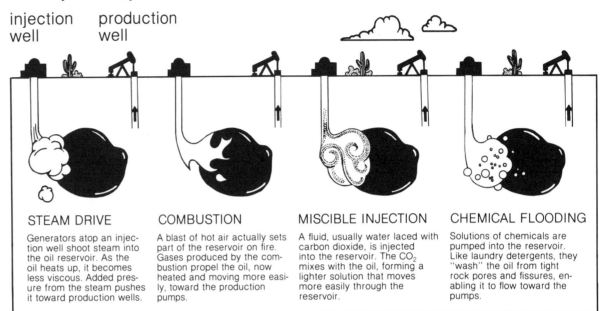

STEAM DRIVE
Generators atop an injection well shoot steam into the oil reservoir. As the oil heats up, it becomes less viscous. Added pressure from the steam pushes it toward production wells.

COMBUSTION
A blast of hot air actually sets part of the reservoir on fire. Gases produced by the combustion propel the oil, now heated and moving more easily, toward the production pumps.

MISCIBLE INJECTION
A fluid, usually water laced with carbon dioxide, is injected into the reservoir. The CO_2 mixes with the oil, forming a lighter solution that moves more easily through the reservoir.

CHEMICAL FLOODING
Solutions of chemicals are pumped into the reservoir. Like laundry detergents, they "wash" the oil from tight rock pores and fissures, enabling it to flow toward the pumps.

Oil: Potential Environmental Impacts Summary[11,12]

	Water	Air	Land	Solid Waste	Noise	Aesthetics
Production	Disposal of brine; oil spills from off-shore production		Oil spills		Fairly high when drilling	Problems due to oil rigs, pipelines, spills, etc.
Upgrading	Thermal pollution; sulfuric acid; spent caustic	Sulfur oxides; hydrocarbons; nitrogen oxides		Spent phosphoric acid catalyst; spent clay		Problems due to stacks, storage tanks, smell, etc.
Transportation	Tanker accidents; tanker dumping	Pipeline spills				
Utilization	Thermal pollution	Auto and diesel exhaust emissions; sulfur oxides; nitrogen oxides			Not considered serious— 30 db at 100 m from power plant	Problems due to chimney stacks, cooling towers, storage tanks, etc.

Advantages

1. Its stored energy is easily released.
2. It is easily mined; a Mid-East oil well can flow at a rate of 10,000 barrels a day at a "cost" of 10¢ per barrel.
3. It is easily transported, stored, and controlled.

Disadvantages

1. Once combusted, it is lost forever.
2. There is a short supply; at present increasing rates of consumption, world oil could be gone by the year 2000.
3. The supply potential is unsure and unstable due to political factors.
4. It is not found in many places around the world.
5. Environmental dangers include:
   (a) *Air Pollution.*[11] About 70,000–160,000 tons/yr. for a 1,000 MW power plant (load factor .75).

Aldehydes	.2 kg/t
CO	.1 kg/t
Hydrocarbons	1.2 kg/t
NO_x	38.6 kg/t
Sulfur	61.7 kg/t
Dust	4.4 kg/t
Radioactivity	nil

   The environmental impact of oil can be reduced through such techniques as hydrogenation to yield sulfur-free gas and emission-control devices for cars and industry.
   (b) *Water Pollution.* About 3,000–6,000 tons/yr. for a l,000 MW power plant. In addition, ocean spills and petroleum-related pollution of lakes and rivers results in more than 6×10^6 metric tons of oil discharged into the seas per year.[3]
   (c) *Thermal Pollution.* Half of the total energy consumed in electric power plants, and much more in transportation, is lost.
   (d) *Land Use.* Between 70 and 85 km² for all operations—oil extraction, refining, power plant.
6. Primary production only removes 25% of the oil in place; secondary removes about 5% more; tertiary recovery is not widely in use yet. The improvement in recovery techniques will probably result in more oil than the discovery of new oil reserves.
7. Oil production is getting more and more expensive; the investment to recover oil and gas from the North Sea will be greater than the cost of putting a man on the moon.

70% of the operating expenses of a supertanker is insurance.

Oil Shale/Tar Sands

Trapped Oil

Oil shale is shale in which a petroleum-like substance is trapped. To get oil, the shale must be either mined (from underground or open strip mines), crushed, and heated to 900° Fahrenheit in a retorting process by which the solid organic material in the oil shale is converted to gas and oil vapors, or, in another process, heated underground and the resulting gas and oil pumped out. A third technique, which offers promise for developing countries, is a labor—instead of capital—intensive method of oil extraction. In this method holes are dug into the oil shale, filled with broken oil shale, and then ignited. Heat then separates the oil from the shale and the oil settles at the bottom of the hole where it is pumped out after the fire is extinguished. Major advantages in addition to the inexpensive set-up include the fact that there are no unsolved technical problems and the process does not need water, hence it can be done in dry areas.[13] The world reserves of oil shale and tar sands amount to about 670 billion barrels of petroleum; about 130 billion are in the U.S. The U.S.S.R., China, and Brazil are the only places where commercial oil shale industries are in operation today. The U.S. will have such a facility in 1983.

Tar sands are similar to oil shale; petroleum is trapped within sand rather than shalc. Oil can be drawn in much the same manner as from oil shale. Another technique for oil extraction under investigation is the uniform heating of the tar sands through microwaves. Preliminary investigation indicates that this technique could be

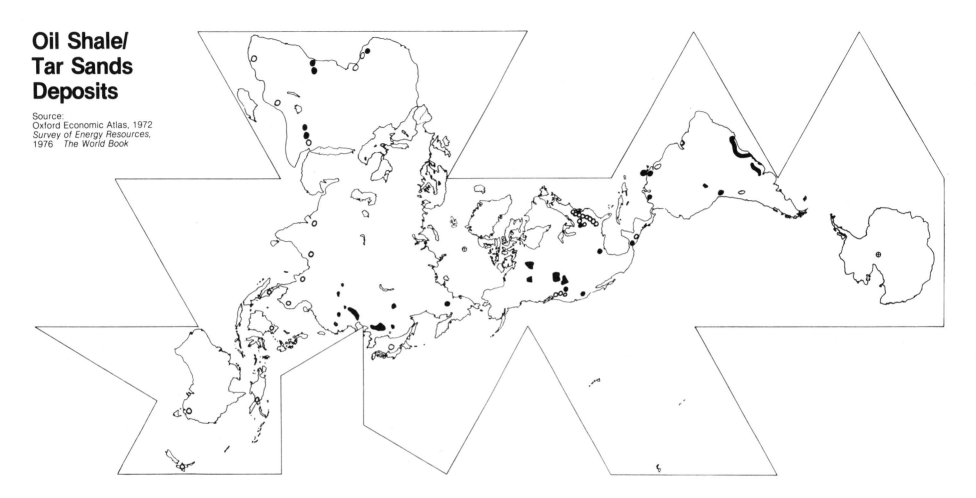

economically attractive and have little environmental impact. The tar sands of Alberta, Canada, have been in limited production (50,000 barrels of oil per day) since 1967, with a scheduled 129,000 barrel-per-day operation in 1978. Other sites where tar sands have been exploited, on a small scale, are Trinidad, Albania, Romania, and the U.S.S.R. World tar sand and oil shale ultimate reserves are approximately one-third as large as those of liquid petroleum. Because it takes so much energy to get oil from shale or tar sands, the net energy reserves are much less.

History

Prehistory	Ute Indians use "rock that burns."
1735	France: Subsurface oil mining from tar sands.
1800's	Appalachia shale used for lamp oil and medicine.
1855	Mormons in Utah produce oil from shale.
1862-1962	Scotland: Shale oil industry.
1918	200 oil shale companies spring up in U.S.
1925	Venezuela: Extra-heavy oil in tar sands discovered in Orinoco Basin (200 billion to 2 trillion barrels worth).
1935-1962	South Africa: 250,000 ton/year oil shale industry.
1940-1952	Australia: 350,000 ton/year oil shale industry.
1940-1966	Sweden: 2 million ton oil shale industry.
1944-1955	Colorado, U.S.: Oil shale mining experimentation.
1945-1970	U.S.S.R.: 13 million ton/year oil shale industry; from 1970 on, production has increased to 30-35 million tons per year.
1953	U.S.: Sinclair Oil tests in situ oil shale recovery.
1967	Alberta, Canada: Plant to obtain oil from tar sands in limited production.
1970's	China, Brazil: Oil shale industries in operation.
1975-1976	U.S.: 10,000 barrel batch of oil shale oil produced for evaluation by Department of Defense.
1978	Alberta, Canada: Second tar sands oil production plant (129,000 barrels per day) built.
1983	Colorado, U.S.: Occidental and Ashland Oil begin operation of $440 million, 57,000 barrel/day commercial oil shale facility.

Advantages

1. Using these sources increases the oil supply.
2. Production of liquid fuels from oil shale is technically simpler and cheaper than from coal.
3. Oil shale water, containing the widely-used industrial chemical ammonium bicarbonate, could be a usable and valuable resource.
4. Valuable industrial chemicals can be produced from shale oil waste—alumina, soda ash, and nahcolite.

Disadvantages

1. High amounts of energy are needed to get it; the net energy may not be enough to warrant extraction.
2. Vast quantities of shale or tar sands must be mined to produce significant quantities of oil.
3. Environmental dangers include:
   a. *Air Pollution.*[11] Total: about 1,000 tons/day for a 1×10^6 bbl./day plant.

NO_x	.09 kg/bbl.
Sulfur	.5 kg/bbl.
Particulates	.4 kg/bbl.

   b. *Water Pollution.*[11] Water requirements are very large—$100\text{-}150 \times 10^6$ m³/yr. for a 1×10^6 barrel per day plant. Pollution includes high salt concentrations, chemical runoff from erosion and leaching of spent shale deposits, and oil spills. In addition, water tables in the vicinity of the mine will be lowered if underground water is used.

   c. *Land Use.*[11] Cumulative land requirements over a 30-year period may reach 250 km² for a 100,000 bbl./day plant. Change in local topography and wildlife are a result of oil shale mining and production.

   d. *Solid Waste.* Shale: The volume of original rock material is increased by as much as one-half; consequently, a place must be found for the overburden and mine tailings. Tar sands: For each 50,000 barrels of oil produced, 33,000 cubic meters of overburden must be removed, and 100,000 tons of tar sands must be mined and disposed of.
4. Revegetating and controlling leaching from spent shale deposit sites is necessary.
5. The resource must be located near a source of water for economic production of liquid fuels; most oil shale is not near water.
6. Tar sands/oil shale exploitation is very expensive; over $2 billion will be spent for the tar sand facility in Alberta, Canada.

Natural Gas

Dinosaur breath . . .

Natural gas is also a fossil fuel, formed as a product of millions of years of pressure on organic materials. It is a mixture of gaseous hydrocarbons, predominantly methane, and is found in nature either associated with oil or alone. Globally, about 60% of natural gas reserves are found unassociated with oil and 40% associated.[1] In addition gas is found in coal seams, hydrates (including permafrost areas and deep ocean substrata), and dissolved both in underground water and stratified water bodies.[2]

Gas is combusted to release its stored chemical energy. It is used as a fuel for water and space heating and cooling, refrigeration, and electric power generation as well as being the major feedstock in the manufacture of chemical fertilizer. If there is no local market for associated natural gas this gas is then often burned as an unwanted by-product. Each year there is enough gas flared-off in the world to produce all the fertilizer the world currently uses.

A major problem of natural gas is transporting it to where it can be used. Gas is transported by pipelines (see map), or by tanker or truck in the form of liquified natural gas (LNG). The major

> *The total energy of the bombs dropped in the 1945 raid on Tokyo, which killed 83,000 people & destroyed more than 250,000 dwellings, is about 1% of the energy in one large liquid natural gas storage tank.*[3]

problems with LNG are that it is very capital intensive, inefficient (25% of the primary energy is lost in processing), and dangerous. A tanker explosion in a port could result in a major catastrophe. On shore, a serious storm, earthquake, or terrorist attack could cause a major rupture in LNG storage tank which could result in large amounts of extremely explosive gas filling nearby sewers and subways and possibly setting off a large series of explosions.[3] The major advantage of LNG is that, as a liquid, natural gas has 600 times less volume than the same amount in a gaseous state. This makes practical the shipping of large quantities of gas. Another advantage is that using LNG increases the supply of natural gas in places that need it. One other way of transporting natural gas is to convert it into liquid methanol; this would greatly simplify and reduce the risk of its transport as well as being useful as a gasoline substitute. The major disadvantage is again efficiency: 40% of the natural gas is lost in processing.

History

900 B.C.	China: Natural gas use mentioned.
400 A.D.	China: Natural gas used to evaporate brine from salt wells.
1659	England: Natural gas discovered.
1670	England: Gas distilled from coal.
1796	Philadelphia, U.S.: First demonstration of gas use in U.S.
1802	Genoa, Italy: First commercial use in Western world, lighting streets.
1821	Fredonia, N.Y., U.S.: First natural gas well.
1821	U.S.: Natural gas pipeline in use.
1829	Lake Erie, U.S.: First natural gas lighthouse.
1915	Ontario, Canada: Gas first stored underground.
1920's	U.S.: Large deposits of natural gas found in southwest and midwest.
1925	U.S.: Seamless, electrically welded steel pipe makes possible economical, long distance transport of oil and gas.
1944	Cleveland, Ohio, U.S.: First LNG accident kills 130 people.
1946	U.S.: Transcontinental gas pipelines.
1952	Gas first stored in water sands.
1965	150 million ft^3 of gas liquified daily for shipment aboard three specially insulated tankers.
1967-69	U.S.: Three underground nuclear explosions used to fracture tight sandstone formations to facilitate recovery of natural gas.
1970's	Deepest drilling up to 10,000 ft (in the 1950's it was 200 ft.).

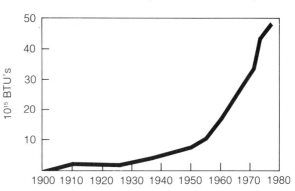

World Natural Gas Consumption (1×10^{15} BTU)[4]

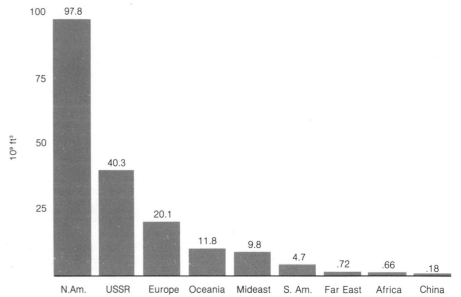

World Natural Gas Consumption 1976: 47.6×10^{12} ft^3 total[4]

Region	10^{12} cubic feet
N. Am.	23.3
USSR	10.4
Europe	9.6
S. Am.	1.5
Mideast	1.2
Far East	.92
Africa	.27
Oceania	.25
China	.15

World Natural Gas Consumption: Per Capita Average 11.8×10^9 ft^3

Region	10^9 ft^3
N.Am.	97.8
USSR	40.3
Europe	20.1
Oceania	11.8
Mideast	9.8
S. Am.	4.7
Far East	.72
Africa	.66
China	.18

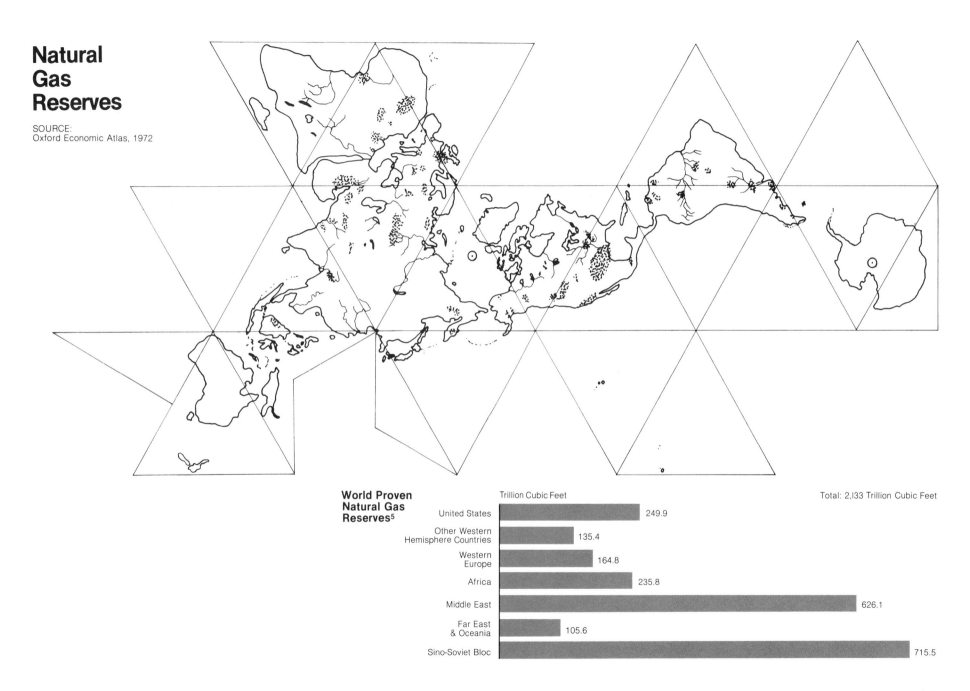

World Natural Gas Production

Source:
World Energy Supplies 1972–1976 (United Nations, 1978).

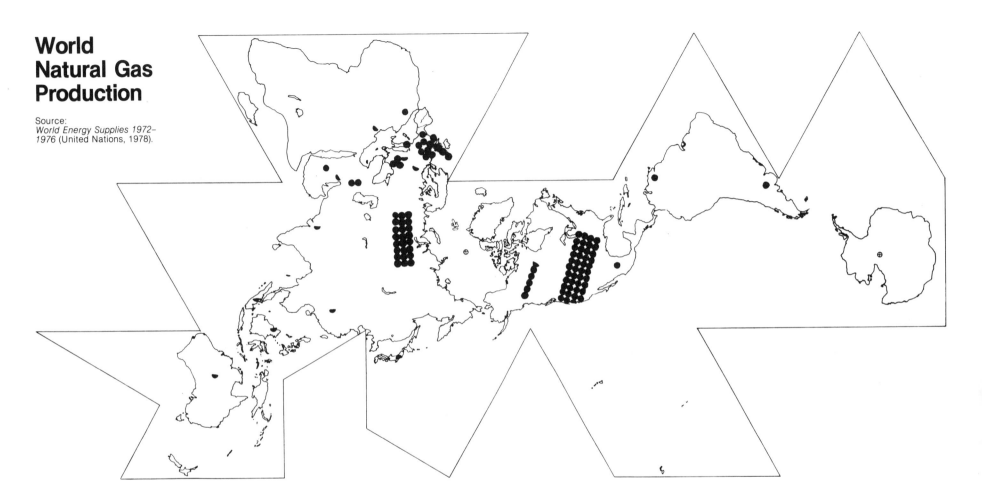

● = 1% of World Natural Gas Production
● = 12.0 × 10⁴ teracalories

Gas and Oil Pipelines

SOURCE:
Pergamon Atlas, 1968

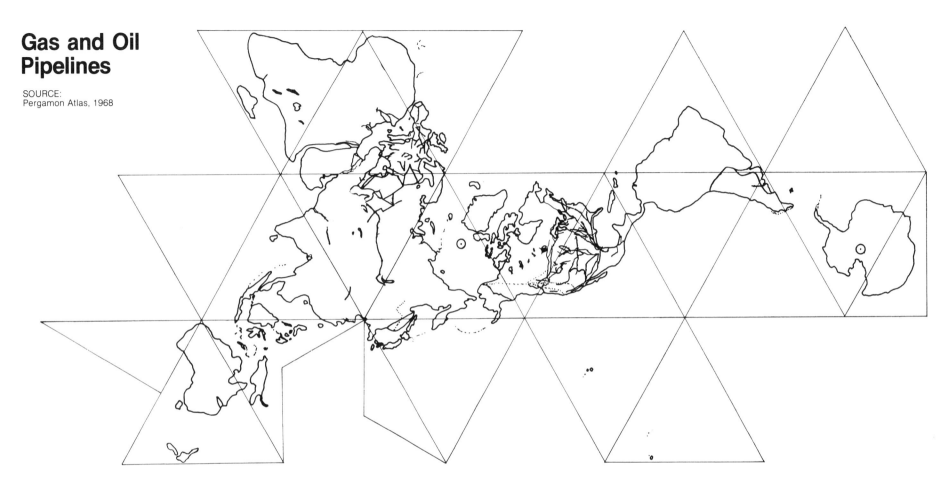

——— Gas and oil pipelines

·········· Proposed pipelines

Pipelines transport about three times as much energy as the electric grid.

Advantages

1. Its sulfur-free combustion makes it the cleanest of all fossil fuels.
2. It is easily stored, transported, and controlled in large volumes.

Disadvantages

1. Once combusted, it is gone forever.
2. It is in short supply; known and available deposits of gas leave only about eleven years of gas left in the United States at current output. The world may run out in 1991. Ways to get more gas (into the United States) would be to import liquified natural gas (LNG) from foreign sources, gasify coal, or extract methane from organic waste, coal, and refuse.
3. Environmental dangers include:
   a) *Air Pollution.*[8] About 24,000 tons/year from 1,000 MW power plant (load factor .75).

Aldehydes	.06 kg/+
NO_x	20.9 kg/+
Sulfur	0.1-.2 kg/+
Particulates	.9 kg/+

   When burned in large power plants, the high temperatures produce large quantities of nitrogen oxides.
   b) *Water Pollution.*[8] About 1,000 tons/yr. for a power plant of 1,000 MW.
   c) *Land Use.*[8] About 85 km² for all operations; about 4 km for a 1,000 MW power plant.
   d) In liquified form, there are risks involved in handling (vapor clouds, fire, and flameless explosions); no fire-fighting technology exists for dealing with LNG transport fires.
4. Gas is expensive to transport from well to consumer. Pipeline networks are so expensive that they are only justified when and where there are large reserves and assured demands.
5. Imported liquified natural gas would: a) make a minimal contribution to U.S. domestic supply; b) need a capital investment of about $11 billion to construct the necessary tankers and receiving terminals; c) run the same risks associated with large oil imports; and d) add about $4 billion annually to the U.S. balance-of-payments outflow.

Natural Gas: Potential Environmental Impacts Summary[6]

	Water	Air	Land	Solid Waste
Production				
Upgrading				
Transportation		Nitrogen oxides at compressor stations		
Utilization	Thermal pollution	Nitrogen oxides		

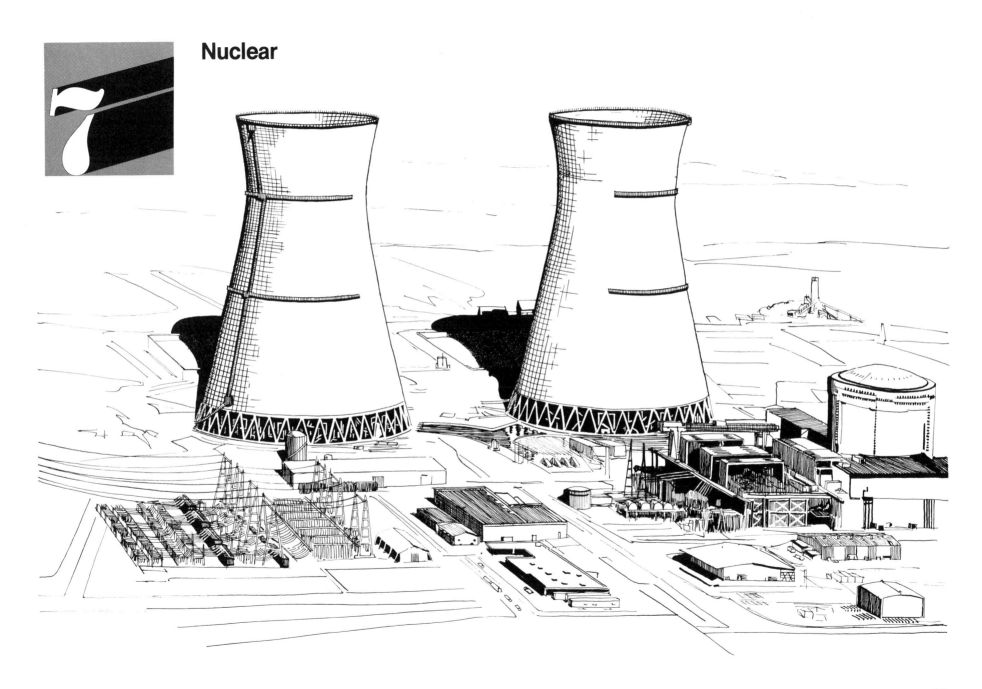

7 Nuclear

U & n (neutron) → ^{239}U → ^{239}Np (Neptunium) → ^{239}Pu (Plutonium)

Electric Bomb

Nuclear fission is the result of bombarding uranium with subatomic particles called neutrons, which split the uranium atoms in two, releasing large amounts of heat in the process. A nuclear reactor is a device for the controlled "burning" of a nuclear fuel—just as the standard coal-fired thermal-electric power plant is a device for the controlled combustion of coal. There are two types of fission reactors under consideration. Conventional reactors consume fissionable materials such as U-235. Radioactive materials produced are considered waste. Breeder reactors produce fuel by neutron bombardment while consuming the original fuel supply. Conventional reactors have the obvious drawback of consuming the Earth's very limited supply of fissionable fuels. At projected use rates, the world will be out of currently economically recoverable uranium in about seventeen years. Breeders solve the depletion resource problem by producing more fissionable fuel than they consume; however, like the conventional reactor, it produces exceedingly toxic radioactive wastes which must be safely disposed of and monitored where stored for many centuries. As events at the Three Mile Island reactor make abundantly clear, there are a number of operational difficulties with both breeder and non-breeder reactors, particularly with cooling systems. Failure of the cooling system will cause the reactor core to melt. Temperatures will then rise rapidly to the point where surrounding structures will also melt, finally producing an atmospheric release of steam and gas which sends a cloud of radioactive material over a wide area. In this "disassembly process" the resulting radioactive release could endanger the lives of millions. Each 1,000 MW breeder would contain 50 metric tons of uranium and plutonium and 40,000 cubic feet of radioactive sodium coolant at temperatures up to 1,000 degrees F, which is hot enough to catch fire if exposed to air.[1]

In addition to land-based nuclear power plants, there are proposals for ocean-based plants. In one variation on this scheme, nuclear plants would be "mass" produced in Florida and then towed to a site off New Jersey where they would be anchored and protected from storms by a surrounding breakwater.

There are many more nuclear "facilities" in the world than is popularly known. Besides the 170-odd commercial nuclear electric power plants located primarily in the U.S., England, Russia, Japan, Germany, and France there are at least 355 other nuclear "research" reactors scattered throughout the world (see map) and nearly 1,500 existing nuclear facilities (waste storage, reprocessing, etc.) in just the U.S. So-called "research" reactors, these facilities nevertheless have a lot of the same problems and dangers that commerical reactors have, including those of weapons proliferation dangers and decommissioning. In addition to these "civilian" nuclear facilities there are over 50,000 nuclear weapons in the world (30,000 in

---

American oil companies own 47% of the U.S.'s known uranium-ore reserves.

A technician monitors the nuclear fuel loading operation at Unit No. 1 of the 880-MW Calvert Cliffs Nuclear Power Plant, Maryland. Each fuel bundle (of a total of 217 bundles) is being positioned by an automatic fuel handling machine, shown extending down to the upper core level.

just the U.S.) with over half the U.S. arsenal on the high seas or in other countries. There are numerous nuclear powered ships as well. The U.S. Navy alone operates 114 nuclear reactors for its 106 nuclear submarines, two aircraft carriers and six cruisers. Lastly, nuclear power sources are in wide use in outer space. Both the U.S. and the Soviet Union have been using nuclear power sources for their spy and

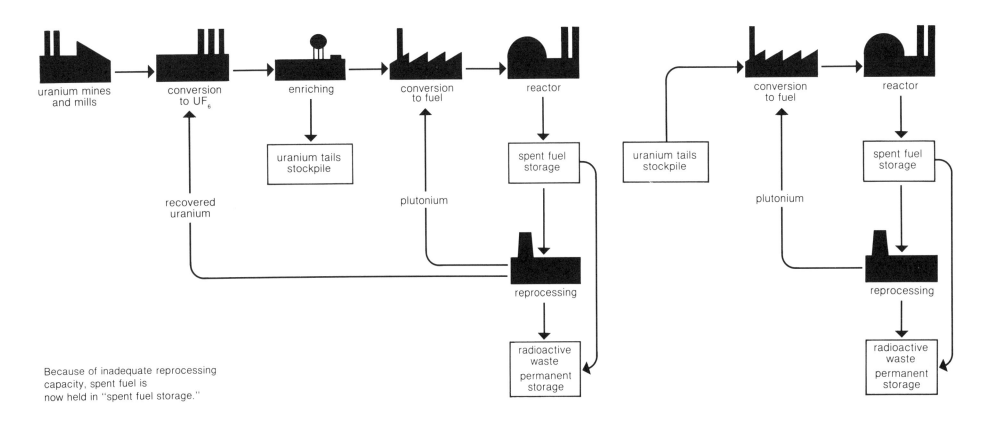

Fuel Cycle of Fast Breeder Reactor[4]

Fuel Cycle of Light Water Reactor[4]

Because of inadequate reprocessing capacity, spent fuel is now held in "spent fuel storage."

navigation satellites for years. So far, there have been three nuclear accidents involving outer-space nuclear power (that the public knows about) that have resulted in radioactive waste being dumped back into the Earth.[2] In addition, there have also been three catastrophic accidents involving the total loss of three nuclear powered submarines, two American and one Russian.[3]

At the moment, dismantling nuclear facilities is costing more than their original construction. One nuclear facility in Minnesota cost $6 million to build and $6.2 million to dismantle; another in New Jersey will cost an estimated $100 million to dismantle even though the original cost was only $65 million. The nuclear industry has said that a nuclear reactor will probably have to be buried underground for 65–110 years before the cobalt-60 in the reactor vessel is sufficiently decayed to permit manual dismantling. There are 300 obsolete nuclear facilities right now in the U.S. (20 of which are power plants); ERDA has estimated that it will cost $25-$30 million annually for 100 years to decommission these facilities alone.[5] To decommission all of the existing facilities just in the U.S. will run about $15 billion.[6] World-wide costs could be $40

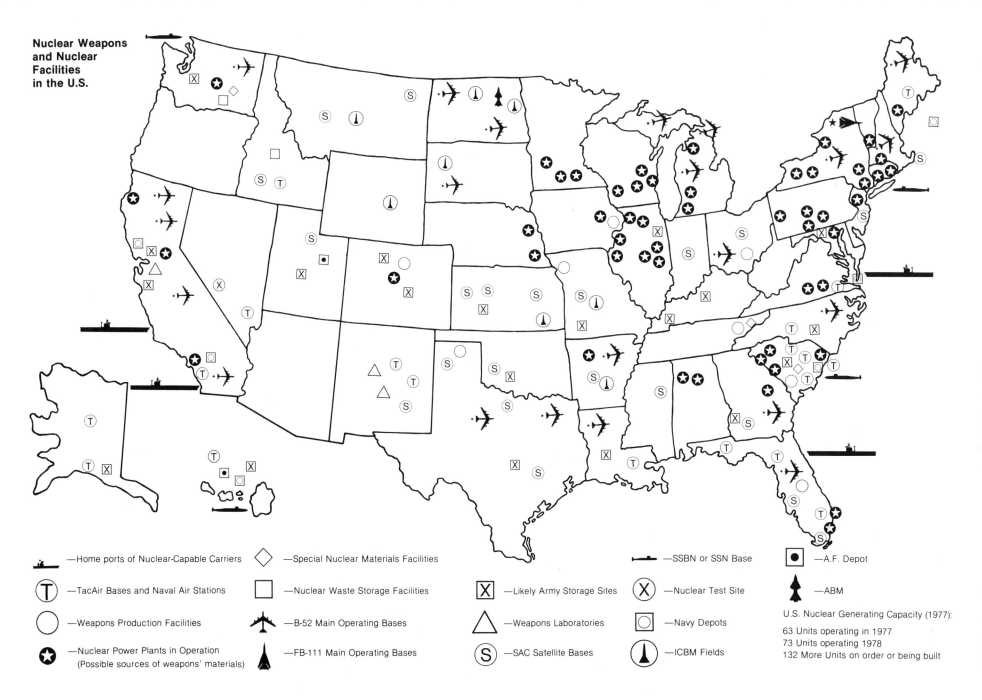

Should utilities post a bond now to pay for cleaning up obsolete reactors later? Or should consumers in the year 2000 pay higher electrical rates to cover the cost of dismantling today's reactors?

---

billion[7] or more. And that is just for *existing* facilities. Nowhere in utility bills are these added costs figured in. The amount of structural and reinforcing steel tied-up by nuclear power plants is also substantial. At least six million tons are currently being made so radioactive by nuclear power plants around the world that they will be unavailable for recycling for centuries. This long-term tie-up of the steel does not occur with other types of power plants. When their useful life is over, the plant is disassembled, sold for scrap—usually at a price in excess of the dismantling costs—and recycled.

In addition, nowhere in the price of nuclear-generated electricity is the cost of storing nuclear waste. Fuel disposal could run as high as $23 billion for the U.S.[8] (World total could reach $66 billion.) There are *already* 75 million gallons of high level military nuclear wastes and about 2,500 metric tons of nuclear waste from the existing commercial nucelar power plants. By 1985, this figure will have grown to 25,000 tons, and if nuclear power goes the way the Nuclear Regulatory Agency projects it, there would be 125,000 tons by the year 2000. Nowhere in the price of nuclear-generated electricity is the cost of guarding nuclear wastes from theft for thousands of years. Plutonium is worth about $10,000 per kilogram, considerably more than heroin or gold. Black market demand could be much higher; and the

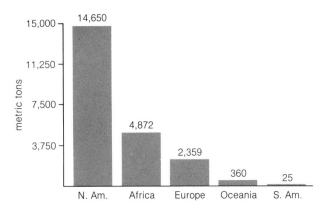

World Nuclear Production[12] 1976: 22,268 metric tons —Uranium Content

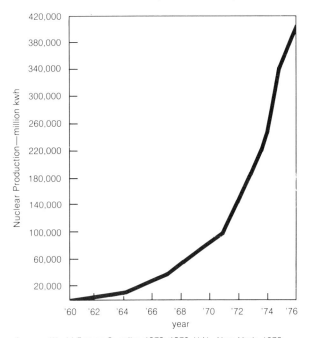

Nuclear Power Production (in million kwh)[12]

Source: *World Energy Supplies 1972–1976*. U.N., New York, 1978.

Between 1969 and 1976, ninety-nine separate incidents of threatened or attempted violence against licensed nuclear facilities were reported in the U.S. alone.[9]

---

more something is worth, the more some people will want to steal and resell it.

About 25 million tons of mill tailings—residual radioactive materials—have accumulated at 22 inactive uranium mills since the 1940's.[10] The cost for cleaning up these sites has yet to be determined. Another unaccounted cost is the immense expense of evacuation in case of emergency. As the Three Mile Island disaster made clear, evacuation plans have not even been adequately drawn-up for over half of the reactors in the U.S., nor have the plans that do exist been tested or communicated to the populations which they affect, nor has it been determined who is to get stuck with the bill for the execution of the plan. Also, there is not reflected anywhere in the price of nuclear-generated electricity the cost of the Nuclear Regulatory Commission ($256 million annual budget; $25.6 billion for the next 100 years), or the over 30 other government agencies or departments that are involved in some form of service or subsidy to the nuclear power industry.[11] The taxpayer, both today and in the future is stuck with the bill (in more ways than one), which amounts to nothing more than corporate socialism. Government subsidy pays for industry profit.

Uses

Nuclear technology is used or can be used for the following:

1. Generation of electricity.
2. Desalination, principally for municipal and industrial uses.
3. Radiation entomology.
4. Pesticide residues and food protection (e.g., sterilization of fruit flies).
5. Plant breeding and genetics.
6. Animal protection and disease control (e.g., parasite irradiation).
7. Food irradiation, to improve storage life.
8. Nuclear medicine.
9. Radiation biology.
10. Radiation dosimetry (i.e., standards of measure for radiation dosage).
11. Hydrology.
12. Weaponry.
13. Industry (e.g., mining of natural gas). Russia has used nuclear blasts to create underground chambers for storing gas and oil and is now planning a major ground-level excavation project to divert water from several northern rivers to the Caspian Sea via a deep 113-kilometer canal.

History

1906 Switzerland: Einstein postulates that $E = Mc^2$.
1938 Germany: Hahn and Stassman observe first fission.
1942 Chicago, U.S.: First atomic fission reactor under the direction of Fermi.
1945 Alamogordo, New Mexico, U.S.: First fission explosion; uranium bomb dropped on Hiroshima, plutonium bomb dropped on Nagasaki, Japan.
1949 U.S.S.R.: First Soviet nuclear bomb explosion.
1952 First British nuclear bomb explosion.
1954 U.S.: First nuclear-powered submarine, *Nautilus*.
1954 U.S.S.R.: First nuclear power plant for civilian use (5MW).
1957 England: First nuclear power plant for civilian use (Calder Hall) in England begins operation.
1958 Pennsylvania, U.S.A.: First civilian nuclear reactor begins operation.
1958 U.S.S.R.: Accidental atomic explosion at nuclear facility; "hundreds" killed.
1960 First French nuclear bomb explosion.
1960-65 U.S. Atomic Energy Commission oversees development of fission reactor technology for electrical power generation.
1963 Nuclear-powered submarine Thresher lost in Atlantic.
1964 First Chinese nuclear bomb explosion.
1964 U.S.: Plutonium nuclear reactor disintegrates over Madagascar on entry into atmosphere after aborted orbital attempt.
1964 Idaho, U.S.: Experimental 20-MW breeder reactor begins operation (still operating in 1979).
1966 U.S., near Detroit: Fermi laboratory near "melt-down" incident in AEC-developed breeder reactor.
1968 Nuclear-powered submarine Scorpion lost in Atlantic.
1970 U.S.: Plutonium nuclear reactor plunges into atmosphere from orbit.
1970 Fission produced electricity becomes "economical" and a large number of plants begin construction.
1975 U.S.A. has 56 operating nuclear power plants (38,000 MW, 8% of U.S. electric power industry total).[13]
1976 U.S.S.R.: 150-MW nuclear breeder begins operation on the shore of Caspian Sea.
1977 World has 204 commerical reactors with a combined capacity of 94,841 MW (U.S. has 64 that furnish 12% of U.S. electricity).
1978 U.S.S.R.: Nuclear-powered spy satellite plunges into Earth atmosphere above Canada.
1978 Israel: Israeli scientist says Israel can produce uranium from phosphates at a cost below world uranium market price.
1979 Pennsylvania, U.S.A.: Near melt-down at Three Mile Island nuclear power plant.
1979 U.S.A.: Shutdown of five commercial power reactors in eastern U.S. for possible dangers posed by earthquake damage to cooling systems.

Uranium Resource Locations

Source:
Oxford Economic Atlas, 1972
Congressional Quarterly,
"Energy Crisis in America."

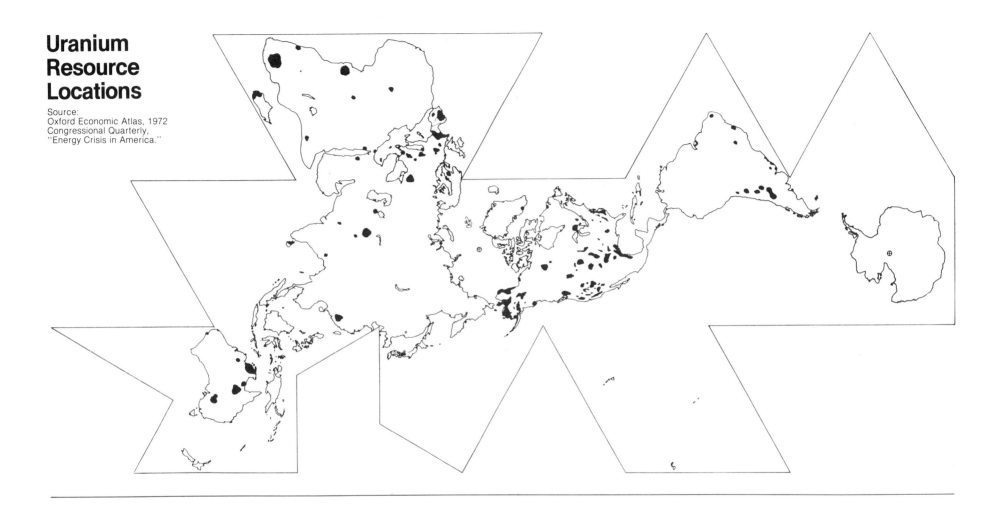

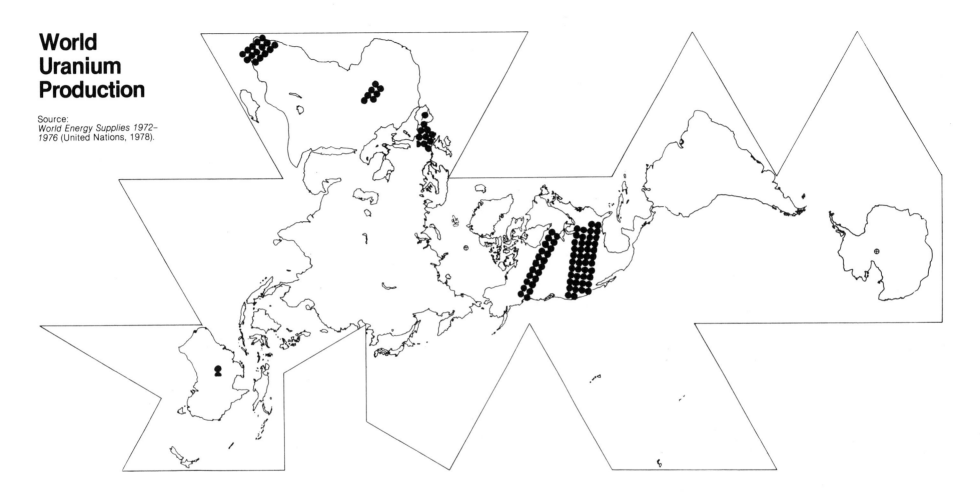

Nuclear Research Reactors In Operation

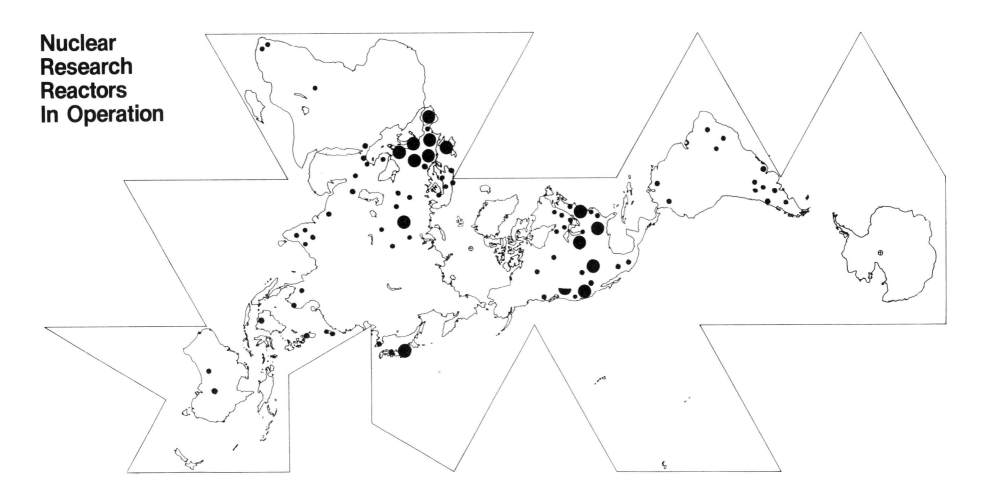

- ● 20 research reactors
- ● 10 research reactors
- • 1 research reactor

Number of Power Reactors (Over 20 MW$_e$)

Source:
Nuclear News, "World History of Nuclear Power Plants," 1977.

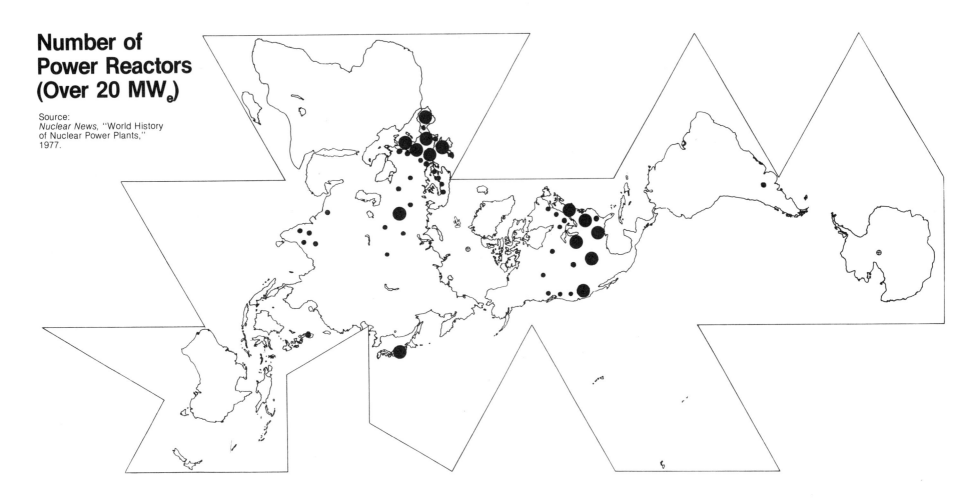

● 10 reactors
• 1 reactor

185 Commercial Reactors operating in 1977; 504 units on order.

Advantages

1. Substitution of locally concentrated waste by-products (radioactive) for the more diffuse emissions of conventional fossil fuel power plants; such as CO_2, SO_2, CO, etc.
2. Uranium ores require less mining, thereby causing less water pollution, land disruption, and human injuries.
3. Its fuel supply is concentrated: 1 ounce of uranium has roughly the same power potential as 100 tons of coal.
4. Transportation costs for nuclear fuels are lower than fossil fuels.
5. The life expectancy of the remaining fuel supply (for breeder reactor) is high.

Disadvantages

1. Environmental dangers include:
   (a) The release of large quantities of waste heat; about 40% more than conventional fossil fuel power plants.[14]
   (b) The production of some of the most toxic substances known to man in the form of nuclear waste, which endangers health by genetic mutation, cancer, and radiation burns. These wastes must be stored for *hundreds of thousands of years* before they become non-toxic to humans,[15] containers for storage last decades while toxicity lasts for millennia. So far over 430,000 gallons of high-level radioactive waste has leaked since 1945.[16] The volumes are not meager: low level waste schedules for production in just the U.S. by the year 2000 will amount to about one billion cubic feet—enough to cover a four-lane highway coast to coast (3,000 miles) one foot deep.
   (c) The steady release of the gas Krypton 85 may alter the electrical properties of the atmosphere, thereby affecting global climate.[17]
   (d) High water use for evaporative cooling can pose ecological problems where water is scarce.
   (e) A recent survey of cancer incidence indicates that residents of regions with nuclear power plants have higher cancer rates than regions without nuclear power plants. Low-level radiation is increasingly becoming a cause for concern. For each two trillion kwh generated by nuclear plants, 20 to 200 deaths will result from occupational radiation exposures; .3 to one death from routine radioactive atmospheric emissions; and 4,000-7,000 nonfatal accidents and illnesses.[18]
2. Large initial capital investments are necessary to build plants. A 1,000 MW power plant costs in excess of $1 billion: this does *not* count fuel costs, regulatory costs, waste storage costs, decommishing, disassembly, and "entombment" costs, or the social costs.
3. There is a short supply of uranium fuel; only 17 years worth of uranium is left (without breeder) at current fuel costs.
4. Its thermal efficiency is low; about 32%.[14]
5. In conventional nuclear reactors large amounts of energy are needed to "enrich" the naturally occurring uranium to the chain reaction sustaining U-235.
6. Accident and sabotage risks are high; humanity's vulnerability to terrorism will be increased.
7. Nuclear materials can be diverted for use in

An ideal site for a nuclear power plant is a place where there is no evidence of any seismic activity over the past milennia, is not subject to hurricanes, tornadoes, floods, or any other "act of God." It should be in an endless expanse of unpopulated desert with an abundant supply of very cold water flowing nowhere and containing no aquatic life. Most importantly, it should be adjacent to a major electrical use center.

> *Protecting ourselves against future terrorism means nothing less than building a nuclear system able to withstand the tactics of future terrorists fighting for a cause that has not yet been born.*[9]

nuclear weapons. Proliferation of commercial reactors will lead to proliferation of nuclear weapons.
8. Huge nuclear facilities will engender a more centralized technocratic and authoritarian society; it has even been said that nuclear power is *inherently* totalitarian, and therefore, by definition, *unconstitutional.*[19]
9. Nuclear power reactors in wartime are vulnerable targets.
10. Nuclear power produces only heat and electricity.

Nuclear Fuels: Potential Environmental Impacts Summary[20]

	Water	Air	Land	Solid Waste	Radiation
Production	Leaching of waste banks	Strip mining	Waste from under-damage	Exposure of ground mining	miners
	Uranium mine water				
Upgrading	Leaching of waste banks	Particulate emission and waste banks		Wastes from ore dressing	Exposure of plant workers
Transportation					
Utilization	Thermal pollution			Waste disposal from fuel processing plants	During generation and disposal of waste

Income
Energy
Sources

Introduction: Old Myths and New Structures

Not Enough

One of the prevailing fallacies about so-called "alternative" energy sources (what is alternative about the Sun?) is that they are, or could be, of insignificant magnitude when compared with the total amount of energy now being used. For example, the contribution that biomass could make *immediately* (medium and long-range contributions aside) to solve the U.S.'s energy problems has been termed "small compared to the 75 quads (10^{15} BTUs) of energy now consumed by the U.S. every year." This comparison ignores a number of fundamental points, not the least of which is the fact that by using biomass for energy on the farm or in industry, where large-biomass wastes are accumulated, the end-use efficiency will be much higher than by using electricity or gasoline that is produced elsewhere and transported to the farm. The contribution of biomass in this instance is thus much larger than what appears in a superficial comparison because one quad of biomass already delivered and appropriately matched in energy quality to end-use needs could be worth two to three quads of energy at the electric plant or oil refinery. In addition, the comparison mentioned above neglects to point out how much of that seventy-five quads is wasted. If we were to make efficient use of our energy we wouldn't need so much of it and then the immediate contribution of biomass (or any other underutilized energy source) would be a much higher percentage. Also, the comparison fails to point out that the amounts of energy available from non-utilized energy sources are cumulative and synergetic; that is, biomass should be counted along with wind, tidal, hydroelectric, solar thermal, solar electric, etc. Very often, closer examination of a particular site and its daily and seasonal variations will reveal that the various regenerative energy sources are complimentary; that is, when the Sun isn't shining, the wind is blowing, or the waves are high, etc.[1]

Last, but not least, it should be noted that the above quote about the small contribution that biomass can make is basically fallacious when viewed from a global perspective. Biomass (just in the form of fuel wood and biogas plants) already produces more energy than nuclear and hydroelectric power combined. In fact, biomass accounts for nearly 10% of the world's total energy use.

Costs

Another argument often used to disparage income energy sources is their reputed costs. When a windmill, solar collector, biomass facility, etc., is priced today, it is usually compared to energy sources that are subsidized by tax policies, past Federal research and development funds (at least $100 billion and possibly as much as $300 billion over the last twenty years), present research and development funds and demonstration grants (more than five times the amount spent on income energy sources), depletion and depreciation allowances (over $40 billion has gone to just the oil industry), socialized insurance risks, institutional arrangements, building codes, government requirements and regulation, utility rate structures, and "external" or "social costs"—such as environmental effects—which are paid for by society. Even under these circumstances, regenerative energy sources often cost less than their capital energy counterparts when compared on a life-cycle cost basis. If compared on the basis of the replacement costs of the present-day fuels, that is, what it would cost to produce one additional unit of the energy source in question, say, to produce additional natural gas or oil with synthetic gas or oil (at at least twice today's prices), then the income energy sources come out cheaper. Adding to this replacement cost all the other costs outlined in the first part of the book under the heading of regenerative accounting systems, then the non-depletables are cheaper by far. Yet another factor that positively affects the cost of our non-depletable energy sources is the fact the the difficulty of acquiring that energy does not increase, as it does with oil extraction for example, as we extract more and more of it. Just the opposite happens. The more solar energy in use the more we learn about extracting it ever more efficiently.

Implicit within the comparison of one energy source vs. the total needed energy is the idea that this one energy source is being considered in a monological way: here is *the* answer. There is no one answer. We need to consider the whole energy system, not just one narrow facet. We need to consider systems, not just fuels. When this is done, it will be readily seen that an orchestration of energy sources to match the galaxy of energy needs in their unique locations is what is needed to solve our energy problems.

In relation to this, the opposite argument has been made for large-scale nuclear and fossil-fuel power. It is said that large-scale (1,000 MW or larger) thermal electric power plants are the only things that will make a significant dent in meeting our growing energy needs. This statement suffers from a number of false assumptions. The first has to do with the already mentioned matching of end-use efficiencies with energy sources. It is not necessary to have a thousand or million-degree fire 100–1,000 miles away from your house to heat your kitchen or your tea to 65° F or 212° F respectively. Large-scale power advocates also assume that one centralized facility is "better" than ten thousand decentralized facilities. It has been pointed out recently that this is not the case, both from a social well-being standpoint and, surprisingly, from an economic standpoint as well.[2] Small-scale decentralized income energy facilities have numerous advantages besides those mentioned above. For a given power output, mass production is today proving to be cheaper than one of a kind production, no matter what size. The cost reduction of "economies of scale" that are often claimed for large-scale power plants is far outweighed by the even larger cost reductions of mass production and the diseconomies of large-scale power—primarily breakdowns that cripple large segments of the power system; transport/transmission costs and losses; and loss of flexibility in dealing with fluctuations in demand and emergencies. For example, a small heat engine can be mass-produced at a cost of what it takes to produce an automobile engine—about $13 per kilowatt. This compares with the $500 to $1,000 per kilowatt for a coal or nuclear power plant. Decentralized power facilities also enable the user of the energy to bear the social or environmental costs (and benefits) rather than some distant nonentity.

Storage

The need for energy storage is another oft-repeated objection applied to income energy sources. But this "need" is in fact a remnant of the conceptual set used for judging centralized power sources, where concentrated sources are systematically diffused over a large area to energy customers. It is not necessary when starting with an already diffuse source to first concentrate it to the level of a centralized power plant and then diffuse it again over the electric grid or gas pipeline. Storing income energy on the scale of its end-use is not difficult. Water tanks, rock beds, fusible salts, flywheels, hydrogen, and compressed air are all available for storing heat, electricity, or mechanical motion on a daily or seasonal basis, and on a household, neighborhood, or industrial level.

When making decisions about energy systems, our evaluation, be it technological, economic, social, or ecological, has to be in terms of whole energy systems, not just conversion systems or fuels; evaluation should be in terms of the needs, possibilities, and compatibilities of the whole Earth and the life forms co-evolving with and on our planet.

8
Hydroelectric

Controlled Rain

Falling water is an income energy source. Hydroelectric power is the conversion of the gravitational pull of the falling water of rivers and controlled release water reservoirs through turbine generators. The amount of power available is determined by the volume of water and by the distance the water falls. A large amount of water dropping a little or a small amount of water dropping a great distance can produce the same amount of energy.

Hydropower provides a clean, efficient means of producing electrical power; it does not alter the energy balance of the Earth and does not release pollutants to the environment. Currently, hydroelectric plants produce 2.3% of the total global energy production (1.46×10^{12} kwh total, 365 kwh per capita[1]), and about 23% of the world's electricity. It is estimated that the undeveloped global hydroelectric potential is 5×10^{12} to 50×10^{12} kilowatt-hours per year.[2] South America, Africa, and Southeast Asia contain, respectively, 20%, 27%, and 16% of the world's theoretical hydro-capacity.[3] Power output from existing hydroelectric facilities could be upgraded by replacing old turbines with newer, more powerful ones and by adding additional turbines at sites where this is possible.[4]

Tidal "lift translator" has two sets of fixed guide vanes that are raised or lowered, depending on the direction of the tide. These improve the efficiency of the moving vanes by optimizing the angle at which water passes them. In a river or canal setting, only one set of fixed vanes is needed.[8]

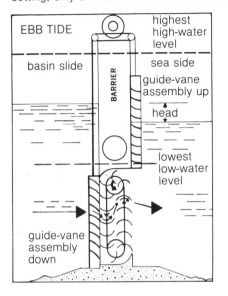
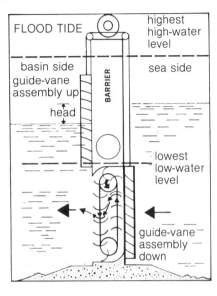

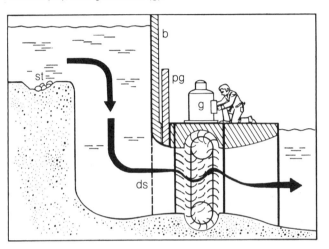

This cross section of low-head dam shows barrier (b), stone and sand trap (st), pressure gate (pg), debris screen (ds), and generator (g).[8]

Small-scale water power is also feasible throughout the world. Such installations could power grain and lumber mills as they have done extensively in the past, or produce electricity in the one-kilowatt to ten-megawatt range. China alone has over 70,000 small hydroelectric facilities in operation, most of which are in the 25-kilowatt or less size range.[5] The U.S. could almost double its hydro-output by simply outfitting each already-built dam with a generator. There are some 48,000 sites for such development[6] that could supply an additional 54,000-MW of power to the U.S.[7]

In addition to conventional small-scale hydro turbines, that are, in effect, merely miniaturizations of Grand Coulee-size hydro turbines, there is a new type of energy-harnessing device that offers promise of being able to tap very small elevation differentials and slow-moving hydro sites, such as rivers, irrigation canals, and tidal locations. Called a "lift translator," the device extracts energy from large cross-sections of slow-moving water (or even air) through a continuous revolving belt of foils—like a vertical conveyor belt—that the passing water lifts, much the same way air flow over an airplane wing lifts the plane off the ground (see illustration).[8] Such "lift translators" could tap the energy of rivers having a slope as shallow as a drop of five feet in ten miles. The potential energy tied up in this source has been estimated to be more than 2.5×10^{12} kwh.[8]

Water Power

SOURCE: Hubbert, Resources and Man, 1969

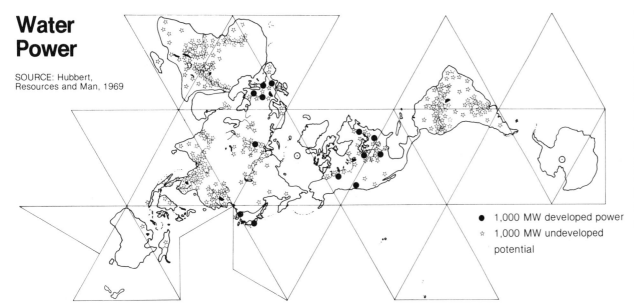

- ● 1,000 MW developed power
- ☆ 1,000 MW undeveloped potential

"The total world capacity is given as 2,857,000 megawatts. Of this, it is significant that the continents of Africa and South America, both of which are deficient in coal, have the highest potential water-power capacities of all the continents. . . . It thus appears that if the world's potential conventional water-power capacity were fully developed it would be of a magnitude comparable to the world's total present rate of energy consumption. From this, it might be inferred that without the present supply of fossil fuels, the world could continue at an industrial level comparable to that of the present on water-power alone."[3]

World Hydroelectric Production

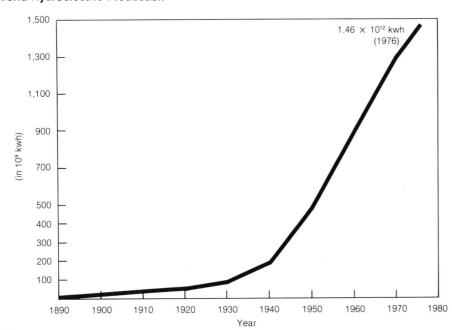

1.46 × 10^{12} kwh (1976)

World Hydroelectric Production 1976 (Total: 1,456 × 10^{12} kwh)

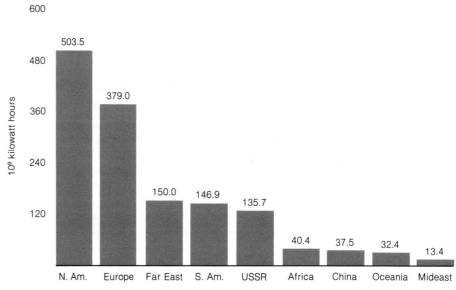

Region	10^9 kilowatt hours
N. Am.	503.5
Europe	379.0
Far East	150.0
S. Am.	146.9
USSR	135.7
Africa	40.4
China	37.5
Oceania	32.4
Mideast	13.4

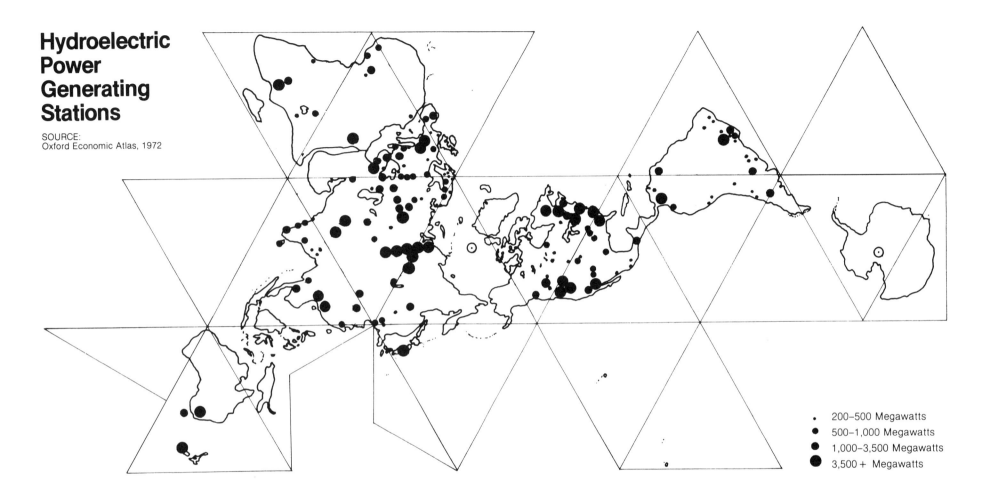

World water-power capacity

Region	Potential (10³ MW)	Percent of total	Percent developed	Development (10³ MW)
North America	313	11	33	104
South America	577	20	5	31.2
Western Europe	158	6	71	113
Africa	780	27	1	9
Middle East	21	1	13	2.8
Far East	497	17	8	40
Oceania	45	2	20	9
U.S.S.R., China and satellites	466	16	18	88
Total	2,857	100		152

Source: M. King Hubbert, 1962, Table 8, p. 99, computed from data summarized by Francis L. Adams, 1961; World Energy Supplies 1971–1975 U.N., N.Y. 1977

Ocean Hydroelectric

Another energy source that is of comparable magnitude to all the world's conventional river hydroelectric potential is the use of suitable areas of the world's oceans, which are isolated (dammed) to produce a water-level differential. This can then be tapped the same way a conventional river hydroelectric plant taps its power.[9] These areas can be considered as sources or sinks for water, so that, if isolated, the level of water inside them will rise or fall continuously. A source area is an area where local precipitation exceeds local evaporation, and a sink area is an area in which the opposite is true. A sink area can be selected anywhere in the arid belts of the world, but five major oceanic areas exist that are almost completely surrounded by land, in addition to being arid. Relatively small gaps connect these areas with the world's oceans, but closing any of them would be a major feat. The vast amounts of waste rock, overburden, and mine tailings from possible shale oil development could possibly be considered as "fill" for a project of this nature.

Another variation of this theme is termed "evaporative hydroelectric." According to this alternative, any place where evaporation greatly exceeds precipitation and that has the right geographical features, could be a potential site. For example, Pacific Ocean water could be tunneled to Death Valley, California, where because of the enormous evaporation rate and an elevation 200 feet below sea level, the flow of water in and subsequent fall could generate about 2,000 MW of electric power through conventional hydroelectric techniques, plus provide needed moisture to the Southwest.[10] As far fetched as this may sound, a team of engineers in Israel have proposed to do something quite similar with the Dead Sea. Their plan calls for a 45 mile (75 km) tunnel running from the Mediterranean to the Dead Sea. Here the water would fall 1,312 feet (400 m) into the Dead Sea and generate 300 MW of power.[11]

Yet another potential source of energy related to hydroelectric is to tap the temperature differential that builds up in the fresh-water reservoirs behind dams. The thermal energy contained in a typical reservoir—even taking into account the difficulties in converting this diffuse energy to useful work—greatly exceeds the gravitational hydroenergy.[12] A temperature differential power plant similar to those currently under development for ocean use (see Temperature Differential section) or an advanced version of the Nitinol engine (see Phase Transformation section) could produce power from the temperature difference between the sun-warmed surface waters and the cold deep waters. Surface waters are generally 5–20° C greater than the water temperature 10 to 20 meters below the surface.[12] It would be possible to greatly increase power output at hydroelectric installations that had such temperature differentials simply by adding such power-producing engines. Some of the advantages of this approach are that hydroelectric plants are already equipped with the necessary equipment for power distribution and transmission lines, there should be no negative environmental impact, and the greatest temperature difference occurs in the summer, which corresponds to the time of peak power demand and also often the time of low power output capabilities at the hydroelectric facility because of low levels of water in the reservoir.

Areas Suitable for Oceanic Hydroelectric-Power Generation

Area	Excess evaporation	Width of entrance and sill depth	Theoretical power potential for a 100 m head and 80% excess evaporation (1 MW)	Rise in world ocean level for a 100 m drop in marginal sea level
Mediterranean	67,000 m³/sec	30 km, 320 m	52,500 MW	67 cm
Red Sea	50,000 m³/sec	20 km, 100 m	39,200 MW	12 cm
Persian Gulf	20,000 m³/sec	95 km, 100 m	15,700 MW	7 cm
Gulf of Calif.	16,000 m³/sec	175 km, 3,000 m	12,500 MW	5 cm
Gulf of Mexico	50,000 m³/sec	225 km, 1,600 m; 220 km, 1,600 m	39,200 MW	41 cm

History

300 B.C.	Water wheels in Greece.
100 B.C.	Horizontal shaft water wheel produces 2 kilowatts power.
100 B.C.	Water wheels used by Romans.
500 A.D.	Water wheels come to Europe.
1000	Water-driven blast furnace.
1500	Water pumping works.
1600	Versailles water works produces 56 kilowatts power.
1800's	Development of the hydraulic reaction turbine.
1827	France: Outward-flow turbine installed.
1882	U.S.A.: Pelton invents Pelton Wheel turbine.
1882	Appleton, Wisconsin, U.S.A.: First hydroelectric plant produces 125 kw of electricity.
1897	Nearly 300 hydroelectric plants in operation throughout the world.
1900	Europe: First pumped storage.
1960's	Rapid growth of pumped storage for hydro-power stations.
1970	U.S.S.R.: 6,100 MW hydroelectric plant, largest in world.
1979	U.S.A. and Colombia: New small-scale hydroelectric units of 1,000 and 1-kilowatt respectively.
1980	New York, U.S.A.: Small-scale hydro produces hydrogen for commercial sale.
1983-89	Brazil: 12,600 MW hydroelectric plant will begin operation; it will be largest in world.
1985	U.S.A.: 9,600 MW hydroelectric power plant at Grand Coulee.

Advantages

1. No fossil fuels are consumed in electrical energy production.
2. Production costs are low: .6 to 1.6 mills per kwh.
3. Conversion of force of water to electrical energy produced can be 90% efficient.
4. Hydroelectric generation produces no chemical or thermal pollution.
5. Hydroelectric plants can be used to produce hydrogen electrolytically.
6. Hydroelectric power requires little maintenance cost.
7. Reservoir lakes can be used for recreation in many, but not all, areas.
8. Reservoirs can provide considerable, but not complete, flood protection to downstream areas.
9. Reservoirs are capable of storing large quantities of water for long periods of time, although not indefinitely; using dams does not usually entail the storage problems that hinder so many other renewable energy sources.
10. The downstream flow can be managed to aid water-quality control, and to level out the extremes of winter-versus-summer stream conditions.
11. Groundwater reserves are increased by recharging from the reservoir.
12. The greatest potential for hydropower is in those areas that are energy poor.
13. Hydropower is one of the most responsive—easy to start and stop—of any electric power-generating source.
14. Pumped storage is one of the best means to store excess electric power.
15. Most hydro systems have an expected life span 2-3 times longer than conventional thermal power plants.
16. Where power from small-scale hydro installations can be used locally, distribution and transmission costs and losses can be reduced.

17. Hydro installations can be used to breed fish and other aquatic products (as is done in China).[5]

Disadvantages

1. Construction cost per installed kilowatt capacity is high: $100 to $600 per kilowatt (although adding hydropower to existing dam sites is less expensive than building new power plants of any kind).
2. Damming of rivers causes changes in the ecological cycles of the rivers and surrounding landscapes; wildlife are displaced or destroyed. Spawning grounds of migratory fish such as salmon are destroyed, etc.
3. Sedimentation and silt accumulation progressively affect rivers' flow and land drainage patterns. (The Chinese have claimed to have developed a method of eliminating silt fill-in or damage caused by hydroelectric dams by providing tunnels along the banks of the dam to run silt off, using desilting pipes and a special turbine that resists friction from mud.)
4. There is a shortage of feasible sites.
5. It causes loss of land suitable for agriculture and/or recreation.
6. Power production may be curtailed or even discontinued in times of severe drought.
7. Some water is lost by evaporation and seepage from the reservoir surface; it has been estimated that as much as 9% of a 21,000 acre reservoir is lost in this way.[13]
8. In coastal areas, the construction of dams prohibits upstream migration of anadromous fish, such as Pacific salmon, unless some arrangement such as a "fish ladder" is installed.
9. Large hydroelectric installations can cause social disruptions by the sudden input of 3 to 4 times the amount of energy than was previously available.
10. Dams are vulnerable to natural forces, human error and acts of war; due to dam failures, hydropower has the highest direct death rate of any currently-used energy source.
11. The weight of the water in a dam could trigger local earthquakes.
12. Rivers no longer deposit fertile silt downstream.
13. The river channel downstream from a dam is more susceptible to erosion.
14. Certain disease vectors (for instance schistosomiasis) are encouraged.
15. Soil quality could be reduced because the water table behind the dam may be raised and thereby raise surface salts and minerals.

Geothermal

Less is known today about the interior of the Earth than is known about outer space and the deep oceans.[1]

Paradiso powered by Inferno.

Geothermal power is a capital energy source of such magnitude that the distinction between "capital" and "income" energy sources becomes clouded. Geothermal energy will be around as long as will all the other income energy sources.

Geothermal energy is the heat from inside the Earth, tapped at shallow depths as dry steam, wet steam, brine, hot dry rock, pressurized liquid, or magma—that is, molten rock, It is generated by the natural fission process occurring in the Earth's crust as uranium, thorium, and potassium decay, and by frictional and tensional processes. On the average, the temperature rises about 120° F per mile (30° C per kilometer) below the surface of the Earth. At 60 miles into the Earth, it is white hot. Power is derived from this stored thermal energy.

Theoretically, any place on Earth, if drilled deep enough, and provided with heat exchangers, could provide geothermal energy. This heat can be used for powering a wide variety of activities. The potential for generating electricity is well established and is undergoing significant development (see History). However, the extent to which geothermal "oceans" underlie the Earth's land masses is a matter of

The heat content of the Earth is 7.5×10^{24} kwh; or five million times the total amount of solar energy the Earth receives each year from the sun (1.5×10^{18} kwh).

Each year the Earth releases about 300×10^{12} kwh of heat from its interior.

considerable speculation. Sources are reported on all continents including Antarctica, even though only unsystematic searches have been undertaken. Prospecting techniques are only beginning to be developed, and it might be speculated that existing estimates are extremely conservative in lieu of the limited data on which they are based. American geothermal exploration companies have leased about 8 million hectares of geothermal lands in the Western states and are researching the vast geothermal potential of the Gulf Coast; which, in addition to vast geothermal resources (enough to generate up to 115,000 MW of electricity—more than ¼ of U.S. installed capacity [2]) may contain as much as $1,400 \times 10^{12}$ cubic feet of dissolved methane[3] (about 14% of the world's total known natural gas reserves and more than five times the U.S. proven reserves).

The technology for converting steam sources is very well developed.[4] It eliminates the procedure of fuel refinement and distribution needed with fossil-fueled thermal plants by directly providing the steam that would otherwise be artificially created from water by combustion of fossil resources. Energy extraction from molten rock located within the upper 10 kilometers of the Earth's surface would furnish virtually a limitless supply of

We should be able to develop about 20,000 megawatts of geothermal power in the next two decades.[5]

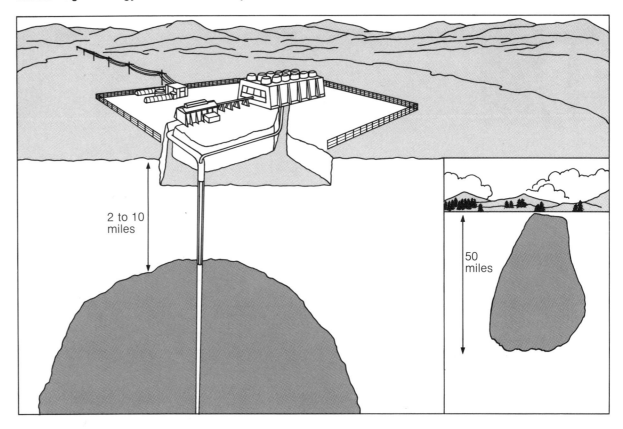

Sandia Magma Energy Power Plant Concept

energy. It has been estimated that the heat in the Earth's molten rock could supply our current level of energy consumption for millions of years without measurable effect on the environment. This energy could be tapped by placing heat exchanges directly in the rock. It would take about 55% of one year's oil and gas drilling capacity to develop the geothermal equivalent of the total installed electrical capacity of the U.S.[7]

Geothermal power generation potential of Japan for the near future is estimated to also be 20,000 MW.[6]

A recent development in harnessing geothermal energy is a scheme for producing hydrogen and methane directly from the reaction of molten rock and water. The hydrogen that is produced could then be used

Model of a high-temperature hot-water geothermal system.

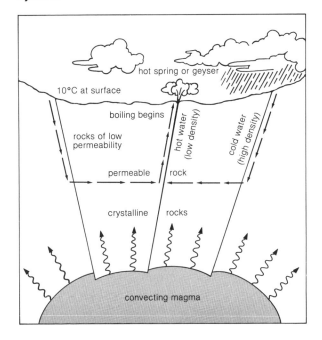

as a fuel in a wide variety of ways (see Hydrogen section). In this design, water, including seawater, is pumped into subsurface bodies of molten rock or magma where a chemical reaction between the water and hot ferrous iron in the magma takes place. Magma, as the highest temperature (600–1300° C) geothermal resource, is the ultimate heat source for most other geothermal energy resources.[8] Some molten rock deposits are located relatively near the surface of the ocean floor and could be reached with current drilling techniques. Ocean water could then be pumped into the magma and the resulting hydrogen extracted. An estimated 2×10^6 mt of hydrogen could be produced per km^3 of magma.[8] In addition to hydrogen, large quantities of steam would be produced that could be used to generate electricity. The amount of hydrogen produced could be doubled or tripled by adding biomass—such as sewage sludge—to the water because the hydrogen content of the biomass would be released during the reaction with the high-temperature magma.

Uses

1. Electric power generation.
2. Heating and cooling for space and industrial processes.
3. Sewage heat treatment.
4. Heating livestock barns.
5. Breeding fish.
6. Greenhouse horticulture.
7. Heating irrigation water.
8. Drying and processing.
9. Mineral water.
10. Medicinal uses; hot springs for bathing.
11. Highway defrosting.
12. Mineral production.
13. Hydrogen production.

History

1904   Laradello, Italy: Electricity generated from a geothermal source. Though capacity was extremely small with the initial installation, capacity is presently 405.6 megawatts, providing the power for much of the nation's railroads.

1913   Italy: Continuous generation begins.

1924   Beppu, Japan: 1-kilowatt steam generator.

1925   Iceland: Large-scale use of geothermally-originated hot water for

System for extracting energy from a dry geothermal reservoir.

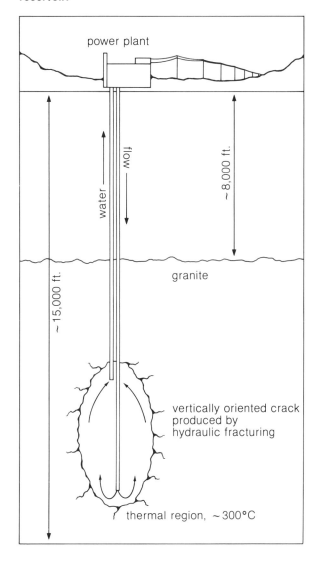

domestic heat. Iceland's fields were estimated in 1954 to have an electric generating capacity of 300 megawatts.

1946 Yunosawa, Japan: 38-kilowatt generator.

1960 Sonoma County, California: "The Geysers" plant built by Pacific Gas and Electric Company, which has been steadily enlarging the plant's capacity to 700 MW. This is the largest geothermal plant in the world.

1966 Otake Field, Japan: 13 MW power plant completed.

1970 China: First small (86 kilowatt) geothermal power plant built.

1973 Onuma Field, Japan: 10 MW power plant completed.

1974 The Soviet Union announces a major plan for exploration of geothermal resources in its eastern regions. Geothermal energy potential of the USSR is greater than all other Soviet energy resources combined.[10]

1974 China: Second geothermal power plant begins operation.

1975 China: Third geothermal power plant (300 kilowatt) begins operation.

1974 Nobeoka, Takamori, Japan: One 25 MW, and two 50 MW power plants under construction.

1975 Ahuachapen, El Salvador: 60 MW plant completed.

1975 Hungary: Geothermal installed capacity reaches 84,569 MW; used almost exclusively for non-electric purposes, such as district heating, industry, agriculture, waterworks, and heating swimming pools.

1977 U.S.: Discovery of potential geothermal energy sources near major urban centers in eastern U.S. areas: Baltimore, MD, Norfolk, VA, Jersey City, NJ, and Savannah, GA.

1977 U.S.: California's geothermal energy potential is estimated at 30,000 MW for electric generation and about 10×10^{12} kwh for non-electric direct-heat applications.

1978 Reno, Nevada, U.S.: Vegetable dehydration plant becomes first U.S. domestic industry to utilize geothermal energy as its primary source of power.

1978 New Jersey, U.S.: First drilling on East coast of U.S. for geothermal energy resources.

1979 France: About 20,000 apartments heated with geothermal; by 1985, 500,000 dwellings are targeted for geothermal use.

1979 U.S.: Warm underground waters being mapped for use with heat pumps to heat homes and commercial buildings.

1979 U.S.: Geothermal energy used to distill alcohol.

1979 Cerro Gordo, Mexico: 150 MW power plant and 70,000 ton per year potassium chloride by product facility operating.

1980 Hawaii, U.S.: 3.5 MW power plant begins operation; steam reservoir estimated at 500 MW at this one site.

1981 U.S.: 1,128 MW of capacity operating at the Geysers power plant in California.

1984 Central America: 465 MW of geothermal generating capacity.

1995 U.S.: Geysers generating capacity at 2,600 MW.

Other geothermal sites:

Wairakei Field (192 megawatts) and Kawaerau, New Zealand; Pauzhetka and Partunka, USSR; Namafjall, Iceland,[4] El Tatio, Chile (15 MW); Tiwi, S. Luzon, Philippines (110 MW, 225 by 1981, 1,600 MW by 1987); Turkey (10 MW).

Advantages

1. No fuel is combusted at a geothermal power installation; geothermal energy already exists as heat.
2. There is a plentiful supply of geothermal energy; the geothermal heat content beneath just the U.S. down to six miles is over 1,300 times the world's supply of coal[11]; and over 1,900 times the world's supply of oil.[4] A .01° C decrease in the mean Earth temperature (probably too small to detect) could supply the world's energy needs for about 25,000 years.[12]
3. Geothermal capital costs are lower than comparable-size fossil or nuclear-fueled power plants. Once a plant is built, there are no fuel costs as in conventional power plants.
4. Geothermal energy is relatively simple and easy to tap; plant set-up time is small, averaging three years from planning to operational stage (nuclear power plants take about ten years).

Solar systems are upward oriented, collecting stellar radiation energy; geothermal systems reach downward, toward planetary thermal energy.[9]

5. The use of geothermal energy can decrease dependence on the price and supply uncertainties of fossil fuels, especially imported oil.
6. Geothermal energy has multiple use capabilities; it is well suited for integrated industrial use—for producing electricity, process steam and heat for a variety of industries and agricultural activities clustered in one region.
7. There are no major economies of scale in geothermal power production; small generating modules can be added one at a time as local development and need progress.
8. Using surface geothermal energy in an environmentally considerate way has the potential of improving the chemical pollution situation caused by the natural release of the same geothermal energy.

Disadvantages

1. Geothermal characteristics change with the area so exploration is not easy; surface does not always indicate what lies below; e.g., Hawaii started drilling in 1,200° C magma at surface and stopped drilling at 160 meters when temperature was down to 137° C.[13]
2. Geothermal turbine efficiency is comparatively low (22%) due to the low temperature and pressure of the steam input. Overall plant efficiency for geothermal electric power production is estimated about 15% less than that of a fossil-fueled plant. A geothermal plant would require 22,000 BTU's to generate one kilowatt hour, while a fossil-fueled plant would require 9,000–10,000 BTU's.
3. There is uncertainty about how long each geothermal well will last; at low exploitation rates it could last almost indefinitely, at projected exploitation rates it may deplete.[9]
4. Environmental dangers of geothermal power include:
   a) possible land subsistence;
   b) disposal of waste water with high mineral content (this problem could be solved by distilling the mineral rich wastes and retrieving these valuable materials or, possibly, by using the exceedingly salty water—three times saltier than sea water—to power salinary gradient power plants [see Water Salination section]);
   c) disposal of hot water produced in the power conversion process (this problem could be solved by pumping these waters back underground to the geothermal source, thereby increasing the expected life of the well);
   d) release of noxious gases, such as hydrogen sulfide, ammonia, and boron, into the atmosphere, as well as large amounts of water vapor;
   e) amount of land taken up (average of 20 km^2 for all operations) and noise given off (steam emerging at high pressure makes a noise sometimes higher than 100 db.);
   f) seismic activity may be generated from fluid withdrawal and/or reinjection.
5. Use of geothermal heat must be near the source; it is not possible to transport geothermal heat very far.
6. The geothermal energy supply is of low quality; that is, it is largely diffuse, like solar energy, rather than concentrated, like coal.

Geothermal Energy: Potential Environmental Impacts Summary

	Water	Air	Land	Solid Waste
Production	Water waste with high mineral content	Noxious fumes, water vapor, noise	land subsistence, seismic activity	salts
Upgrading	thermal pollution			

Currently Utilized Geothermal Sources

Source:
Saint and Jasso, Cal State, Fullerton.

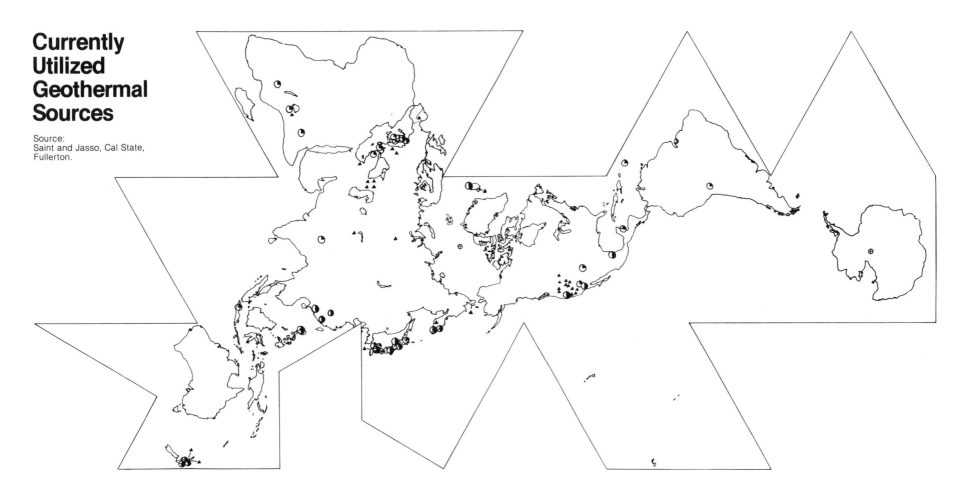

Three quarters of the geothermal energy now being used by humanity goes into non-electrical uses.[9]

◐ existing geothermal power product
◔ geothermal plants under test or construction
▲ non-electrical use of geothermal

In addition to these locations of geothermal energy, offshore geothermal may be a significant resource which has not yet been adequately surveyed.

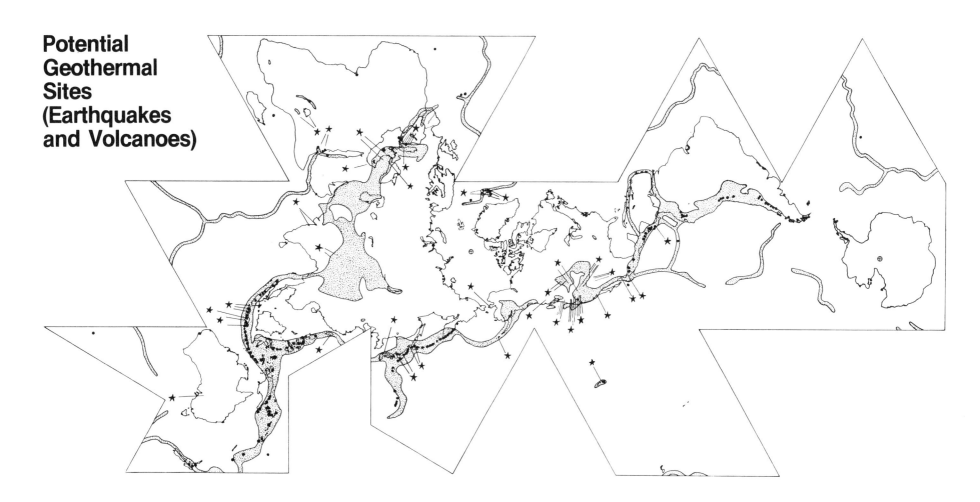

10 Wind

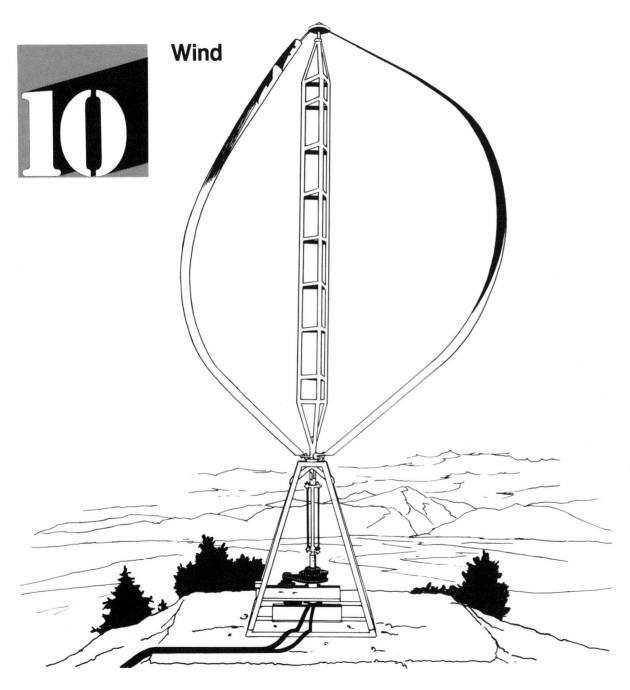

> *... in any reasonably windy region the amount of wind energy available on an annual basis is larger than the average energy flux of sunlight.*[1]

Solar Storage Battery

Wind is continually regenerating in the atmosphere under the influence of radiant energy from the Sun. The atmosphere is, in fact, a huge storage battery for the Sun's energy. About 2% of all solar radiation to the Earth is converted to kinetic energy in the atmosphere. This is equivalent to about 30,000 trillion kwh per year or more than 500 times the total energy consumption of the world. About 35% of this energy (11,000 $\times$ 10^{12} kwh/year) is available within the boundary layer at the Earth's surface.[2] The amount of energy that is available for performing useful work for humanity may be more than this figure because as energy is removed from the winds close to the ground, kinetic energy is transferred downward from higher altitudes.[3]

Wind power turns vanes, blades, or propellers attached to a shaft. The revolving shaft spins the rotor of a generator which produces electricity, or does other useful work. The effectiveness of a windmill increases rapidly with the wind velocity. When wind velocity doubles, power output is multiplied by eight.

Wind is an attractive power source because it does not impose an extra heat burden on the environment, as does energy extracted from fossil and nuclear fuels. Unlike hydro-power, the

> *Without a doubt, wind energy is the cheapest form of solar electricity available today.*[1]

Storage 1.4 Wind Machines Taxonomy

HORIZONTAL AXIS

Single-Bladed

Double-Bladed

Three-Bladed

U.S. Farm Windmill Multi-Bladed

Bicycle Multi-Bladed

Up-Wind

Down-Wind

Enfield-Andreau

Sail Wing

Multi-Rotor

Counter-Rotating Blades

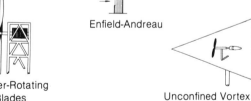

Unconfined Vortex

Cross-wind Savonius

Cross-Wind Paddles

Diffuser

Concentrator

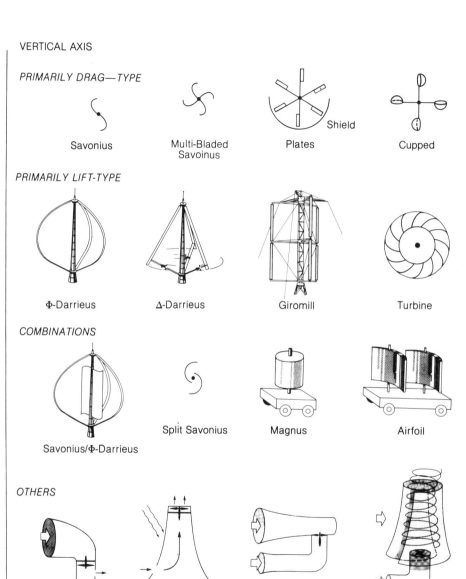

VERTICAL AXIS

PRIMARILY DRAG—TYPE

Savonius

Multi-Bladed Savoinus

Plates Shield

Cupped

PRIMARILY LIFT-TYPE

Φ-Darrieus

Δ-Darrieus

Giromill

Turbine

COMBINATIONS

Savonius/Φ-Darrieus

Split Savonius

Magnus

Airfoil

OTHERS

Deflector

Sunlight

Venturi

Confined Vortex

119

wind is available almost everywhere, but, like hydro-power, there are high concentrations of wind power that are provided by the natural geography of the countryside that may double the wind's intensity. The wind machine taxonomy chart on the preceding page illustrates the wide variety of wind-harnessing devices.

At the moment, the most exotic wind-harnessing devices are the Confined Vortex (or "Tornado" wind turbine) and the Magnus effect rotor. The former is a tall cylindrical tower with an open top, slotted side openings, and guide vanes that create a swirling tornado-like vortex that spins the turbine within the cylinder. Air is pulled into the base of the tower and then upward by a low-pressure zone that develops within the tower by the passage of air into and over the tower. The wind flow within the tower would turn turbines that would generate electricity. It has been estimated that such a wind-harnessing device would be about 100–1,000 times as efficient as that of a conventional wind turbine of the same diameter. It appears that what a focusing or concentrating collector is to solar energy, the confined vortex is to wind energy.

The Magnus effect rotor takes advantage of the phenomenon in which air flowing over a spinning vertical cylinder develops a strong lateral propulsive force. The spinning cylinders would propel vehicles around a circular track, and axle-mounted generators would produce power.[4]

In addition, there are many different backup and storage systems for use with these harnessing devices. Such systems are usually necessary because the wind does not blow at a steady rate. Storage batteries are one system used to store electricity for use when the winds have diminished and power requirements are high; other storage systems include compressed air storage (possibly using depleted natural gas fields for the storage reservoir), electrolysis of water to produce hydrogen, which can be burned as a fuel or used to produce electricity in a fuel cell (see Hydrogen section), pumped storage at hydro-electric sites, and the "synchronous inverter." This last item is basically a device that converts the electrical output of a wind machine to the same current as that of the electric grid and then puts the windmill's output into the grid—in effect running the electric meter backwards. The grid then becomes the storage medium for electric energy. The electric company buys the windmill's power at wholesale prices and sells it back at retail. An additional advantage of this type of storage and backup system is that it helps decentralize the electric utility.

There are many ways of tapping wind energy. Illustration 2 points out some major modes of operation: alone, and as part of a grid. The wind-powered mill, water pump, and electric generator are all situated on top of a tower, as in drawings A, B, and C of Illustration 3. The greater the velocity of the wind, the higher the power output. Because wind velocity goes up with altitude, the higher the tower the better. A major part of the planet's potential wind energy is 30,000 to 40,000 feet above the surface of the Earth in the 200 mph jet streams. Some ways of tapping higher-velocity winds at high

> *We have gone to the moon since the last time someone seriously looked at windmill design and engineering.*

Illustration 2: Two Modes of Utilizing Wind Machines[4]

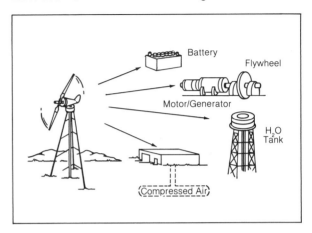

1. Operating Alone. Wind turbines of any size can be used alone to generate power for a variety of tasks, such as producing electricity or heat, or pumping water. However, if a continuous source of power is needed, some form of energy storage may be required, since the winds blow intermittently in most locations.

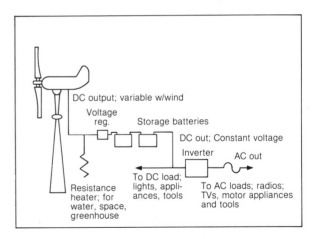

Mode 1. Simple, on-site domestic unit (two to five kw).

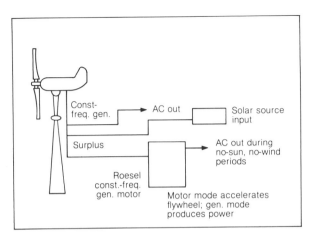

Mode 1. Simple system with mechanical storage (can couple with solar input).

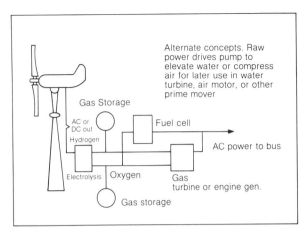

Mode 1 or 2. Complex, intermediate energy conversion: medium power range.

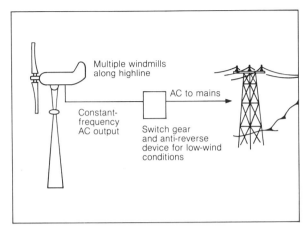

Mode 2. Simple, direct feed to main bus: megawatt range.

---

2. **Operating in a Grid.** Large wind-turbines driving electrical generators of MW-scale (the size now being built by Boeing, General Electric, and Hamilton Standard in the U.S.) can be tied into public utility networks and used to save the fuel of the regular generating plants when the wind is blowing. Energy storage for large-scale wind power can be handled by existing hydroelectric dams as "pumped storage"; where wind energy is used to pump water up into a hydroelectric reservoir for later use by the hydroelectric facility, and in underground formations that are suitable for compressed air storage (both options are abundant in many windy areas).[1]

Wind turbines distributed to take advantage of prevailing winds.

Wind turbines integrated with an existing hydroelectric system.

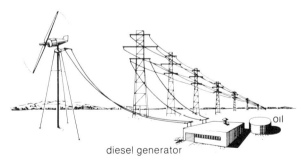

Wind turbine integrated with existing diesel electric system.

altitudes are by using windward sides of mountains, high-rise skyscrapers, kites, blimps, and balloons (Illustration 3, drawings D, E, F, and G). Another way of tapping wind energy would be to use the thermal current around existing buildings (drawings H, I, and J) of cities, or better yet, to design our buildings or cities so as to create low-pressure zones with resultant high-velocity winds that can be harnessed (K), or air scoops utilizing the roof for smaller units (L). Another recent design, the Lebost Wind Turbine, is also suited for urban use. This vertical axis wind turbine is encased in a shell that, in effect, magnifies the wind. Because of its low speed and high torque it can be used for heat production via friction devices that convert mechanical energy into heat.[5] Such a combination may make the system ideal for heating and cooling city structures.

To get an idea of the vast wind power potential available in just the U.S., it is useful to examine the quantities of energy that could theoretically be generated by outfitting each of the U.S.'s high-tension electric transmission line towers with the largest current technology windmills and placing these towers in good wind sites. The U.S. has $3,500 \times 10^{12}$ kwh of potential wind energy[6] and about 180,000 miles of high tension wires.[7] The tallest wind turbine *planned* by the DOE would only be slightly higher than long distance electrical transmission towers. Outfitting each tower and assuming a .50 load factor would generate more than five and one-half times the total electric output of the U.S. (M).[8] There are also proposals to hang wind generators from high-tension lines above the Great Plains, as well as using the air-space over railroad right-of-ways for locating wind power plants (N).

Illustration 3: Wind Power Possibilities

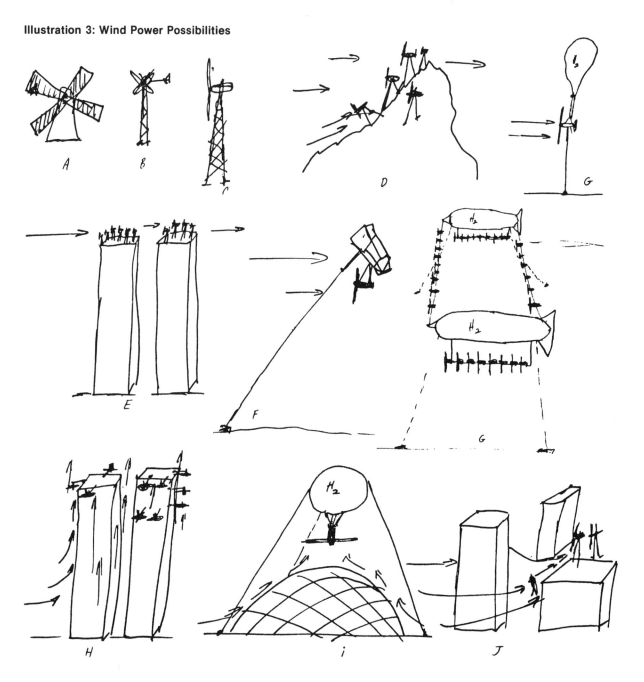

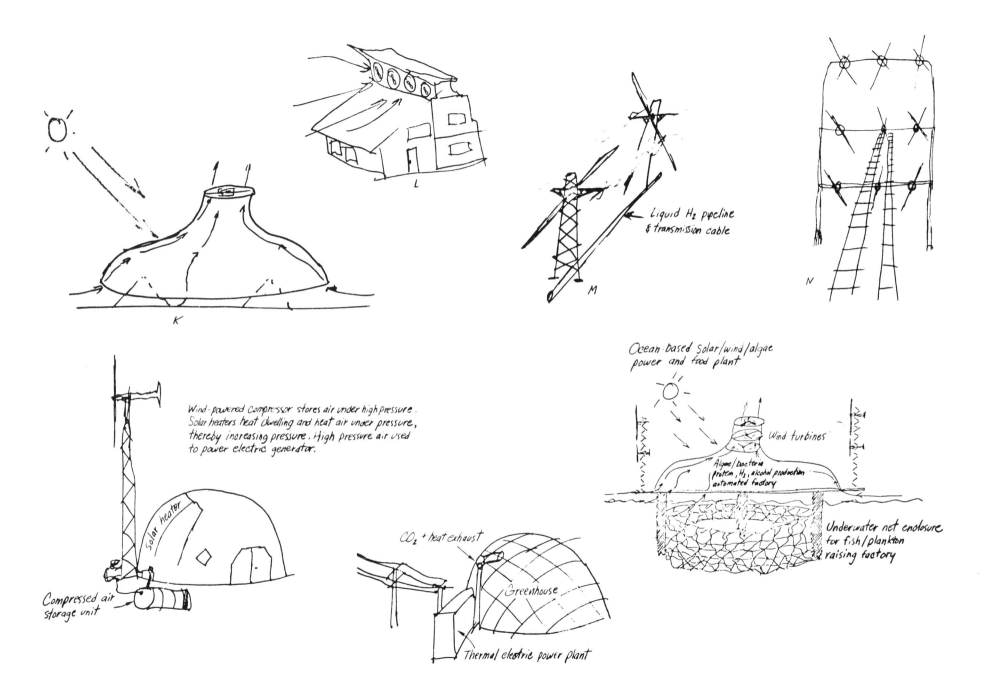

Another excellent example of wind power potential is illustrated by comparing the ongoing U.S. national program to store a billion barrels of imported oil underground by 1985. By spending the same amount on large wind generators as would be spent on the oil and its storage, more than a billion barrels of oil that would otherwise be burned to produce electricty would be saved.[9] "If oil were imported at the same rate as it would be in the absence of a wind generator program, domestic oil fields would, after ten years, contain a billion barrels more petroleum than they otherwise would have. Moreover, the wind generators, which have an expected life much greater than ten years, would be delivering energy at a rate equivalent to 175 million barrels of oil per year."[9]

A large-scale application for harnessing wind energy would be "energy farms," where the land would be covered with a grid of mass-produced, wind-powered generating units—a "forest" of wind generators. The land could be purchased or the wind power rights merely leased, since a grid of well designed wind generators would be entirely compatible with high-yield farming. Power densities of 45,000 kw per square mile are possible in the Midwest U.S. with this approach. By far, though, the most concentrated and intense wind power site would be in the Antarctic (see map). The vast wind power potential of this continent could power much of the world (see Storage section, *Ho-ping: Food for Everyone*).

"The intermittent nature of wind energy has been a matter of great concern to some people. I believe the problem is overrated for the case of utility power generation. If wind power is introduced into multi-regional power grids at

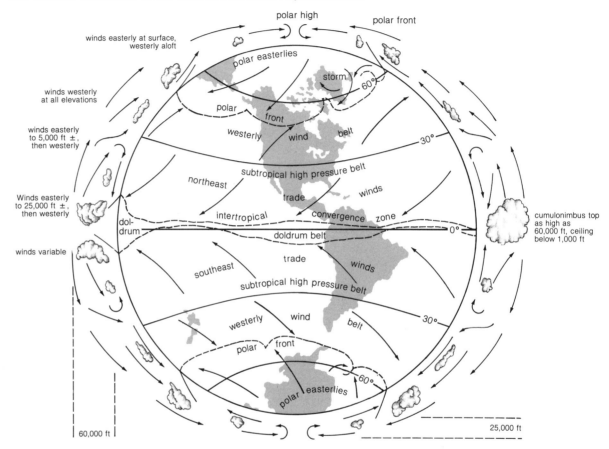

The General Circulation of the Wind in Schematic Representation.

base load capacity, the emergency fill-in and peaking can be accomplished by existing fossil fuel units (or hydrogen-powered units). The key to this approach is to cover a sufficiently broad area so that the wind is sure to be blowing in some parts of the sub-grids at all times."[10] One study of potential wind systems in a wide grid in West Germany found that a single wind turbine might be idle 35% of the time, but the entire network would be idle less than 1% of the time.

Small-scale applications of wind power are also very promising. A ten-foot rotor would recharge a small urban car overnight; a 25-foot rotor would provide enough energy for a single family home in many parts of the U.S.[10]

Using off-the-shelf components (a 14-meter diameter blade, a two-square-meter hydrogen storage facility, and a nine-horsepower lawn-

Wind

Source:
Wind Machines, Mitre Corp.,
Oct. 1975.

- ■ over 5000 kwh/kw
- ▨ 3750–5000
- ▦ 2250–3750
- ⸬ 750–2250
- ☐ less than 750

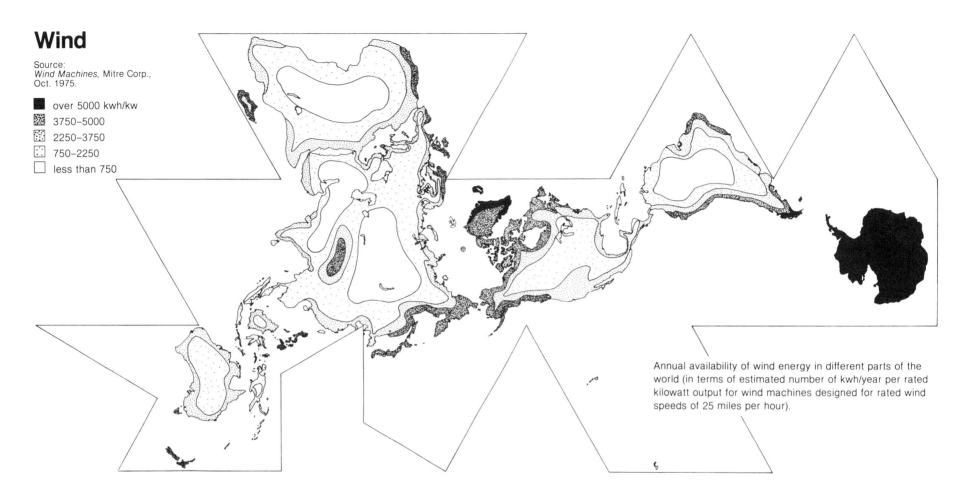

Annual availability of wind energy in different parts of the world (in terms of estimated number of kwh/year per rated kilowatt output for wind machines designed for rated wind speeds of 25 miles per hour).

mower engine), and using hydrogen as a storage medium, it is possible today to have a simple, mass-produced wind-powered, self-contained, non-polluting residential power system for about $5,000, which at current rates can be amortized in about ten years.[6]

By mass producing wind turbines in lots of 100–1,000 a year and locating these turbines at sites having average winds of 12–18 mph. wind energy costs would be in the range of 1.5 cents to 7 cents per kwh.[12]

One last wind energy application is in transportation. Sail power is a tried and proven technology for moving bulk cargoes around the world. Increasing attention is being paid to the development of modern sailing ships that could move very large cargoes at a fraction of the energy cost of present-day fossil-fueled ships. Fuel expenditures make up 20 to 30% of the total cost of operating a modern vessel; the world's non-communist commercial fleet of 25,000 ships greater than 1,000 deadweight tons consumes 5 to 8% of the world's oil.[13] Given these figures it is readily apparent that the new advances in naval architecture, such as synthetic fabric sails that have a strength approaching steel, are mildew-proof, and minimize maintenance as well as new hull and rigging design, could make a significant impact on reducing energy consumption. One recent analysis postulates that as much as 75% of the world's ocean transport requirement could be met by sails, thereby saving an estimated 5 to 22 billion dollars per year.[13]

Uses

1. Electrical power generation.
2. Pumping water or air.
3. Mechanical power for grinding, milling, etc.
4. Electrolysis.
5. Heating and cooling.

History

3000 B.C.	Egypt: Wind-powered sailing ships.
2000 B.C.	Babylon: Hammurabi used windmills for irrigation.
200 B.C.	Hero of Alexandria uses windmill to provide air for an organ.
644 A.D.	Persian windmill reported in Seistan.
1100 A.D.	Middle East: Windmills used extensively; windmill appears in Europe (coming from Islamic world via Persia).
1150	First horizontal-axis windmill.
1200	China: Windmills reported in Northern China.
1300	Netherlands: Windmills used extensively to drain and maintain submerged lands.
1400	Pope Celestine III claims the ownership of the wind; windmills could use his air at a price.[14]
1582	Netherlands: First wind-powered mill built for pressing oil from seeds.
1586	Netherlands: First wind-powered paper mill built to meet demands for paper resulting from invention of the printing press.
1600	Wind-powered sawmills.
1622	U.S.: Dutch settlers introduce first windmills to America at New Amsterdam (now New York).[14]
1800	About 9,000 windmills in use in the Netherlands for a wide variety of purposes.
1854	U.S.: Halladay introduces multibladed American windmill.[14]
1860	Use of windmills in America equivalent to 1.4 billion horsepower hours of work, or 1.04 billion kwh. Since 1850 more than 6 million small windmills have been built and used in the U.S. to pump water and generate electricity. More than 150,000 are still in operation.[4]
1883	U.S.: Steel windmill fabricated by Perry.
1890	Denmark: La Cour develops four-vaned windmill for electric power generation.
1894	Wind used for electric generation by the Arctic explorer Narsen.
1900	Denmark: 3,000 industrial windmills and about 30,000 other types used for homes and farms supply power of about 200 MW.
1910	Denmark: Several hundred systems built consisting of 80 ft. tower, 75 ft. diameter, four-bladed rotor generating 5 to 25 kw.
1920's	U.S.: Madaras invents rotor power plant.
1927	France: Darrieus invents new vertical axis wind-powered rotor.
1927	Netherlands: Dekker adds airfoil to the leading edges of mill sails.
1929	France: Electric wind turbine, 20 m. in diameter, capacity not known, built at Bourget.
1931	U.S.S.R., near Yalta: Electric wind turbine 20 m. in diameter, 100-kilowatt capacity. Annual output: 280,000 kwh.
1941	U.S., Grandpa's Knob, Vermont: Electric wind turbine, 175 ft. in

Wind in the Antarctic

SOURCES:
1. National Geographic Atlas of the World, 1963
2. Science News, January, 1974
3. The Antarctic, 1969
4. Physical Geography: Earth Systems, 1974
5. Lewis, R. S. & Smith, P. M., editors, *Frozen Future*. Quadrangle Books, 1973, NY

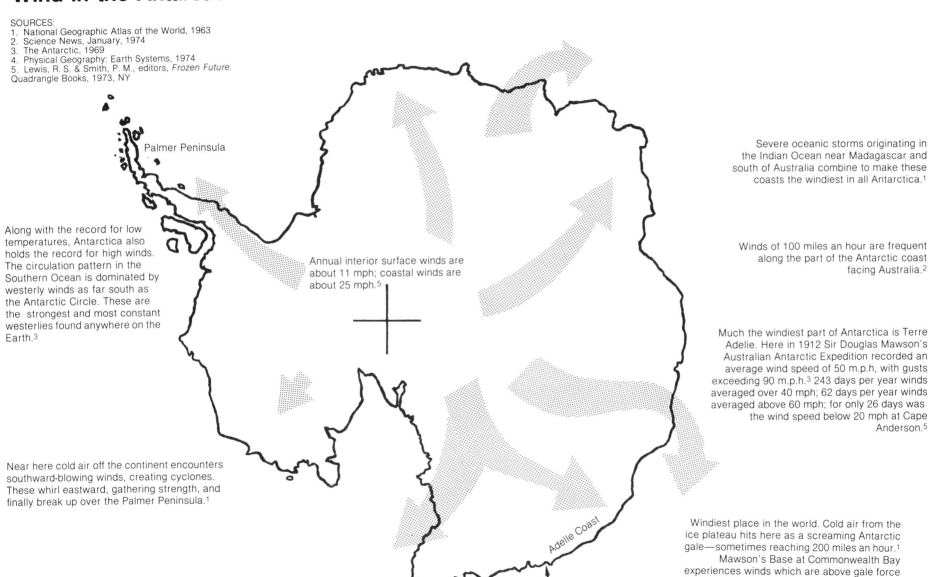

Severe oceanic storms originating in the Indian Ocean near Madagascar and south of Australia combine to make these coasts the windiest in all Antarctica.[1]

Winds of 100 miles an hour are frequent along the part of the Antarctic coast facing Australia.[2]

Much the windiest part of Antarctica is Terre Adelie. Here in 1912 Sir Douglas Mawson's Australian Antarctic Expedition recorded an average wind speed of 50 m.p.h, with gusts exceeding 90 m.p.h.[3] 243 days per year winds averaged over 40 mph; 62 days per year winds averaged above 60 mph; for only 26 days was the wind speed below 20 mph at Cape Anderson.[5]

Windiest place in the world. Cold air from the ice plateau hits here as a screaming Antarctic gale—sometimes reaching 200 miles an hour.[1]
Mawson's Base at Commonwealth Bay experiences winds which are above gale force (44 kph or 28 mph) more than 340 days a year.[4]

Along with the record for low temperatures, Antarctica also holds the record for high winds. The circulation pattern in the Southern Ocean is dominated by westerly winds as far south as the Antarctic Circle. These are the strongest and most constant westerlies found anywhere on the Earth.[3]

Near here cold air off the continent encounters southward-blowing winds, creating cyclones. These whirl eastward, gathering strength, and finally break up over the Palmer Peninsula.[1]

Annual interior surface winds are about 11 mph; coastal winds are about 25 mph.[5]

1946   diameter, 1,250 kw capacity, constructed by the S. Morgan Smith Co. and designed by P. C. Putnam.
1946   Denmark: Three experimental systems of 12, 45, and 200 kw built and tied into existing public utility networks.
1950   England: Enfield Cable Co. builds 100 kw wind turbine at St. Albans, England.
1951   Thomas 6,500 kw wind generator designed; featured two 200-foot blades atop a 475-foot tower.
1954   The U.S.S.R. revealed the number of windpower plants operating in the country as 29,500, with an aggregate capacity of 1.1×10^9 kwh.
1955   England: 100 kw wind turbine system built on Cape Costa in the Orkney Islands by John Brown Co.
1957   Denmark: Gedser Windmill, a fully automated 200 kw unit with three blades 45 feet long and mounted on a 75-foot prestressed concrete tower. By 1961, the unit had produced over 400,000 kwh per year for the Danish Public Power System. Mass-produced, it would cost $40,000.
1958   France: 800 kw, wind turbine built and operated near Paris.
1960   France: 132 kw, wind turbine built in Southern France.
Germany: Hutter-Allgaier 100 kw wind generator built.
Denmark: 600 kw Gedser wind generator designed.
1970   U.S.: Heronemus proposed Shoreham plan: two networks of floating windpower stations off Long Island. Each station would support three towers, each having two 200-foot wind turbines. Costs competitive with future costs of conventional systems.
1973   Schonball 70 kw wind generator (power supply for five families). Estimated cost of mass production: $32,200.
1974   U.S.: Lewis Research Center: NASA testing 100 kw generator; initial phase of program that could produce generators of megawatt capacity.
1974   U.S.: NSF and NASA state that a major U.S. development progress in wind power could result in an annual yield of 1.5×10^{12} kwh, the amount of electricity consumed in 1970.
1975   U.S.: Yen designs Vortex generator with power output 100 to 1,000 times that of conventional wind turbine of same diameter.
100 kw wind generator built by NASA; 125-foot diameter, two-bladed propeller, and a 125-foot tower; prototype for much larger units (1,000 + kw).
1977   U.S.: Boeing selected to build largest windmill in history; 2,500 kw (2.5 MW) system with 300-foot diameter blades.
1977   Denmark: Tvind college completes a 2 MW wind turbine (world's largest) at a cost of $360 per kilowatt. Privately financed and designed by students and faculty, wind turbine was completed in 3 years.
1977   Canada: 200 kw Darrieus wind turbine begins feeding grid; largest vertical axis wind machine in existence.
1978   U.S.: Largest operating windmill (250 kw) in U.S. installed in Pennsylvania by Mehrkam, private inventor, at less than half the cost of government windmills.
1978   U.S.S.R.: Construction begins on 4,500 MW network of wind power stations, to be completed by 1990.
1979   U.S.: 2 MW wind power system begins operation near Boone, North Carolina.
1979   U.S.: 1–2 kw, 8 kw, 15 kw, and 40 kw wind energy systems prototyped and tested by DOE.
1978-80   U.S.: Five 200 kw wind turbines installed and begin operation at Clayton, New Mexico; Culebra Island, Puerto Rico; Block Island, Rhode Island; Cuttyhunk Island, Rhode Island; and Oahu, Hawaii.

Advantages

1. Wind is everywhere, free, renewable, and plentiful; the wind power available in the conterminous United States is 13 kw/hectare (about $3,405 \times 10^{12}$ kwh)[15]. The average wind power in the U.S. Great Plains over the course of a year is over 200 watts per square meter.[1]

High Voltage Electric Grid

▨ primary and secondary electric power trunk lines

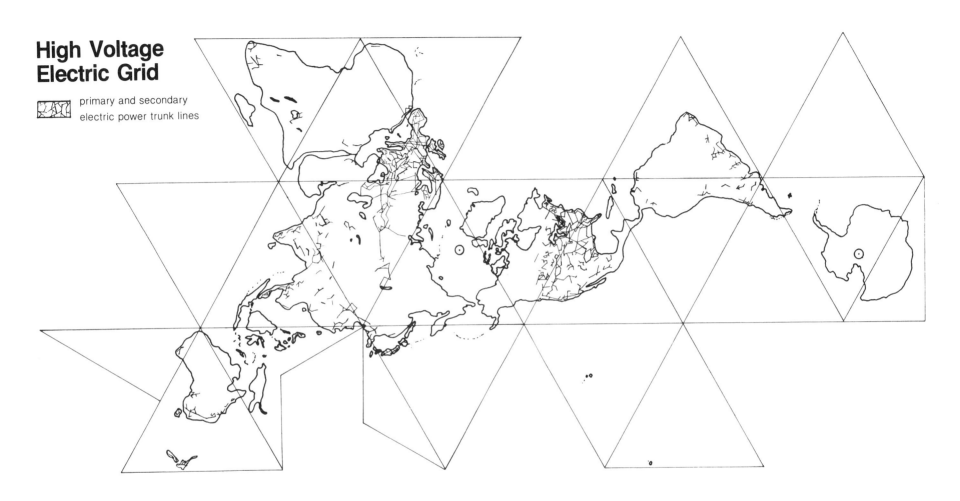

SOURCES:
Power Today, World Power Conference, 1955, 1956
Electrical World Magazine, 1965, 1967, 1968
Electrification & Power Construction in USSR, P. S. Neporozhnji
Electric Power Industry Survey of Taiwan, Vol. 1
Korea Electric Co: "Electric Power in Korea"
Thailand Electric Power Study, 1966
Yanhee Electric Authority, Annual Report, 1965
4th Annual Electric Power Survey of India, 1960
International Bank for Reconstruction & Development, 1960
The Israel Electric Corp., Ltd.: 39th Annual Report. 1961/1962
U. N. Publication: Latin America Survey Document
U. N. Publication: Situation, Trends & Prospects of Electric Power Supply in Africa
Edison Electric Co.: Report on Coordination Practices of European Electric Power Systems, 1967
Edison Electric Co.: Coordinated Planning & Operation of the Electric Power Systems in Japan, 1967

About 1,450 sq. miles (3,766 km^2) of land are taken up in just the U.S. by high voltage power lines.

> *Climatological studies indicate that a potential installed capacity well over 100 gigawatts (equivalent to over 100 1,000 MW power plants) is possible in the 17 Western states, and many times this much in the Arctic region of North America.*[16]

2. It causes no damage to surrounding environment; no air, water, or thermal discharge, etc.
3. It does not add to the thermal burden of the Earth.
4. Technology for harnessing the wind is already developed.
5. Wind speed is generally higher during the middle of the day when the use of electric power is also at its highest point.
6. In high latitudes, the wind blows mainly in the winter when heat needs are greatest and sunlight scarcest.
7. Local geography can provide natural concentrations for wind energy that can double its power potential.
8. Wind turbines are twice as effective as direct solar systems in extracting work from the natural medium.[1]
9. Wind that is "lost," that is, not captured and delivered, is renewable.
10. Wind turbines are ready to be mass produced, resulting in great savings to the consumer.
11. Manufacturers project sales price of large wind systems to decrease 15% with each doubling of the number of units produced.[16]

Disadvantages

1. Most locations produce relatively small power outputs.
2. There are variations in power-plant output due to flux in duration and intensity of wind which necessitates storage facilities.
3. Wind must move at speeds greater than seven mph to be usable in most cases.
4. Rotation of windmill blades can interfere with home T.V. reception. The higher the broadcast frequency the more pronounced the interference, and the larger the blades, the farther away will pictures be disrupted. (Interference results because the rotation speed of large wind systems is near the synchronization speed of U.S. television—30 cycles per second. This is not a problem with small machines or with large systems further than a mile from the site; in addition, fiberglass rather than metal blades cut interference in half. In addition use of cable TV for the area affected eliminates the problem.)[1]
5. Capital costs are high.
6. Mechanical failures of wind turbines could be dangerous.
7. Navigation hazards to aircraft and ships if facility is on mountaintops or offshore.
8. Restriction of further building that might block winds.
9. Environmental effects include:
   a) possible tree removal
   b) hill alteration to promote winds
   c) hazards to birds
   Siting of wind systems is much more critical than solar systems because wind power increases as the third power of the wind speed. Available energy from each site, even sites in the same area, will therefore greatly differ (factors of ten or more).

11

Solar Energy

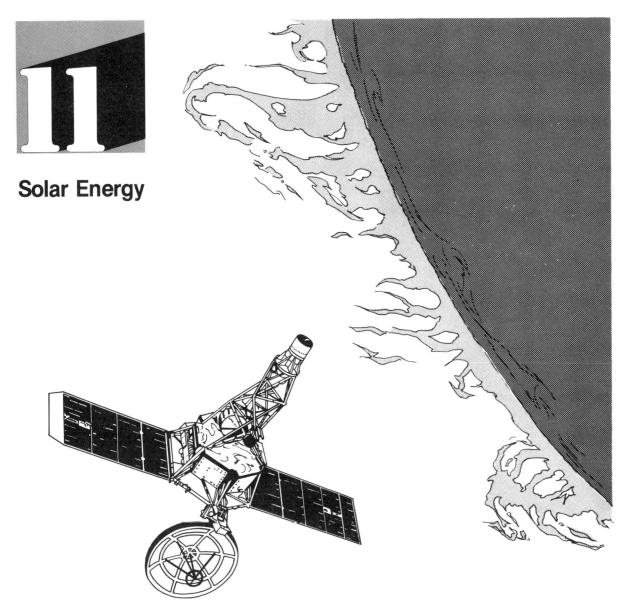

Terrestrial

Universal Engine

Radiation from the sun is the Earth's primary source of energy. More than 99% of the processes that are happening on Earth—including all of humanity's—are energized by the Sun either directly or indirectly (see Chart 1). The Sun is a thermonuclear reaction taking place 93 million miles away whose energy products are received free of charge on Earth. A fraction of this solar radiation is impounded in physical and chemical processes occurring naturally in the Earth's atmosphere, hydrosphere, and lithosphere. As the chart on page 000 shows, about 1,500,000 trillion kilowatt hours of solar energy arrive at the outer atmosphere of the Earth each year. This is more than 25,000 times larger than humanity's current (controlled) energy consumption. Most of the energy is either reflected back to space (35%), converted directly to heat (43%), or powers the global evaporation, precipitation, etc. processes. All of it is necessary to maintain the Earth as we know it. The miniscule amount of energy humanity controls and valves in its preferred directions is not enough to maintain even a small fraction of humanity's actual life-support needs of food, warmth, and shelter. The world's total remaining supply of all fossil and nuclear fission fuels would be enough to power the Earth system for two weeks. Humanity's total annual oil production would "operate" the

---

In a different way than in the past, man will have to return to the idea that his existence is a free gift of the sun.[1]

> *In the early 1970's, the surprising truism of solar energy was that residential systems were ready for use. The largely unappreciated truism of the late 1970's is that the key components of industrial and commercial systems are now ready for wider use.*[3]

world's rainfall system for about thirty-five hours.[2] All the benefits imparted to society by nature (for "free"), including the absorption and breakdown of our pollutants, the cycling of nutrients, the building of soil, the degradation of organic waste, the maintenance of a balance of gases in the air, the regulation of the radiation balance and climate, and photosynthesis are all driven by our solar "engine." These processes "maintain clean air, pure water, a green earth, and a balance of creatures;....(they) enable humans to obtain the food, fiber energy, and other material needs for survival."[4]

Humanity's technological capabilities for collecting, storing, and converting solar radiation to perform useful work is at a high level. There are many different types of solar energy harnessing systems. They can be roughly characterized as either "active" or "passive." Active solar systems use pumps and/or fans to move heat around. Passive systems do not use any additional energy, but rather use the structure of the building as the collector and storage medium. The basic generic types are illustrated in Charts 2 and 3.

> *We think we heat our homes with fossil fuels, forgetting that without the sun those homes would be −240° C when we turn on our furnaces.*[5]

Chart 1. Solar Energy Use in Perspective

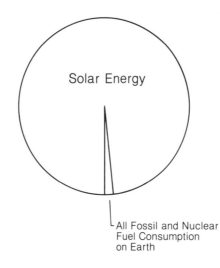

The Sun, by any measure, is the largest single energy input into the world's economy.

The most widespread active solar energy harnessing device in use is the flat plate collector (Fig. 1). Solar energy strikes a black metal absorber that is encased in an insulated box with a glass or plastic cover. Collected heat is transferred to a carrying medium, usually air or a liquid such as water, and then is transported to either a storage tank of rocks or a liquid or used directly as space heat or hot water. Different flat plate collector designs make use of variations in the type of metal absorber used, the type of treatment given to the metal to enhance its light-absorbing and heat-retaining properties, the materials of the box, and the amount of insulation and glazing used.[6]

Figure 1. Flat Plate Collector

A variation of the flat plate collector, but significantly less expensive, is a plastic flat plate collector currently under development that consists of multiple layers of thin, flexible plastic films with air-filled pockets at the top and bottom providing insulation and a center channel for water to flow through. Another design, even simpler than the flat plate collector, is tubular plastic or synthetic rubber collectors for heating swimming pools.

A relatively new design is evacuated tube collectors. These devices consist of an inner glass cylinder that has been blackened to absorb sunlight and that is encased within an outer clear glass tube. The space between the two tubes is pumped to a vacuum to provide

Chart 2. Solar Energy Harnessing Artifacts Taxonomy

Flat Plate collector for water & air heating

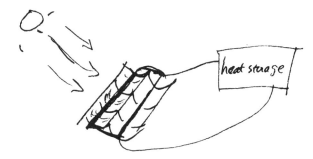

Focused collector for water & air heating

Focused collector for industrial processes

Photovoltaic cells for electricity production

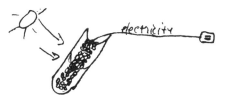

Focused Photovoltaic cells for electricity production

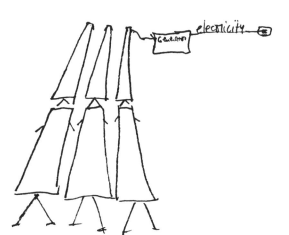

Flat plate solar power plant

Solar ponds

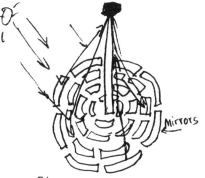

Solar Tower Power plant

Chart 3. Solar Energy Harnessing Systems Taxonomy: Passive

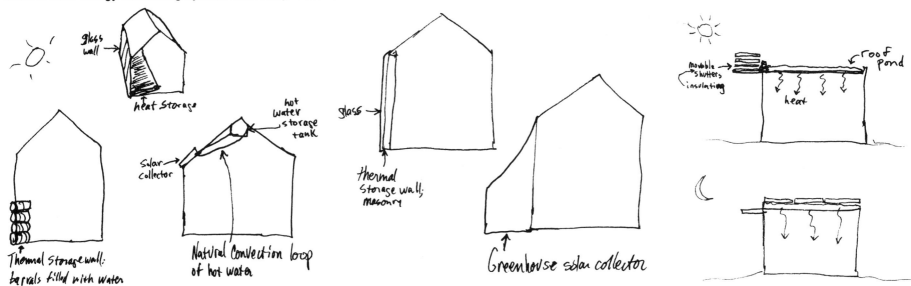

insulation. Evacuated tubes offer the possibility of being mass-produced at relatively low cost in similar fashion to fluorescent light bulb manufacture, as well as being able to absorb light coming from any direction, being immune to the effects of high winds and cold weather, and operating at higher efficiency (40–50%) and thus higher temperature than regular flat plate collectors.

Concentrating collectors increase the amount of solar energy entering any particular collector, thereby increasing the temperatures that can be reached. Economic advantage occurs because mirrors or lenses that concentrate sunlight are usually less expensive than a comparable area of flat plate collectors.

There are three different types of concentrating collectors: nonfocusing concentrators, trough focusing concentrators that track the Sun by rotating along one axis, and two-axis tracking concentrators.[6] These designs produce low, intermediate, and high degrees of concentration respectively. High temperature solar concentrators are capable of producing temperatures in the range of 300–500° C, but temperatures as high as 3,300° C (6,000° F) have been produced at a solar concentrator "furnace" in France.

"Solar ponds" are yet another way of producing power from the Sun. In this relatively

Another use for concentrator collectors is in the production of electricity. A number of relatively small concentrator collectors that produce electricity for the home with small turbines have been built and tested. One design even freezes water in the summer time for air conditioning use.

simple technique, shallow ponds are heated by the Sun and the heat extracted to do work. High temperatures are achieved because the hot water at the bottom of the pond is kept from circulating to the surface and releasing its warmth to the atmosphere. This is accomplished by having the pond's salinity increase with depth. The Sun passes through the lighter surface layer of fresh water, heats the pond's heavily saline lower layer, and because of the density gradient, no convection takes place, thereby allowing the temperature to build up almost to the boiling point. Such ponds do not require direct sunlight: they can collect heat with hazy light and can store their heat for long periods of time, depending on their size. A small pond can hold heat for a few days, a square kilometer pond could hold its heat for an entire season. A 100 square meter solar pond in

Solar furnace, located near Odeillo in the Pyrenees Mountains of southern France, was built for the testing of materials under extremely high temperatures. In the foreground, an array of 63 mirrors (heliostats), each measuring 6 meters by 7½ meters, reflect sunlight onto the curved mirror surface of the office building in the background. This in turn focuses the sunlight on an aperture in the tower at center, where temperatures of 6,000 degrees F. can be produced—enough heat to melt any known material.

The 100-megawatt (electric) heliostat power plant concept. The tower (260 meters high) near the center of the field has a boiler on top. About 20,000 heliostats (6.4 by 6.4 meters) would be required, spread over an area of about 3.5 square kilometers. A 10-megawatt (electric) pilot plant is under development by DOE. Water is pumped up to the receiver and the steam is brought down to the conventional steam-electric generator usually employed by utilities.

Israel is coupled to a closed-cycle turbine that produces 5 kw of electricity. A square kilometer solar pond of similar design could produce 5 MW.

Captured heat from all types of active solar systems is available for immediate conversion to work or for storage in such mediums as large containers of water, rocks, or euteric salts. Some storage, called "annual cycle energy systems," are capable of storing heat in summer for use in winter. These neighborhood scale storage systems could reduce costs and dramatically improve the reliability of solar heating.[6]

There are at least six distinct techniques used in modern passive solar systems, either alone or in combination. Clustering south-facing windows with interior walls and/or floors that provide storage for the collected heat, thermal storage walls made of either masonry or water in drums, roof ponds, natural convection loops,

> *If sunbeams were weapons of war, we would have had solar energy centuries ago.*[7]

greenhouses attached to the south side of a structure, and moveable insulation for night-time window coverage. Passive solar systems seem to be able to equal or excel the performance of flat plate collectors as well as being less expensive.

Large-scale solar harnessing devices have been proposed, and some are now being built. The "solar tower," in which a series of large mirrors focus the Sun's energy onto a boiler on top of a tower, is one such device. Another is the "solar farm" in which rows of flatplate or concentrator solar collectors would be deployed to harness the Sun's energy.

Another large-scale solar facility that has been proposed is a huge geodesic-tensegrity sphere that would "float" in the upper Earth atmosphere similar to a high altitude hot air balloon.[8] Such a craft would be a mile in diameter, and would stay aloft indefinitely.

Roof-type Solar Still

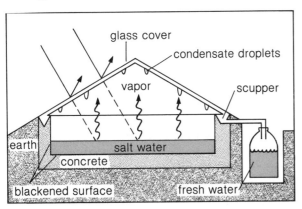

Concentrated sunlight from 1,775 mirror facets is shown striking the steel target mounted 114 feet up on the 200-foot-high "power tower" at the U.S. Department of Energy's Solar Thermal Test Facility at Sandia Laboratories, causing molten steel to drip from the target. In the first major focusing test of 71 heliostat arrays, a 2 × 3-foot hole was burned in the one-quarter of an inch thick steel plate in less than two minutes.

Buoyancy would be maintained by solar heated confined air. Preliminary analysis indicates a 5–21,000 MW net surplus of energy.[9] The same microwave sending and receiving technology

Simple Solar Heating System using water to store the heat and forced air to distribute it.

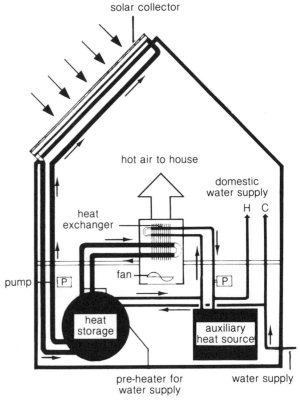

that has been proposed for orbiting solar power stations could be used in this floating sphere power station. Besides use as a power station, such a floating sphere could be used for stratospheric and microbiological monitoring, radio, optical, infrared, and high-energy particle astronomy, Earth resources and meteorological measurements, peace-keeping surveillance, and as a long-range communications relay.[9]

Mean Annual Solar Radiation

SOURCE:
Physical Geography: Earth Systems, 1974

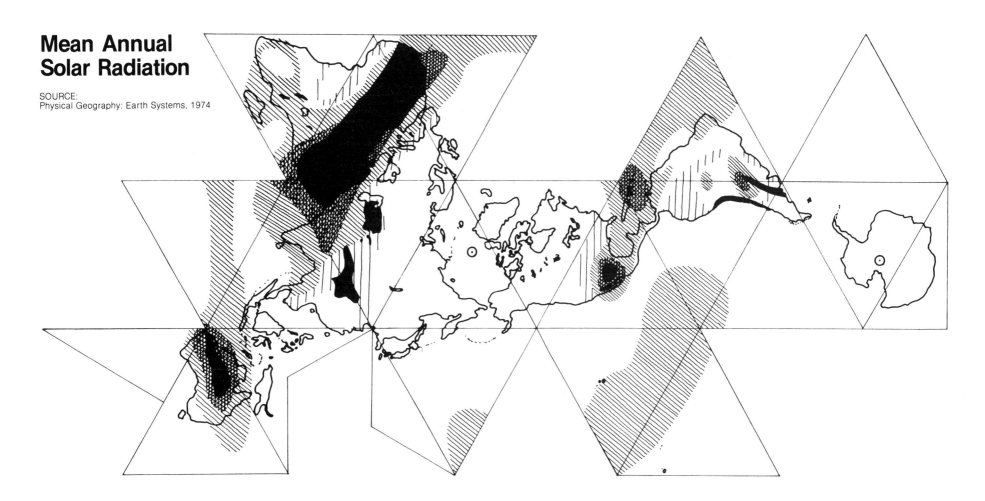

Kilowatt per square meter per year

- ■ 465+ kw/m²/yr
- ▦ 415 kw/m²/yr
- ▨ 370 kw/m²/yr
- ▥ 325 kw/m²/yr

Solar radiation is incident at the top of the atmosphere at 1.4 kw per square meter, and rarely exceeds 1 kw per square meter at the surface of the Earth. Radiation temperature of the sun is 5,900°K. (10,000°F.).

A Naturally Air-Conditioned Building (Designer: Harold Hay)

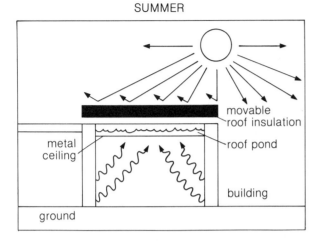
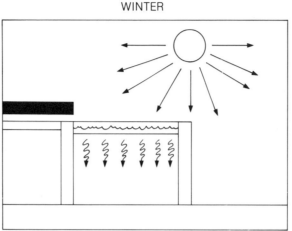
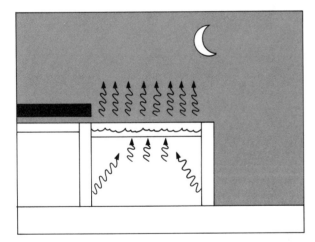
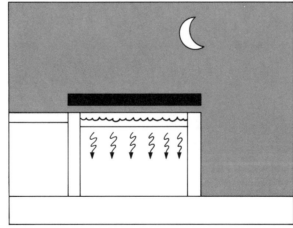

The Power Mountain Concept is a possible approach to economic collection of solar energy for large-scale use. It is estimated that, at a comparatively-reasonable construction cost, large quantities of air could be heated to around 350° C to operate a Rankine-cycle turbine via a heat exchanger.[10]

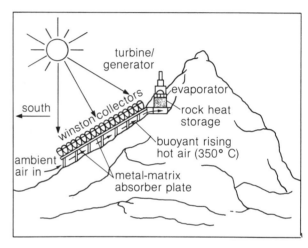

The illustration outlines yet another design for a large-scale solar energy facility. In this design, the heated air rises without the aid of mechanical pumps, ducts, or pipes, but through its own buoyancy, because of the sloped arrangement made possible by mounting the collection field on the mountain-side.[10] This drastically reduces the cost of transporting heat that other ground-based collectors have. The concentrator collectors are of such a design that they do not need the precision tracking of the Sun that the power tower, for instance, requires.

Photovoltaic cells ("photo" meaning light, "voltaic" meaning electricity) are still another way of harnessing the Sun. Photovoltaic solar cells convert sunlight directly into electricity.

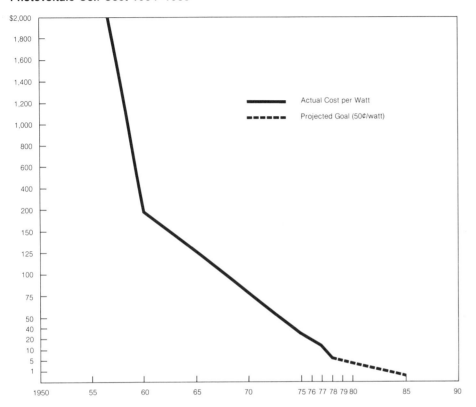

Photovoltaic Cell Cost 1954–1985

Actual Cost per Watt
Projected Goal (50¢/watt)

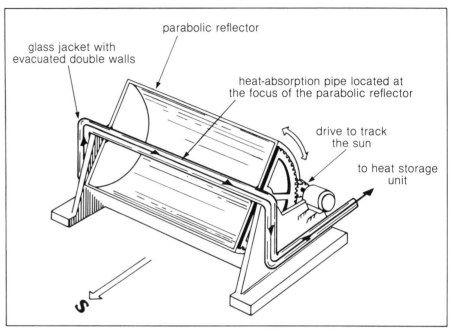

Schematic diagram of a solar-energy collector which focuses the incident radiant energy on one axis. Evacuated double walls are used to minimize convective heat transfer.

They are not new; there have been over twenty years of reliable proven experience with solar cells.

The basic patents for solar cells were first established in the early 1940's, but at that time the cells only existed on paper. The necessary breakthroughs for their manufacture occurred in 1954 in the Bell Laboratories; by 1956 solar cells were being produced. Their first applications were to power space satellites. The Vanguard TV-4, launched in early 1958, utilized

In the climates of North America, Europe, Japan, Africa, South America, China, Oceania, the Middle East, and the Far East, enough solar energy falls on the roof of the average house to supply all of the energy needs.

a solar cell system which provided only 0.1 watts of power. Solar cells were later utilized in the more sophisticated Telstar, Pioneer, Mariner, and communications satellites as well as in the manned Apollo and Skylab missions.

From this development in the mid-1950's to 1973, the primary application for solar cells was in space, though some cells, initially produced for the space program, but not meeting NASA's rigid peformance specifications, were used in remote terrestrial situations in the Arctic, Antarctic, and isolated mountain and desert regions. Presently, solar cells are used for powering everything from wristwatches to buoys, railroads signals, road signs, emergency telephones, irrigation pumps, and household devices. Nearly everything that runs on electricity can be powered by the right combination of photovoltaic cells.

The numerous advantages of photovoltaic cells include:

1. no moving parts;
2. no fuel consumption;
3. they are pollution-free and noiseless; they are the most environmentally benign source of electricty yet conceived;
4. they operate at environmental temperature;
5. there is no prior conversion of sunlight to heat, as in solar thermal electric power plants;
6. they have long lifetimes;
7. they require little maintenance;
8. they can be fashioned from silicon, the second most abundant element in the Earth's crust;
9. their modularity makes them useful for a wide spectrum of applications; modules of a few watts can power educational television sets in remote rural locations while huge integrated complexes of hundreds or thousands of megawatts can be used for urban or industrial power requirements;
10. they are net energy producers in less than 2 years (with the possibility of this being cut to a matter of weeks with more efficient production processes); and
11. they can be used in tandem with flatplate or concentrator collectors to harness the 80% or more of the Sun's energy that the solar cells do not convert into electricity.
12. they are ready for mass production; given the market, one or more manufacturing plants, each capable of producing as much

---

$.50 per watt can be achieved "if dedicated processing plants large enough to turn out 50 megawatts of generating capacity are built."[6]

---

as 100 MW of solar cell capacity per year, could be built within two years.[6]

If solar cells are so good, why are they not in widespread use? One of the answers is simple: cost. The major disadvantages of solar cells are their high costs and low power outputs per cell. Because the cells can be used just as effectively in tandem as alone, this latter disadvantage is easily overcome. Costs are a little more difficult.

In 1973 the first photovoltaic cells strictly for Earthside use were manufactured. With their advent, prices tumbled even further and faster than they had in the preceding twenty years. Costs for solar cell arrays in the 1950's reached as high as $2,000 per watt. From 1973 to 1977 the price dropped from a high of about $500 to about $11/watt; a 45-fold decrease (see chart). Since 1977, prices have dropped even further. One of the latest requests for proposals from the Department of Energy for concentrating photovoltaic systems called for systems producing electricity at $2/watt. Nine companies responded saying they could meet this price level by the end of 1979.

In 1973, estimates called for the achievement of practical cost-effective solar cells by 1985. Because advances have been increasingly rapid and productive this deadline has been moved up. Price goals for silicon solar cells are 50¢ per watt. There are no technological reasons for this goal not being achieved; in addition, another type of solar cell, those made from cadium sulfide, are projected to cost half this amount by 1985. Again, in 1973, industry experts predicted that it would not be until 1983 that silicon solar cells could be developed that would be 10% efficient. Two years later, in 1975, two companies succeeded in making such cells. The increasing solar cell efficiency has the obvious importance of reducing the costs: you only need one half the number of 10% efficient cells as you would need of 5% efficient cells.

Major cost reductions can be obtained simply through the mass production of solar cells. The last major price decrease (50%) was brought about by the U.S. government placing a large order. A billion dollar order (what a 1,000 MW power plant costs or half the cost of the Clinch River breeder reactor) would lower the cost of photovoltaic cells to a fraction of current costs. In fact, based on the assumption that doubling the output will lead to a 30% cost reduction (a conservative figure based on recent experiences of photovoltaics and electrical components industries), a $1 billion investment in solar cells would result in over 4,000 MW of power capacity. When solar cells cost $1 per watt they will be cheaper than nuclear power (and the least expensive way of powering street, highway, parking lot, emergency, airport, and marine lighting, and a host of other decentralized energy needs); when the price goes to 50¢ per watt, 1,000 MW of solar cells will cost half what a nuclear power plant costs in the 1970's (and be the least expensive way by far of powering the residential section of the U.S.). At 25¢ per watt, 1,000 MW of solar cells will cost one-quarter what a nuclear power plant costs. High cost ($20.00/watt) solar cells can already compete with batteries and small

Figure 1.[12] **Two Light Shafts** for efficient illumination with Sunlight are shown. On the left, a vertical shaft for a six story building incorporates sized windows; on the right, a curved shaft directs light around a corner.

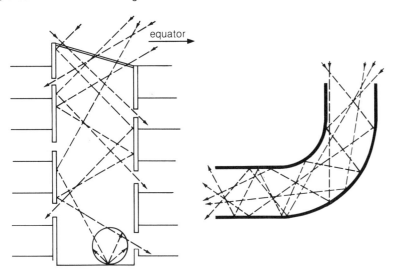

Figure 2.[12] **A Sun Tracker** concentrates light that can be "piped" a considerable distance to an interior room.

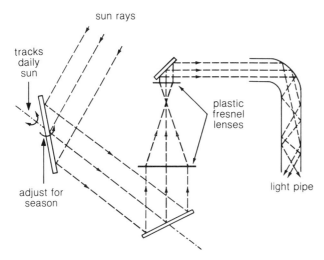

Figure 3.[12] **Light Output from Pipes** can be controlled by using converging and diverging pipes.

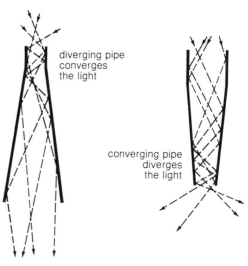

Figure 4.[12] **Two Types of Diffusers** illustrate alternate ways of lighting rooms with solar light.

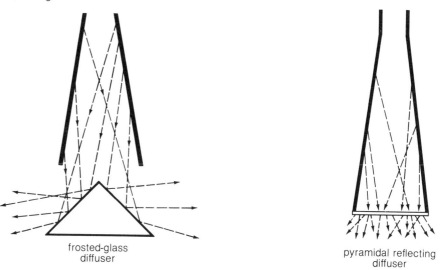

gasoline generators in developing regions and at remote military installations.

One of the more intriguing concepts to come out of the solar energy field in the last few years is the idea of the "Solar Breeder." In this design, the energy needed to manufacture solar cells is produced by solar cells. It has been shown that it is possible to "breed" solar cells in this manner, growing or producing a larger crop each year.[11]

Yet another use for solar energy, one so ubiquitous that we usually forget solar energy presently supplies over 99% of it, is for lighting. Interior daytime lighting of large office and factory buildings consumes a lot of energy, more than is commonly realized, because for every 100 watts of lighting a building requires, 20–40 watts of air conditioning is needed to remove the heat that is generated.[12] There are many optical arrangements for transmitting free, outdoor, solar-produced light into building interiors, even "piping" sunlight into remote areas using tubes with reflecting surfaces on the inside of the pipe.

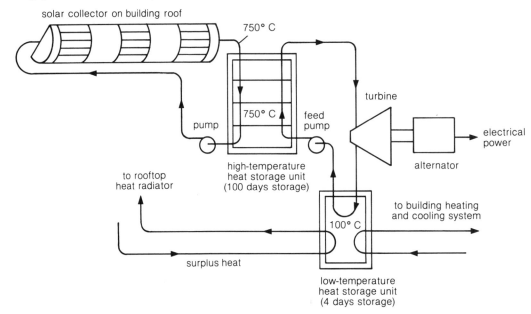

This solar-powered total-energy system is projected for use in suburban shopping centers or industrial plants where space for solar collection is available. One-axis-steerable concentrators would be situated on the roof; high-pressure dry nitrogen, steam, or sodium chloride would transfer heat at 750° C. to a high-temperature storage unit of insulated rocks or molten salt. Heat drawn from the storage unit would drive a turbine to produce electrical power and then, held in a low-temperature storage unit, would be available for space heating or air conditioning. (From Morrow, 1973.)

Uses

1. Low-quality heat for water heating.
2. Intermediate-temperature/quality heat for heat engines and steam for industrial processes.
3. High-quality heat (3,000° F) for some industrial processes in solar furnace.
4. Space heating and cooling.
5. Electric power generation.
6. Agricultural drying.
7. Refrigeration.
8. Distillation, desalination.
9. Cooking.
10. Warming sewage for methane production.
11. Pumping water for irrigation.

History

B.C.	"Passive" solar houses built in the Indus Valley and the U.S. Southwest. Solar heat for distillation of liquids and drying agricultural products. Crops grown with solar energy.
200 B.C.	Greece: Archimedes reportedly concentrates sun with reflecting mirrors to set fire to attacking fleet of ships.
1700's	Switzerland: de Saussure invents solar flatplate collector.
1860's	France: Meurhot developed ½ h.p. solar steam engine.
1890's	Chile: Solar desalination plant produces up to 6,000 gallons of fresh water per day.
1891	Baltimore, U.S.: Kemp patents first commercial solar water heater.
1897	Pasadena, U.S.: 30% of the homes in Pasadena have solar water heaters.

1912	Egypt: Shuman constructs 50 h.p. solar engine to pump irrigation water.
1938	U.S.: First "modern" solar house built in U.S. by Hottel at M.I.T.
1940's	U.S.: Basic patents for solar cells.
1941	Florida, U.S.: At least 40,000 solar water heaters installed by 1941.
1950's	France: 50 kw solar furnace built.
1954	U.S.: First use of solar cells to convert sunlight into energy.
1958	U.S.: First outer-space use of solar cells on Vanguard TV-4.
1960	Use of solar water heaters in Florida (25,000 dating from 1920's), Japan (400,000), Israel, and Australia for individual homes.
1965	Genoa, Italy: 100 kw central receiver solar tower thermal power plant.
1967	Japan: Use of silicon solar cells for isolated radio repeater stations; 170 sets have been installed with 4,000 watts of power.
1969	France (Pyrenees Mountains): Large-scale mirrored solar furnace built for the scientific study of solar energy; produces over 1 MW of power per day and reaches temperatures of 6,000° F.
1971-72	U.S., University of Arizona: Studies conducted by Meinels indicate feasibility of mass solar energy conversion to electricity.
1974	Breakthrough in mass production of photovoltaic cells (Tyco Mfg. Co., Waltham, Mass.).
1975	Connecticut builds forty-unit solar-heated housing project.
1976	U.S.: Thirty-seven states pass solar incentive legislation.
1977	France: Electricity for grid produced by solar tower prototype. Atlanta, Georgia, U.S.: 400 kw solar thermal test facility. Arizona, U.S.: First and largest (38 kw, 10.6 million gallons of water per day) solar-powered irrigation pump. Japan: More than 2.5 million solar hot water heaters sold in last thirty years. U.S.: Los Alamos, New Mexico, becomes first city to prohibit erection of buildings that shadow an existing solar collector between 9 A.M. and 4 P.M.
1978	U.S.: Mass-produced 10 kw solar irrigation systems available. U.S.: Solar-generated industrial process steam for pasteurization, steam bleaching, laundry operation, and can washing developed. Albuquerque, New Mexico, U.S.: 5 MW solar thermal test facility. Israel: 30 kw electricity generating power plant using heat of a solar pond begins operation.
1979	Spain: 50 kw solar farm built by German truck and bus manufacturing corporation. Arkansas, U.S.: 250 kw concentrating photovoltaic system for community college.
1980	San Diego, California, U.S.: Ordinance enacted requiring all new residential buildings to use solar heating systems.
1981	Barstow, California, U.S.: 10 MW solar tower scheduled to begin operation. Japan: Two 1 MW solar towers scheduled to begin operation. France: 5 MW solar tower scheduled to begin operation. Georgia, U.S.: Total solar energy system for factory will produce 200 kw of electricity and 1.5 MW of thermal power for hot water, heating, cooling, and process steam.
1985	Southwest U.S.: 100 MW solar thermal commercial facility.

Advantages

1. It is a clean source of inexhaustible energy.
2. Earth's thermal burden is not increased from terrestrial conversion of solar energy.
3. Solar power plants are relatively easy and quick to construct.
4. Solar power is safe.
5. The supply of solar energy is politically dependable. The enormous geographic and political diversity of equally sunny locations permits global dispersion of production capacities thereby decreasing possibilities of embargo by any one bloc of nations.
6. Solar energy could eliminate the need for energy transportation and distribution networks thereby reducing waste. Up to 70% of the cost of providing electricity in the U.S. is in the distribution of it.[5]
7. Solar collectors have a high energy yield; a conventional collector will produce enough energy in less than a year to pay back the energy used in its manufacture.[5]
8. Solar energy that is "lost," that is, not captured or delivered, is renewable.
9. Large solar farms in desert regions could increase land productivity by shading areas

Solar power in general has several unique implications which do not arise from its obvious advantages. For example, it could help to redress the severe energy imbalance between temperate and tropical zones; its diffuseness is a spur to decentralization and increased self-sufficiency of population; and as the least sophisticated major technology it could greatly reduce world tensions resulting from uneven distribution of fuels and from limited transfer of technology.[13]

that could then be used for rangeland.
10. Large solar energy facilities can produce electricity and fuels with smaller units without economic penalties (unlike nuclear facilities).
11. Many solar power systems do not require sophisticated, complex organizations for installation and operation.
12. Small-scale solar energy devices can be mass produced for great savings to the consumer.

Disadvantages

1. Earth-based solar converters work only when the Sun is shining, i.e., the energy supply is intermittent.
2. Solar energy is diffuse and of low quality.
3. Solar energy is not storable in its primary form (photons).
4. Solar energy systems have a low (5–15%) efficiency in extracting work from the natural medium.[14]
5. Effects of weather are often unpredictable.
6. Environmental impacts include alterations of hydrology in non-arid regions and surface albedo and roughness.
7. Wide use of solar energy could create land use problems.
8. Wide use of solar energy could accentuate electric peak demand in poor weather.
9. Widespread use of rooftop flat plate collectors would use large amounts of steel, copper, and glass compared to central power plants.[15]
10. Only about 65% of existing residential units can be retrofitted with solar space and/or water heating systems.
11. Solar energy is currently disadvantaged by numerous institutional or economic barriers—the cost of borrowing money (it's 20% more expensive to save a kw through solar energy than to add a kw of capacity through utilities), increased property taxes, building codes, lack of skills to install and maintain solar systems, legal uncertainty, the attitude and role of utilities (utility rate structure can discourage conversion to solar), and the difficulty of making economic comparisons with conventional energy systems.[16]

Solar Tower Disadvantages

1. Solar towers and farms will probably only be feasible in semi-arid regions with few cloudy days and little pollution.
2. Solar towers and farms will probably redistribute heat from desert to city.
3. Current power tower design would use as much water for cooling as a fossil fuel plant of comparable size.[15]
4. Power tower would require immense amounts of steel and concrete—500 tons of steel and 2,800 tons of concrete for every megawatt the plant produces (about 15 times the construction material required for a nuclear plant and 35 times the amount needed by a coal plant).[15]

Solar Energy— Extraterrestrial

With the advent of the space shuttle, it will be possible to put an orbiting solar power plant in stationary orbit 24,000 miles from the Earth that would collect solar energy almost continuously and convert this energy either directly to electricity via photovoltaic cells or indirectly with flatplate or focused collectors that would boil a carrying medium to produce steam that would drive a turbine that then in turn would generate electricity. This electricity would then be microwaved to Earth for transmission over the existing electric grid. Present technology for solar conversion to electricity, microwave focusing transmission, receiving, and conversion are all sufficient to accomplish this task.

Advantages

1. An orbiting solar collector is in direct sunlight continuously, with no climatic or weather-induced variations of solar intensity, thereby eliminating the need for storage, and providing base load capability.
2. Space environment is relatively benign in that it is free of physical and chemical factors such as rain, wind, earthquakes, corrosion, and gravity with which terrestrial systems have to deal.
3. Power production can be increased by adding to the orbiting array without increasing the size of either the transmitter or the collector.
4. Sunlight is eight times as intense in outer space as it is on the surface of the Earth.

Schematic diagram illustrating the SSPS concept with photovoltaic power conversion in a synchronous orbit, microwave transmission to Earth, and conversion to electricty of the received microwave radiation.

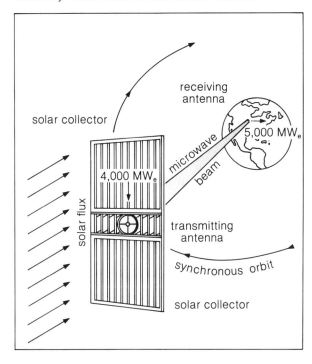

An artist's concept depicting one of the Solar Power Satellite microwave transmission antennas being assembled at the outer edge of the structure.

5. The conversion of microwave energy to direct current would be about 90% efficient, thereby releasing only 1/10 as much energy as it would deliver for use. Fossil or nuclear fueled plants release about 1.5 times the energy which they deliver to the power lines.[17]
6. Land use requirements for the ground antenna array would be only 1/10 to 1/100 as great as for direct photovoltaic energy conversion of sunlight.[17]

Disadvantages

1. Extraterrestrial solar energy increases the heat burden of the biosphere.
2. Ground receiving stations will have to be at least five miles square and the transmitting antenna about one mile in diameter.
3. Microwave transmission could be harmful to life that is exposed to it, as well as effecting inadvertant weather modification.
4. Microwave power transmission from satellites could interfere with microwave telephone transmission.
5. There will be large initial costs of putting the power station into orbit.
6. Solar power satellites would increase centralized power production, and would be a high military sabotage risk.

12 Bioconversion

The Poor Man's Oil

Bioconversion refers to the tapping of the solar energy that has been captured and stored by nature through photosynthesis. There are numerous methods for doing this, ranging from the relatively straightforward combustion of wood or dung for cooking food to the more complicated production of algae to produce methane via anaerobic digestion. There are bioconversion processes for producing solid fuels (wood and charcoal), liquid fuels (oil and alcohol), gaseous fuels (methane and hydrogen), and electricity.

Fossil fuels were once plants. Bioconversion dramatically shortens the time that it takes to make plants available for energy sources. We would have to wait millions of years if we wanted coal or oil, but only six months or less if we used the plants directly.

There are two primary sources of organic materials for conversion into energy products: organic waste and refuse that has been

---

The biosphere annually produces 150 billion tons of organic matter (150 pounds per day for each person on Earth). This is equivalent to 61,500 × 10^{12} kwh (42 × 10^3 kwh per person per day).

Gas producing prototype unit at the University of California, Davis, that uses farm and forest residues to produce methane gas. The methane from this unit fires a boiler that heats and air conditions one of the buildings on campus.

generated as a by-product of human activities (such as garbage, sewage, and livestock waste), and energy crops grown specifically for producing energy.

Much of the solar and chemical energy that went into the production of plants and manufactured goods still remains in organic waste and refuse when it is "disposed" of as no longer useful (approximately 4,500–7,500 BTU/lb. of municipal refuse).[1] Reclaiming some of this energy would greatly increase our overall energy-use efficiency plus reduce the littering of the landscape with garbage dumps or smudging the sky with incinerator smoke. By combusting the 100,000 tons of burnable trash the U.S. produces annually, about 350 million kwhs of electricity could be saved.

Besides combusting this municipal refuse directly in power plants (as St. Louis, Baltimore, and other cities do) there are other techniques to convert the available waste into fuel. One such technique is destructive/distillation or pyrolysis. Pyrolysis has long been used to produce methanol from wood,[3] and it has been used to produce methane from cow manure and urban refuse.[4,5] The major disadvantage of pyrolysis is that it works at high temperature and pressure and thereby needs a significant portion of energy to sustain itself. It is therefore an expensive and inefficient means to produce fuel.[3]

Hydrogasification is the chemical reduction of organic wastes by treatment with hydrogen at elevated temperatures and pressures. The same disadvantages apply to it as to pyrolysis.

Anaerobic digestion is the production of methane and sludge (that can be used as a fertilizer) by the use of anaerobic bacteria to decompose organic wastes.[6] This process has been in use for years in the treatment of sewage. Los Angeles has been treating its sewage at the Hyperion Sewage Treatment Plant in this manner since 1950, using the methane that is produced to run the plant, and

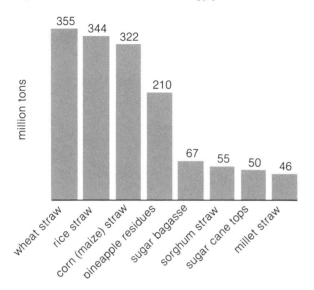

Agricultural waste useful for energy production[2]

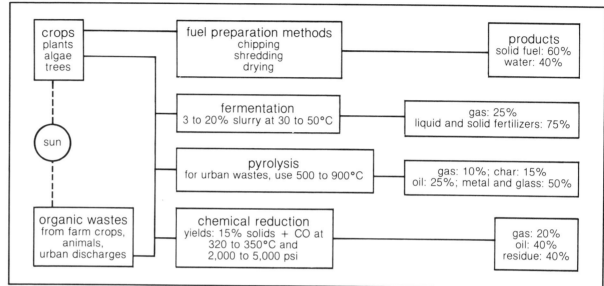

Schematic diagram illustrating methods of fuel production by the use of solar energy[11]

selling the rest to a municipal electrical generating plant. Two problems are partially solved with this approach: energy is produced and waste is disposed of. Recent analysis, though, points out that in-house, non-water-carrying disposal systems, i.e., dry toilets, installed in almost all houses would cost less (in terms of energy and dollars) than modernizing most municipal sewage treatment plants. From 1950–1970, over 2,500 biogas plants were built in India to process cow dung into methane for cooking and heating.[8] Since 1973, close to 45,000 have been produced, with 25,000 sold in 1976 alone.[9] (By burning cow dung directly, more than 90% of the potential heat and all the nutrients are lost; biogas is burned in stoves that are ten times as efficient as a dung fire.) China has over 7 million biogas plants in operation.[10] Some of the reasons for this widespread use are the simplicity of the process and its efficiency. Almost anyone can build a workable methane generator for small farm use following this method. No high temperatures or pressure are needed, only an airtight tank where the anaerobic bacteria can grow without exposure to oxygen and can be kept at about 90–95° F.[6]

In addition, "biogas is not suitable only for household cooking (the process is faster, easier, and less expensive than with wood or straw) and lighting (a gas lamp brighter than a 100-watt incandescent bulb can be used) but it can also generate electricity and power water pumps and crop processing machinery. Moreover, the benefits go beyond the availability of a clean and versatile fuel: savings of fossil fuels and reduction of fuel expenditures offer a significant economic advantage; conservation of forests and grasses has favorable ecological implications; elimination of many insect pests and diseases markedly improves hygienic conditions of rural areas; burning of biogas largely eliminates the tedious, every day gathering of firewood or grasses and lightens household labor; finally, methanogenic fermentation yields an excellent organic fertilizer, an essential ingredient of the Chinese farming. Even in relatively cold Honan, where fermentation efficiencies are lower than in the south, sludge removed from each 10 m³ digester provides annually an equivalent of 30 kilograms of ammonia."[10]

Another bioconversion energy source is algae. Under laboratory conditions, a reproduction rate has been achieved

Algae Production

SOURCE:
UNESCO, Arid Zone Hydrology, 1953

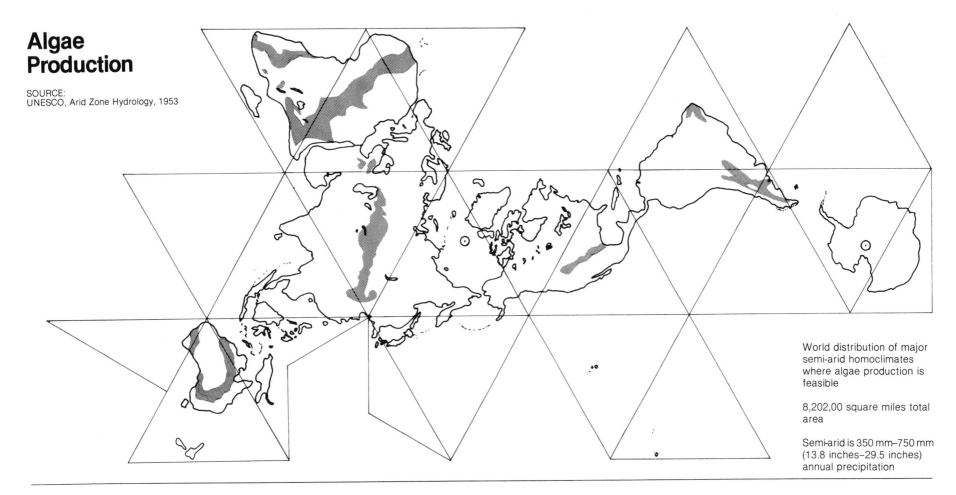

World distribution of major semi-arid homoclimates where algae production is feasible

8,202,00 square miles total area

Semi-arid is 350 mm–750 mm (13.8 inches–29.5 inches) annual precipitation

corresponding to an annual acre-yield of some 75 tons dry weight for the algae chlorella.[7] In another test, blue-green algae grown on sewage yielded 16 to 32 tons dry weight per year. With a heat value of between 9,000 and 13,600 BTU per pound, chlorella can be burned directly as a fuel, made into methane by anaerobic digestion, or converted to alcohol. In addition it is rich in protein so it could possibly be used as a food source. Almost 550,000 kilowatt hours-per acre-per year could be produced in this manner. One hundred six million acres (400 × 400 miles) of algae would need to be produced to obtain all the energy presently used by the entire world. Because algae can be grown hydroponically, it does not have to use prime farm land, but can be grown on waste or semi-arid land, or even the oceans. Also, an acre does not have to be horizontal, but could be somewhat vertical; multi-storied automated greenhouses could contain many acres of algae production on an acre of land. Utilizing just the semi-arid regions that would be suitable for chlorella production, $1,248 \times 10^{12}$ kilowatt-hours annual gross energy could be obtained.

Due to the growing recognition of the importance of hydrogen in our future energy economy, there has been increased interest in the area of direct hydrogen production by photosynthesis. There are currently three approaches to this: artificial photosynthesis systems composed of chlorphyll-containing elements extracted from plant cells, artificial photochemical membranes that would be biochemical analogs of photovoltaic cells (see "The Purple Membrane" in the Exotic Energy Sources section), and hydrogen production from algae. Although it has been known since the 1940's that certain kinds of microorganisms will photo-produce molecular hydrogen,[7] thus far all work in this area is still on the experimental level with no prototypes for practical application yet produced. Recent research promises that this technique may be able to produce sufficient hydrogen to meet the energy needs of an entire household, using an area of only a few square meters for algae growth.[12]

Another way in which the photosynthetic cycle can be utilized as an energy (and material) source is in the production of sugar from the sugar cane or sugar beet. Sugar from these highly efficient plants can be burned (it has the same caloric content as sugar) or converted to alcohol. In this process the thermal efficiency is very good, with practically no loss in going from sugar to alcohol. It takes 12.9 pounds of sugar to make one gallon of alcohol, that is, 64¢ worth of sugar at 1971–72 prices of sugar cane to make one gallon of alcohol. It costs about 20¢ to convert the sugar, making a total of 84¢ per gallon by fermentation. "If the

---

Collectable crop residues and feedlot wastes in just the U.S. contain more energy than used by all the U.S. farmers.

Nearly 30% of Brazil's energy comes from burning wood and sugarcane bagasse.

---

sugar planters in Hawaii, whose gasoline is now rationed, would convert about one-third of their molasses directly into fuel alcohol, they would not have to purchase the 15 million gallons of petroleum which they now do to run their agricultural machinery. In Nebraska, which has about 7 million bushels of spoiled grain per year, this should yield more than 20 million gallons of alcohol."[13]

Despite the lack of large government support, gasohol—a mixture of 10% alcohol and 90% gasoline—is being increasingly used throughout the world. The U.S. has over a thousand service stations selling gasohol. Its advantages include higher octane, better performance and mileage (for older cars), and cleaner burning. Disadvantages include the fact that gasohol (or a pure alcohol fuel) would require a larger fuel tank to obtain the same range because the energy content per gallon is less than gasoline, and a pure alcohol fuel would necessitate some new materials for gas tanks, fuel lines, and carburetors because it is more corrosive than gasoline. Additionally, pure alcohol fuel would have to be preheated before entering the engine because it does not vaporize as easily as gasoline.

Another use of the photosynthetic cycle would be as a source for hydrocarbons for use in chemicals and materials. The Hevea rubber plant, first found wild in Brazil and now grown almost exclusively in plantations in Malaysia and Indonesia, is a source of hydrocarbons.[13]

Yet another biomass resource is aquatic plants such as the water hyacinth or kelp. The water hyacinth, for instance, produces about 134 metric tons per hectare in sewage-spoiled waters. This could be converted to about 18,630 cubic meters of methane gas. About 6 million hectares of water hyacinths would be needed to produce all the natural gas currently used in the world.

The seaweed kelp could be grown in the open ocean by pumping up deep, nutrient-laden waters for nourishment. Kelp is attractive because it is efficient in converting sunlight into stored energy (2%), and because land and water are removed as constraints to production. A small prototype ocean kelp farm is currently operating off the coast of California (see illustration). Current schemes include one design for a 470 square mile ocean kelp farm that would produce as much gas as the U.S. currently consumes.

A bioconversion approach that is receiving more and more attention is the "integrated processing facility," or what has been called the "biomass refinery."[15] In this approach, energy and raw material for the food and fiber industries are produced. For example, corn could be used for producing alcohol, and corn husks and stalks for fiber for paper production or for fuel for drying crops. The high-protein silage left over from the alcohol production could be used as animal feed.

If one assumes a transport sector with three times today's average efficiency—a reasonable estimate for early in the next century—then the whole of the transport needs could be met by organic conversion.[14]

149

History

1200–1850 Wood is the major energy source of the world.

1776 Italy: Voltz discovers that a combustible gas is generated from decaying vegetable matter.

1871 London, England: Concept of obtaining gas from human, vegetable, and farm waste was first demonstrated.

1895 Exeter, England: Methane from sewage sludge used for street lighting.

1905 Bombay, India: First large-scale plant to produce both gas and fertilizer was installed.

1925 Paris, Copenhagen: Garbage is mixed with coal and burned in power plants.

During W.W. II Germany: Biomass convertors used for both gas and fertilizer when petroleum was in short supply.
Sweden: Almost all fuel derived from wood.

1950 India: Patel develops Gobar Gas plants.

1970's Algeria, South Africa, Korea, France, Hungary, China, India, and many other countries produce gas from waste.

1975 Brazil: National Alcohol Program started for the purpose of increasing ethyl alcohol production so that it can be used to replace automotive gasoline, diesel oil, and several synthetic products.

1977 France: 18,000 ton/day waste processing plant fuels water-heating system for 2,500 dwellings.

1978 Brazil: 1.5 billion liters of ethyl alcohol produced from sugarcane to blend with gasoline for automotive fuel (8% alcohol content of total gasoline supply; to be 20% by 1980); 200 alcohol production plants being built.

1978 Pompano Beach, FL, U.S.: 25 ton/day methane-producing, waste reduction via anaerobic digestion plant begins operation.

1978 U.S.: Wood products industry derives 40% of its total energy needs from burning wastes.

1978 U.S.: Peoples Gas Company in Chicago buys methane generated by manure from large cattle feed lots in Oklahoma.

1978 California, U.S.: Diamond/Sunsweet walnut factory produces low BTU gas from walnut shell waste at 80–85% efficiency.

1979 U.S.: About 1 million wood stoves sold in recent years; ½ million more being sold each year.

1979 Vermont, U.S.: Burlington Electric Company building 40×10^6 electric power plant powered by wood chips.

1979 U.S.: One-quarter acre marine kelp biomass energy conversion system begins testing.

1979 U.S.: About 1,000 gas stations, mostly in midwest, offer gasohol. U.S. production capacity at 200 million gallons (757 million liters) per year.

1980 U.S.: Regional test biomass energy farms.

Advantages

1. Biomass is readily available and in large supply.
2. Biomass can be produced almost anywhere and so is not very susceptible to international political pressures; it is constantly being replenished. Many potential sources of biomass can be grown on marginal lands or on bodies of water that are unsuitable for food production.
3. Human-generated wastes are readily, abundantly, and regeneratively available wherever humanity is found.
4. Bioconversion fuels are easily used by present energy-using artifacts.
5. Bioconversion can produce storable solid, liquid, or gaseous fuels.
6. Energy content is high; wood and dry crop wastes have an energy content of 14–18 million BTU per ton—comparable to Western U.S. coal.
7. Biomass contains almost no sulfur, little ash, and will not produce more CO_2 than it removes through photosynthesis.
8. Biomass can be burned or gasified as easily as coal, and liquified easier than coal.
9. Bioconversion is technologically easy for man to use (agriculture has used it for millenia).
10. Using unproductive brushlands for energy production could be environmentally advantageous if coupled with ecologically considerate production methods and fire management programs. In addition, harvesting these unproductive lands on a regular basis could allow these lands to support more wildlife, be suitable for recreational purposes or grazing, in addition to the primary purpose of energy production.
11. Biomass-produced methane can be substituted for natural gas; it provides a convenient and efficient way of storing solar energy; it is non-polluting and the organic residue left from its production makes an excellent fertilizer.

Disadvantages

1. One-third to two-thirds of the energy in biomass is lost in most conversion processes.
2. There is a high initial cost for the conversion and collection facilities.
3. The use of crop residues and animal wastes for energy production essentially robs the soil of the cellulose bulk necessary to maintain soil tilth, microbial activity, and water-retention capacity.
4. Burning of vegetation is the world's largest source of carbon monoxide.
5. There are some difficulties in handling raw biomass.
6. Substantial amounts of land would be required if biomass fuels are to be used on a large scale.
7. Widespread development of biomass plantations could alter surface reflectivity, aerodynamic roughness, and moisture-transfer properties over very large areas.
8. Biomass plantations would reduce regional biological diversity because of the large areas devoted to monoculture.

Wood

Wood is constantly being replenished as forests continue to impound the Sun's radiation. In 1850, 90% of the fuel burned in the world was wood; by 1945, it had decreased to 5% in the U.S., and, until very recently was still diminishing. In the last few years, burning wood has had a minor resurgence in the U.S. (6 million in the U.S. in 1979; 5 million sold since 1974). Fuel wood remains the main source of energy in many developing regions of the world; in some southeast Asian and African countries, over 90% of the total fuel consumption is in the form of wood.[16] In fact, in terms of the total quantity of energy involved worldwide, fuel wood is the world's fifth largest energy source—behind coal, oil, natural gas, and biogas, and ahead of hydroelectric and nuclear.[17]

The widespread use of wood lasted until the easy access forests were eliminated and energy demands rose above what could be provided by wood. Recent uses of wood as an industrial fuel have included its use as a source of "producers" or "wood" gas during World War II to power automobiles and trucks. Individual vehicles were fitted with special ovens in which incomplete oxidation plus dry distillation of the wood would take place. By mixing this gas and air an explosive gas resulted that could be used in ordinary internal combustion vehicles,

Seventy-five percent of the people in Maine use at least some wood to heat their homes.[15]

replacing the usual mixture of air and gasoline vapor. Unfortunately, the energy content of this fuel is not very high. Besides its limited use as a fuel, wood is in high demand as a resource for building and paper.

Another recent concept being proposed involves the widespread utilization of "energy plantations" where certain hybrid poplar, sycamore, or other fast-growing tree species would be grown on short harvest cycles (3–5 years) in dense plantings. Such a utilization would produce about 120 million BTU (35,000 kwh) per acre per year at a cost of between $1.25 and $1.45 per million BTU on a four to six harvests per planting schedule. This cost range, which can be achieved broadly in the U.S., is lower than present fuel oil prices.

To supply the fuel (on a regenerative basis) for all the electrical generating capacity presently installed in the U.S. by this method would require that less than 160 million acres be devoted to such energy plantations, about a third of the land potentially suitable for this purpose. Such areas include lands that have little regular use at present, such as those having slopes too steep or soil too stony to permit more intensive farming, forest grazing land, and grassland range, but would not include land suitable for dirt farming, tree farming, or that included in national parks, forests, or wilderness areas. Present unemployment could be synergetically harnessed in WPA-style plantings for such energy plantations.

In addition to the above method for producing energy from wood, there is also the possibility of using the wastes from traditional logging and milling operations. In the U.S. these wastes amount to some 24 million and 84 million tons per year, respectively.[15] Electric power plants fueled with logging debris could be brought on line much quicker than coal or nuclear-powered facilities—five years compared to ten to twelve years. In addition, using logging debris for fuel would reduce forest fire hazard and insect infestation, as well as improve forest management access and esthetic appearance.

The large increase in residential wood burning heaters and the know-how and experience that has accompanied this new development should make possible another development: a small wood-fueled cogeneration unit that could provide central heat and electricity simultaneously.

World Fuelwood Production[18] 1976: 1,201.5 × 10⁶ Cubic Meters

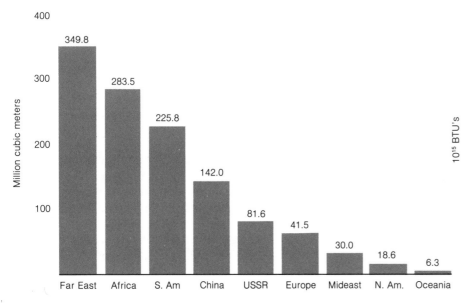

World Fuelwood Production[18] 1976: 6,628.0 × 10¹² BTU's (1.94 × 10¹² kwh); 1.6 × 10⁶ BTU per capita (468 kwh).

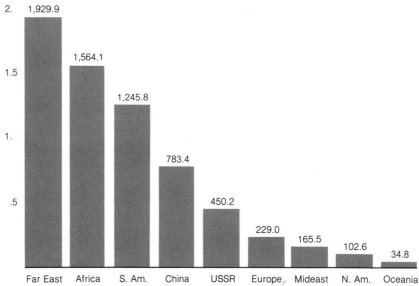

Advantages/Disadvantages

In addition to all those listed previously under "Bioconversion," there are the added advantages that would result from reforestation such as soil erosion control, retardation of dam siltation, and improved air quality; additionally, the wood ash from combustion in power plants would provide a valuable fertilizer.

Disadvantages include the fact that deforesting will increase the global excess CO_2 problem in that forests are highly responsible (over 50%) for removing CO_2 from the atmosphere, and that extensive wood burning can significantly increase air pollution.

World Fuelwood Production

Source:
World Energy Supplies 1972–1976 (United Nations, 1978).

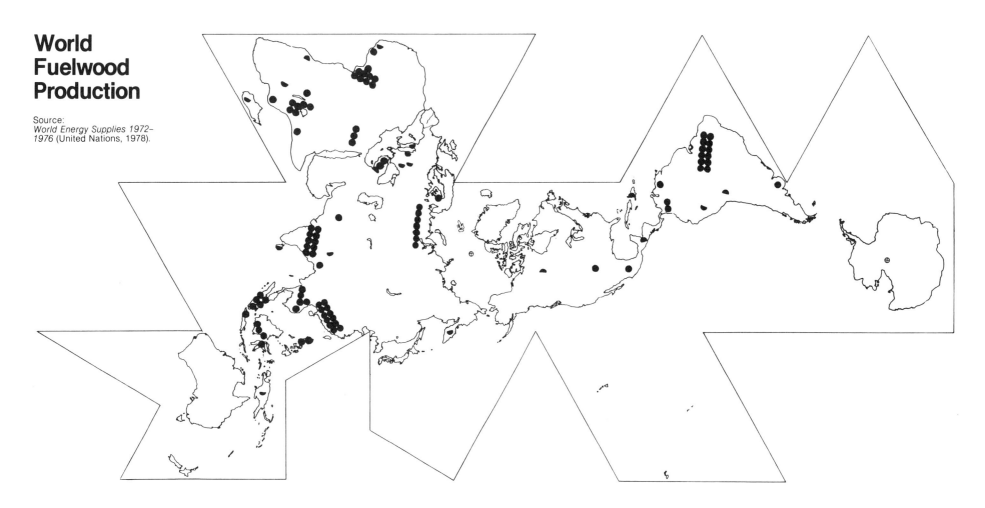

● = 1% of World Fuelwood Production
● = 12.0 × 10² million cubic meters

Forests

Source:
Goode's World Atlas, 1975

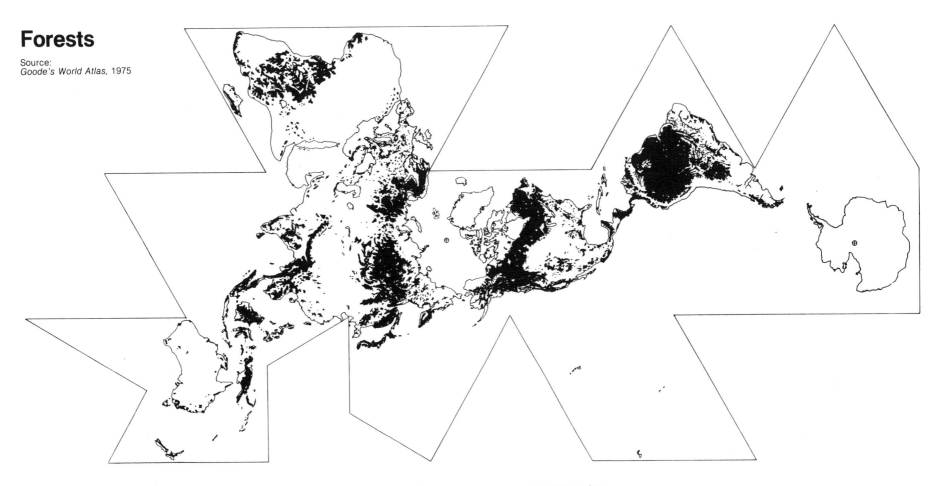

Lands having vegetation dominated by trees cover approximately 30% of the world's land area today; of this amount, 77% is in public ownership. The main types of forest are: tropical rainforests characterized by hardwoods, luxuriant undergrowth, and poor soil; mid-latitute hardwood (deciduous) forests quickly being depleted by exploitation and urbanization; arid Mediterranean or scrub forests with small trees and grassy areas; and coniferous (soft-wood) boreal forests of the northern latitudes with long, cold winters and sparse population. The ecosystem of the forests operates as an influence on the climate and water regime, providing shelter for wildlife, and potential utilization and/or exploitation for recreation, hunting and fishing, and the lumber industry.

Forests comprise about half of the Earth's captured biomass energy.

14 The Tides

Three quarters of the Earth is covered with water. The ocean's tides, waves, and currents are three distinct phenomena that can be harnessed to produce power from this vast source. The tides are caused primarily by the gravitational pull of the moon; the waves are caused primarily by the winds; and the currents are caused by the rotation of the Earth. Each demands its own unique energy-harnessing techniques.

The tides are an income energy source that is continually regenerated through the combined kinetic and potential energy of the Earth-Moon-Sun system. The twice daily rise and fall of the sea causes an oscillatory flow of water in the filling and emptying of partially enclosed coastal basins. The tides can be utilized to produce hydroelectric power by damming the basins and regulating the tidal flow through gates to run turbines in much the same way as falling river water generates electricity on land. Two-way turbines have been developed that can be activated as the tide flows in either direction. The turbines can also act as pumps and the basin can be used to store water to be released during peak periods of energy usage.

The amount of energy derived from a tidal power plant may be large, but there are only a limited number of viable sites. Viability is determined by the height of the tidal range (present technology calls for a minimum range of ten feet for large power systems; lower ranges are suitable for small scale tidal systems), the size of the basin area, its depth opening and shape, the quality of the foundation soil for the barrage, the probability of silting, rate of sedimentation, and the length of barrage necessary to enclose the basin. There are approximately one hundred sites in the world

Tidal Sites

- ● Source A: Hubbert, *Resources and Man*, 1969.
- ■ Source B: *New Sources of Energy and Economic Development*, U.N. Department of Economic and Social Affairs, 1957.
- ▲ Source C: *Business Week*, 11-9-74.

Coast of France
Aber-Benoit
Aber-Wrach
Arguenon
Larcieux
La Rance
Frenaye
Rotheneuf
Mont Saint Michel (6000 mw)
Somme

Bay of Fundy
(World's highest tidal rise and fall: 53 ft or 16 m)
Passamaquoddy
Cobscook
Annapolis
Minas-Cubequid
Amhearst Point
Shepody
Cumberland
Petitcodiac
Memramcook

Severn Estuary (4000 mw)

USSR
Kislaya Islet
Lumbovski Bay
White Sea
Mezen Estuary (6000 mw)

Maraca Island

San Jose

Puerto Santa Cruz
Puerto Gallegos
Cape Virgenes

Gulf of Cambay

Fitzroy River

Cook Inlet (Knik Arm)

Mexico
Rio Colorado

USSR
Sea of Okhotsk (10,000 mw)

Seoul River/Inchon Bay (2000 mw potential)

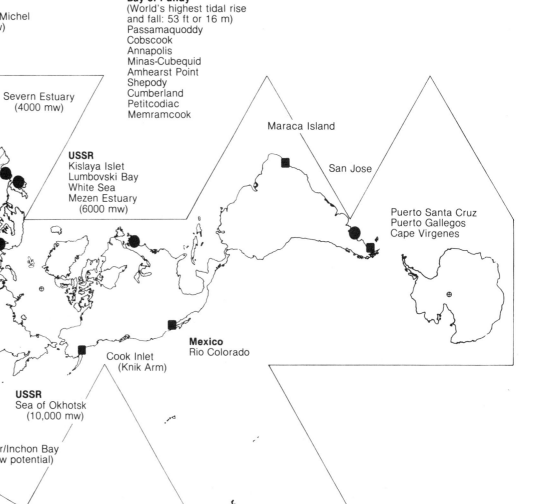

that are suitable for large tidal plants (see map). Some of these plants could be coupled for greater efficiency and productivity.

Besides these large-scale projects, there are innumerable sites for small-scale tidal installations, ranging in size from less than one to ten megawatts. As a cursory look at their history will disclose, tides have been harnessed on a small scale for centuries; nothing new is involved. China currently has planned or built about 120 very small tidal plants with a total capacity of 7,600 kw. Many of the small-scale tidal power sites are located on the coasts of areas lacking in power sources.[1] The power from such installations could be used for electricity generation or mechanical work such as pumping or milling. Electrical energy could be fed into a distribution grid if available, or used to produce hydrogen that could be used to power the machinery of the area or used as a feedstock in the production of fertilizer. The ultimate potential amount of energy from the tides around the world is 36×10^{12} kilowatt-hours.[2] Much less can be harnessed currently.

The 240 megawatt Rance River tidal plant in France, the largest in the world, was built in six years. The life of the plant is determined by how long it takes for silt to fill the reservoir. The Rance plant is expected to last fifty to seventy-five years. Great Britain has studied the projected construction of an 800 megawatt tidal plant for the Severn River at Bristol Channel

---

Each cycle of the Bay of Fundy tide dissipates close to 500×10^6 kwh of energy, a quantity nearly equal in magnitude to the consumption of the entire Canadian electrical network.[3]

Schematic diagram of a one-way, single-basin tidal-power installation.[4]

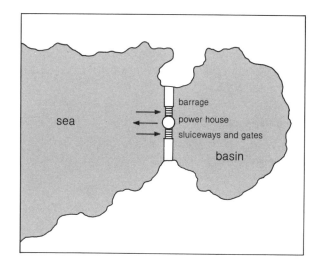

Schematic diagram of a linked-basin tidal-power installation. The directions of water flows are indicated by arrows.[4]

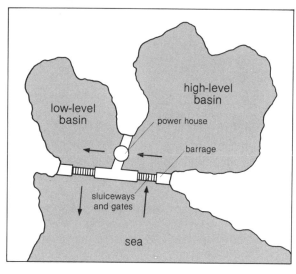

three times, in 1918, 1933, and 1945. If the 1933 or 1945 projects had been carried out, they would have paid for themselves within ten years.

Of the one hundred sites in the world suitable for large tidal plants, only some are near populated areas that would be able to absorb the power produced by the plant. A strategy has to be devised at each location to determine the most efficient way to use the maximum amounts of potential energy available; that is, where and how many barrages are to be built, how many turbines can be supported, etc. A strategy may be worked out for a given site that would provide electrical energy during peak periods. It may be desirable during off-peak periods to switch from generation of electricity into a power grid system, to production of hydrogen which is more easily stored than electriciy, or to pumped hydro-storage in estuarial reservoirs.

Uses

1. Electric power generation.
2. Hydrogen production.

History

Ancient Egypt	Tide powered paddle wheels used to irrigate croplands.
1100 A.D.	Dover, England: Tidal mill, also on Atlantic coast of France.
1100	Suffolk, England: Tide grain mill.
1170	England: Tide grain mill (this mill functioned successfully for 800 years).[6]
1500	Netherlands: Tide mills in operation.
1580	England: 20 ft. tidal water wheels installed under arches of

The wheel for raising Thames water at London Bridge as it was in 1749. 20 feet (6.5 m) in diameter, it was driven by the flow of water between either side of the piers of the bridge as the tide flowed in and out.[8]

	London Bridge to pump water; in use until 1824.
1635	Salem, MA, U.S.: First tide mill pump in America.
1600's	New York, U.S.: Dutch colonists build tidal mills.
1700's	Rhode Island, U.S.: Large tide mill constructed.
1856-1939	Two hundred patents relating to the utilization of tidal energy are registered.
1880	Germany: Tidal pump used for sewage pumping in Hamburg.
1898	Santa Cruz, U.S.: Tide powered motor built by city.
1966	France: World's first large-scale tidal power plant for producing electricity built; it consists of a series of 24 10-megawatt turbine generators.
1969	U.S.S.R.: 400 kilowatt tidal power plant at Kislaga Guba on White Sea north of Murmansk; prototype for 1,500 kilowatt plant. (U.S.S.R.'s tidal power potential: 210×10^9 kwh/yr.[7])

Advantages

1. It causes no waste heat, discharge, or serious bio/geochemical damage or danger to marine ecology.
2. Transportation links can be established across barrages to connect coastal areas otherwise unconnected.
3. Tidal power is not hampered by drought as is hydro-electric because it relies on the relatively unchanging lunar cycle. There is a maximum of only 5% variation from year to year.[5]
4. Tidal power plants are rarely "down"; Rance River tidal plant generates electricity 91% of the time (500 MW hours out of a maximum possible 544).
5. Tidal power is "inflation proof" in that its "fuel" is free.
6. Improvement of navigation to estuary ports.

Disadvantages

1. The costs of construction have not in the past generally been competitive with conventional energy plants due to problems of construction on water-covered sites.
2. While tidal plants produce no pollution, they would disturb the natural water flow patterns and might create or aggravate a problem or build-up of river pollution that would normally dissipate out of the basin.
3. There are relatively few sites available for large power applications.
4. Tides vary, so power production will vary; in addition, variation occurs at different times of day.
5. Because of the marine salt water environment, corrosion can be a problem.
6. Major producing sites are often not near major consuming areas.
7. Tidal power plants cannot be built up piecemeal, but must be fully constructed before any power output (and hence financial return).

15 Waves

Waves are a clean, non-depleting energy source of large magnitude. They are caused by winds blowing across large bodies of water, and, as such, they are effective collectors and storers of wind energy. A mile-long, four-foot wave contains about 20,000 kw of power, or about 18.3 million kwh per year. A ten-foot wave contains 1,200 million kwh per mile per year. One recent British estimate of the annual world wave power potential of only *shore based* wave-harnessing power plants was put at 40 trillion kwh.[1] Other floating devices for harnessing wave power would greatly increase this figure.

Many different proposals have been put forward to harness wave power (see History). Basically, all the various devices exploit one or more of the following characteristics of waves: the oscillating vertical motion of the wave; the circular motion of the water particles within each wave; the varying distances between water surface and sea floor and the associated pressure changes; and the breaking of waves on the shore or breakwaters.[3]

To date, the only wave-power generators in use are in Japan, where there are over three hundred small 70-watt units which are used as power sources for marker and meteorological buoys and lighthouses. These units have very high reliability, even in snow and stormy weather, a fact confirmed by their ten years of

---

Wave power was chosen as the most promising source of alternative energy for Britain in a recent study by the Central Policy Review Staff. "British shores average about 80 kw for each meter of frontage; the potential for the entire 1,500 kw coastline is equal to 120,000 MW, twice the installed electrical capacity of Britain."[2]

> *Japan's wave power potential is three times its current electrical generating capacity.*[4]

practical use. Power is generated by a small air-turbine generator driven by air pressure fluctuation which results from the difference caused by the periodical change of the wave surface around and inside the buoy (see Illustration 1). Larger wave power plants—as large as 3,000 MW—are possible using this method. A 2 MW wave power plant of this design is currently under development. The total conversion efficiency (from wave to air to electricity) is about 28%.[5]

Wave power has been used successfully before—in 1909 at Huntington Beach, California and near Bordeaux, France where a 1 kw wave energy generating device was also in use in the early part of this century. The Huntington Beach device was operated by the California Wave Power Company which built Mr. Alva Reynold's "Ideal Wave Motor" that lighted lamps on the wharf until a storm washed the machine away. The motor used panels moving underwater to transfer the energy of the waves to an electric generator.

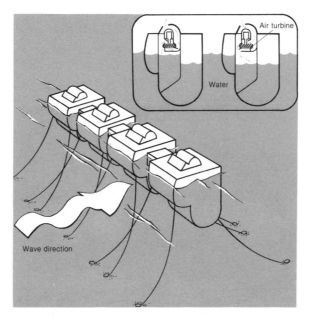

Oscillating water column.[6] Inset: Sectional views of two air pressure ring units.

Oscillating Water Column

This device resembles an empty beer can held with its open end under water. Incoming waves set up oscillations of the water column trapped in the upturned can. As the device rocks back and forth with the movement of the waves, the water within the buoy is distributed from one compartment to another and then back again. The changing distribution of water forces the air inside the buoy through the air turbine which in turn generates power.

Oscillating water column devices are in use in Japan and are under development by the British National Engineering Laboratory, and the Japanese.

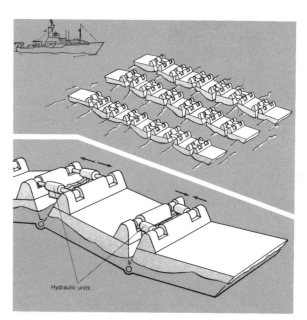

Contouring Rafts.

Contouring Rafts

This device consists of a series of rafts that would "contour" with the waves—that is, they would follow the shape of the waves—so that adjacent rafts would rise and fall relative to each other. Hydraulic motors or pumps between each raft would convert this rise and fall motion into hydraulic pulses which could drive a turbine or store the pressurized fluid, either within the body of the rafts or in separate storage vessels.

Wave Rectifier

This device is divided into box-like compartments set at right angles to wave direction. The compartments are fitted with one-way valves designed so that waves drive seawater into the alternating high level reservoir and empty the low level reservoirs. This creates a "head" between the two reservoirs which can drive a water turbine.

The front face of each high level reservoir has vertical non-return flap valves which open inwards; low level reservoirs have the same arrangement except the flap valves open outward. Waves drive sea water into the high level reservoir and extract water from the low level reservoir through the automatic opening and closing of the flap valves due to pressure differences across them.[7]

Source: Hydraulics Research Station, Wallingford.

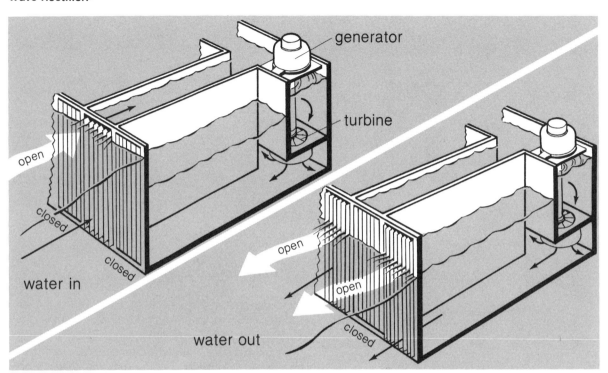

Wave Rectifier.[6]

Pliable Strips

This device incorporates pliable rubber-like strips filled with hydraulic fluid that are firmly secured in concrete troughs submerged along the coastline. The changing pressure from each wave level on the strips forces the hydraulic fluid through a conduit line where it accumulates under pressure. This pressure in turn is made to run a turbine and generate power.[9,10]

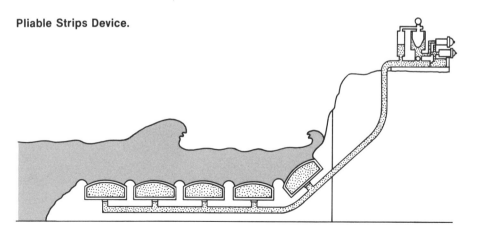

Pliable Strips Device.

Siphon Effect Device.

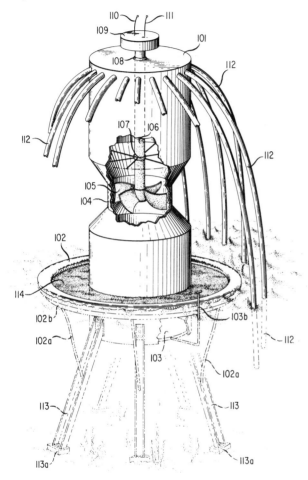

Siphon Effect

This device pulls water into the top of the tank through a siphon effect that is enhanced by the position pressure of waves. The water then flows out the bottom of the tank, throught the turbine, and generates power.[8]

Floating Tubes Device.

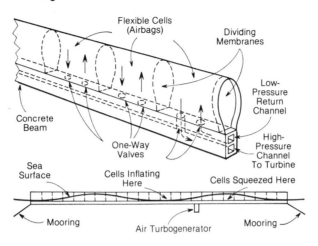

Floating Tubes

This device consists of an array of giant tubes that are moored to the ocean floor at fifty yard intervals. Composed of individual flexible cells—air bags—these tubes would bob in the path of oncoming waves and be squeezed by the rolling, breaking action of the waves. A stream of air would be forced out of each air bag by the wave pressure and then flow through submerged pipes to power a generator. One-way valves would direct the closed circuit flow of air. As wave pressure deflated each tube in turn, thereby forcing air into the high pressure channel, exhaust air from the turbine would be subsequently returned to the bags via a low pressure channel to reinflate the bags in wave troughs.

Nodding Ducks

This device consists of a string of vanes which bob in and out extracting energy from the waves. The vanes would be mounted on a floating boom and would be at right angles to the waves. The circular water particle motion of the waves would be harnessed in this device. Each vane is shaped so as to extract the maximum amount of energy from the waves; the front being flat and the rear surface smoothly rounded so that the nodding movement of the vane does not displace enough water to generate waves there.

Wave Pumps

These devices would have each incoming wave force water, by means of valves and pressure chambers, into tanks above sea level. From here it would flow back to the sea, running a turbine and generating power on its way.

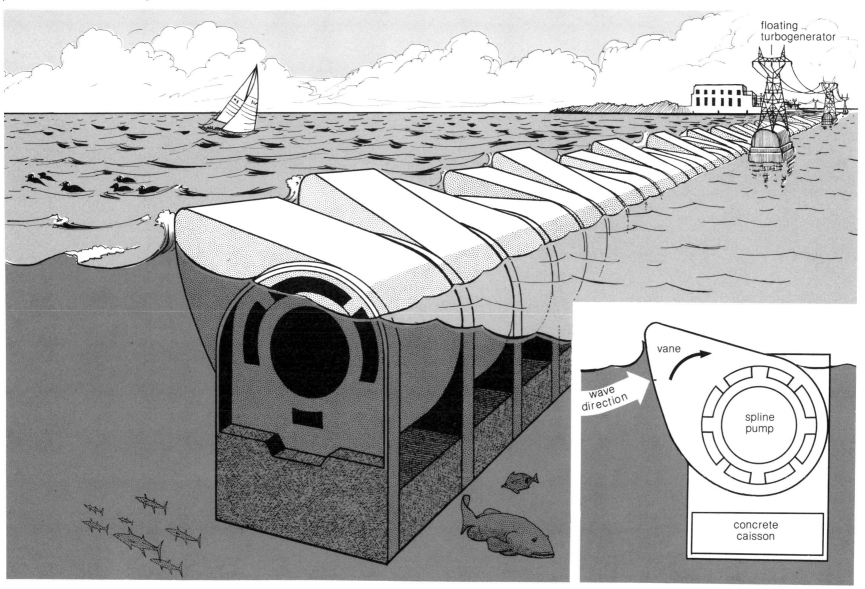

Nodding Ducks. Proposed sea-wave power plant would consist of boom, seen here in cross-section, shaped to present a vane to oncoming waves and mounted in a floating concrete box.

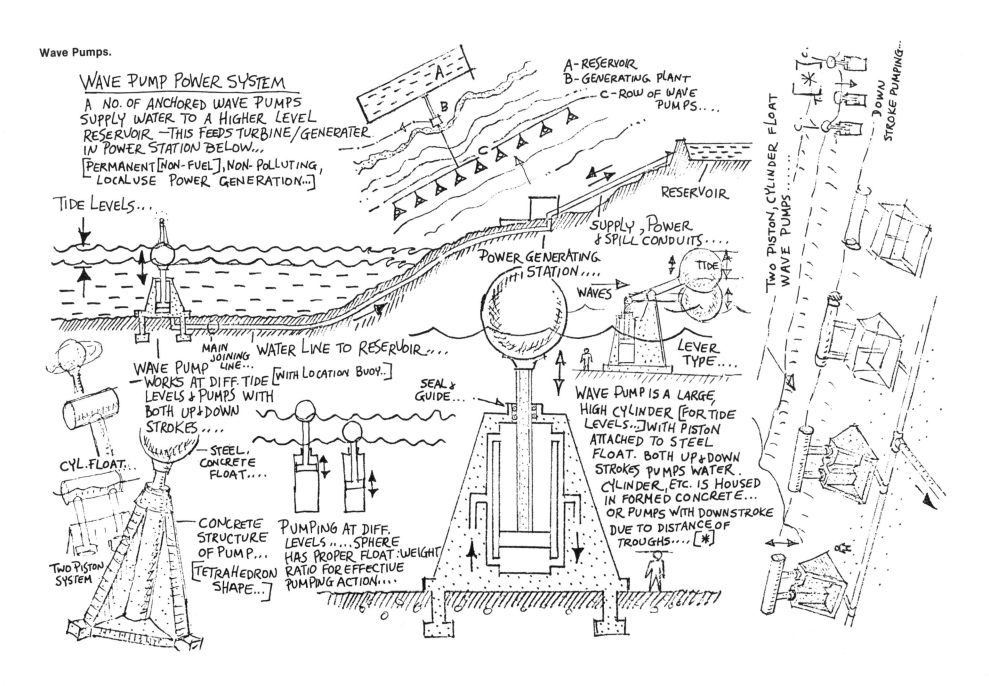

Floating Breakwater

These devices would simultaneously protect the shoreline and generate power through the angular valving of incoming waves. The waves would hit the floating water filled breakwater broadside, thereby contracting it across its girth and forcing water out both ends. Turbines in ring structures of the breakwater would be run by the lateral movement of water in and out of the breakwater.[11]

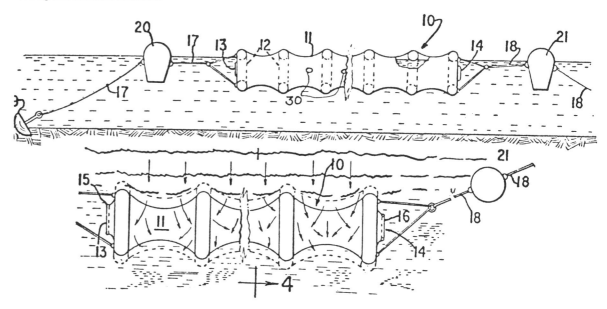

Floating Breakwater Devices.

Shoaling Waves

This device would utilize the momentum transport of shoaling waves to move sea water over a suitably inclined impounding wall, large holding tank, or reservoir. From there, the water would flow back to the sea through a low head turbine. Power generation at high tide would be enhanced through the increased height and volume of the waves at these times. Land-based power plants such as this would be much less prone to damage than other ocean-based designs. Yet another wave-powered device is a battery of floats that is mounted along the shore, each float being connected with the shore by a long boom. The boom would turn a generator through the oscillating motion of the waves and generate power.

Power generation utilizing momentum transport of shoaling waves.

Annual Wave Energy in Specific Areas

in MWh/meter/yr

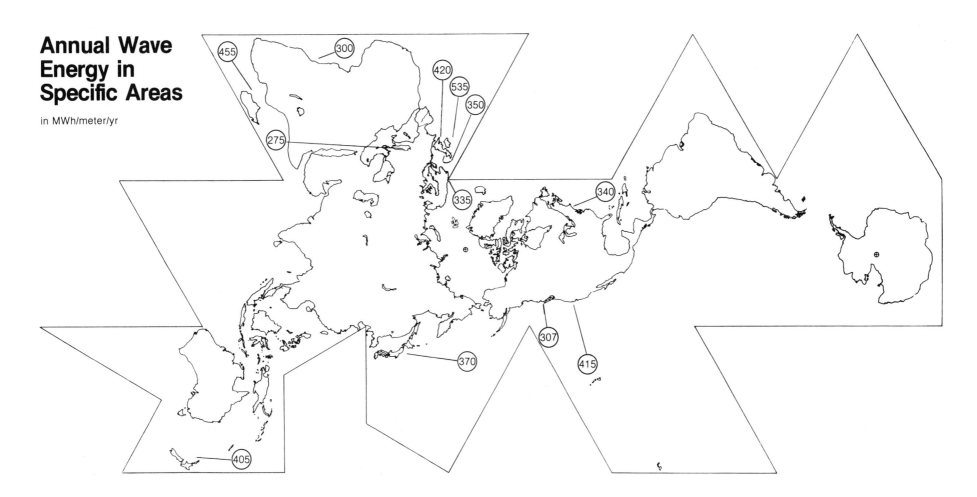

Sources:
The World Energy Book, Crabbe & McBride. Nichols Publishing Co. 1978. "Estimates of the Power of Wind Generated Water Waves at Some Canadian Coastal Locations," Baird, W. F., and Mogridge, G. R. National Research Council Canada, Ottawa, 8-76.

History

1909 Huntington Beach, CA, U.S.: California Wave Power Co. produced electricity for lamps on wharf.
1913 U.S.: Hemmenway obtains patent (#1,082,796) for wave motor consisting of a rotatable cylinder rolling upon a cable by the action of the wind, tides, or waves.
1917 U.S.: Hulden obtains patent (#1,123,104) for similar reciprocating and oscillatory movement.
1923 U.S.: Webb obtains patent (#1,454,801) for wave motor consisting of buoyant globe and counterbalancing mechanism.
1928 U.S.: Hegge (#1,667,152) utilizes vertical motion of floats attached to shafts and gearings.
1941 U.S.: Quinte (#2,242,598); similar to Hegge.
1952 U.S.: Smurr (#2,613,868); similar to Hegge.
1955 U.S.: Searey (#2,707,077); similar to Hegge.
1956 U.S.: Salzer (#2,749,085); similar to Hegge.
1957 U.S.: Caloia (#2,848,189); plurality of floats to actuate water pumps.
1962 U.S.: Corbett (#3,064,137); rising liquid to compress air in buoy.
1964 Japan: First wave-powered buoy goes into operation.
1965 U.S.: Masuda (#3,200,255) utilizes vertical motion of a buoy to compress air and rocking action to get mechanical output.
1966 U.S.: Semp (#3,353,787) utilizes elongated tubes with flexible surfaces arranged in parallel to crest line of waves and below the surface of the water.
1967 Ashika Island, Japan: 40 w wave power plant begins operation.
1970 Japan: 500 watt wave power device on exhibition at Osaka Fair.
1974 England: Salter design for floating structure with vanes that will roll with waves.
1975 U.S.: Fuller (#3,863,455); floating breakwater.
1978 U.S.: Peterson (#4,086,775); large buoy which siphons water to above water line and then generates power through the flow of this water through a turbine mounted in a Venturi tube.
U.S.: Ricafranca and Donato (#4,078,382); converging dams funnel and amplify wave energy into a pressure chamber.
U.S.: Perkins (#4,078,871); channels open to the sea receive deep ocean waves and direct them up a ramp to pressure chamber.
U.S.: Hagan (#4,077,213); plurality of different sized floats are connected to an array through nonlinear interface so their relative motions drive a hydraulic pump.
1980 Japan: Test of 2 MW wave power device.
1980–81 Norway: Pilot wave power plant in operation.

Advantages

1. Waves are a continuous income energy source.
2. Waves are clean, have no by-products, and cause no increase in thermal burden.
3. Wave energy is of large magnitude; about 80 kw per meter of water frontage is available on an annual average basis in Britain. A 1 km unit would have an output of 50 MW. About 15,000 km of wave generators of this capacity would be needed to produce all the electrical energy the world currently uses.
4. Waves are more persistent than the wind and are "self-healing": that is, the energy that is removed is soon restored.
5. The best shore-based wave power sites are mainly waste land now.
6. When wave power devices are built near land in use they would help protect the adjacent coast by causing the wave to do useful instead of destructive work.
7. Coastal wave-powered devices can take advantage of the fact that of all parts of the Earth, the coastline is the best-charted, and that reliable climactic and meteorological data are available for all coasts.

Disadvantages

1. Large-scale use is only at a prototype or a drawing-board stage.
2. Power output in most designs will be variable.
3. Wave generators could be obstacles to navigation and coastal fisheries.
4. Large-scale wave-powered devices might be hard to maintain in the hostile and corrosive environment of the ocean.
5. Social impact could be significant because wave generators would probably be located in remote coastal regions and would inevitably have a strong effect on local communities.[12]

16 Ocean Currents

The ocean currents are constantly being regenerated through the revolving of the Earth. Ocean currents have a high overall energy content, but their energy density is low. Among the predominant ocean currents in the world, twelve have present economic potential for power generation.[1] Total output that could be generated from the twelve currents with present-day know-how is about 1.75×10^{12} kwh. Among the methods under consideration is one in which gigantic cylinders constructed like a Venturi tube are submerged and fixed parallel to the ocean current (Illustration 1). Another design (Illustration 2) calls for the use of a moored ship with a string of parachutes on a conveyor-belt-like structure. With the current, the chutes are open; against it, they close, thereby greatly reducing drag. Power is generated on board the ship, where the slow but powerful motion of the moving conveyor turns an electric generator. A preliminary design that calls for as many as 250 75 MW giant turbines, each 500 feet in diameter and positioned in a single array about 15–20 miles offshore Florida to tap the Gulf Stream, has recently been funded by the U.S. Department of Energy.

In addition to harnessing *ocean* currents for power production, there is the possibility of harnessing lake current. Lake Superior, for example, has currents which could be tapped.[2]

Illustration 1: Underwater Vanes for Tapping Ocean Currents

Illustration 2: Ocean Current Tapping System

Uses

1. Electric power generation
2. Hydrogen production.

Advantages

There is a continuous supply of this income energy source. The Florida Current, a major component of the Gulf Stream, carries more than fifty times the total flow of all the fresh water rivers of the world. Total energy of motion of the current could produce about 25,000 megawatts if all the energy could be harnessed.

Disadvantages

1. Ocean current energy is difficult to harness.
2. The technology is only at the drawingboard stage.
3. The ecological effects are unknown, but could be extensive if large sections of a current were harnessed.

Ocean Currents in January

SOURCE:
The Times Atlas of the World, 1971.

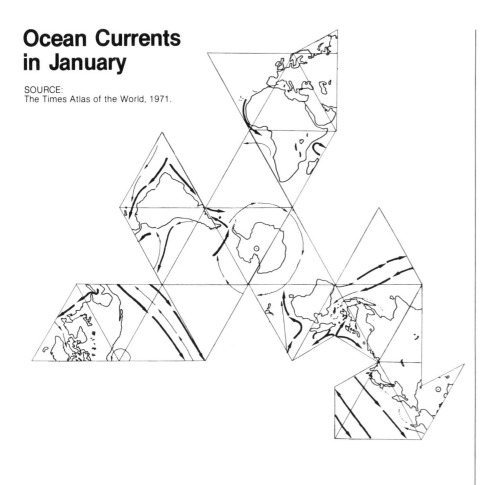

→ primary currents
→ secondary currents

Ocean Currents in July

SOURCE:
The Times Atlas of the World, 1971.

→ primary currents
→ secondary currents

17 Temperature Differential

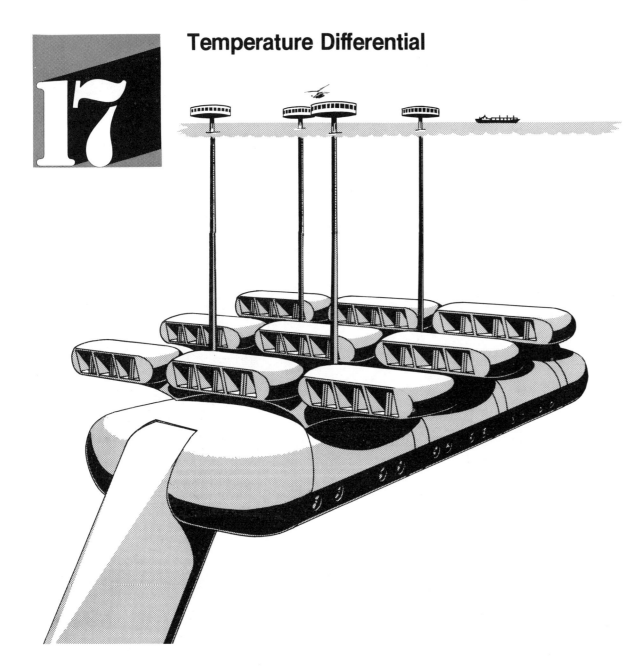

The total amount of solar heat stored in the oceans is about 2.23×10^{20} kwh (7.4×10^{23} BTU), or about 146 times the total amount of energy the Earth receives each year from the Sun.

Temperature differential power can be derived from the thermal gradient that exists between a hot reservoir and a cold reservoir via a Carnot cycle heat engine. Heat will flow spontaneously from a hot region to a cold region. By channeling the flow through a heat engine it is possible to redirect a fraction of the heat energy as useful work. The minimum useful temperature differential is approximately 20°C.[1] Thermal gradients of this magnitude are common in the tropical oceans between the cold, deep waters and the warmer, surface waters. The solar energy impounded in these gradients represents an enormous potential energy supply.[2]

Temperature differential power plants can be built on land where hot and cold ocean currents converge near the coast, or as floating power plants that are capable of operating in the deep ocean.[3] Temperature differential power plants could generate either electricity or hydrogen fuel. Another use would be for a power-intensive metallurgical industry, which always develops wherever there is cheap electric power.[1]

The technology required to harness these thermal gradients exists. The most serious problems to be solved involve the mooring of a

... the world's basic heat source (the sun) and its largest heat sink (the ocean) are being used to drive what is basically a 1920's-vintage ammonia refrigeration plant.[5]

Fig. 1. The OTEC Power Plant Using Hardwire Power Transmission from Offshore Site to Consumer.

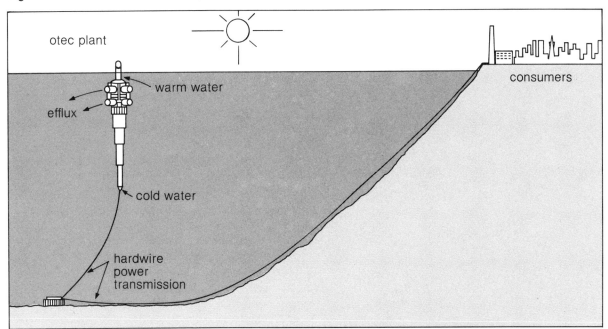

Closed Rankine cycle, ocean thermal energy conversion system.[8]

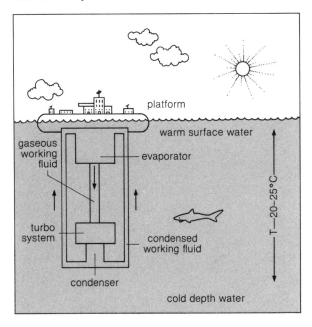

large buoyant structure below the water in the deep ocean, the stabilization of long undersea intake lines, and the transmission of electricity or transport of hydrogen from ocean plants. All of these problems are being or have been dealt with in related contexts.[1,4]

Other forms of "temperature power" are possible. For example, the temperature differences between the warm surface waters and the cold deep waters of hydroelectric reservoirs, the atmosphere and the ocean, mountain tops and warmer valleys, a flowing river and a cold atmosphere, a warm subsurface earth and cold atmosphere, and hot nuclear storage sites and cool atmosphere/earth or water are all potential energy sources for powering heat pumps. Icebergs have even been proposed as energy sources through the harnessing of the thermal and salinity gradients associated with them and the surrounding seas (see Salinity Power).

History

1881	France: D'Arsonval publishes theory that temperature differential engines are possible.
1931	Cuba: 22 kw land-based power plant built on coast by French engineer Georges Claude.
1938	Brazil: Claude tests temperature differential concept aboard the ship Tunisia off the coast of Brazil.
1950's	7 MW land-based power plant built by a French corporation at Abidjan, Ivory Coast.
1971	Project Sea Grant, St. Croix, Virgin Islands: Mariculture plant in operation; fresh water production and power generation projects are planned.
1978	U.S.: Contracts to design 5 MW ocean thermal pilot power plants awarded.
1979	U.S.: First floating ocean thermal test platform begins testing; 15 kw per day net electric power produced.
1979	France: Two studies funded for 100 MW ocean thermal power plant; construction slated to begin in 1980, with completion scheduled for 1984.
1982	U.S.: 25 MW test power plant.
1984	U.S.: 100 MW commercial prototype scheduled for installation.

Ocean Temperature Differentials

(temperature difference between surface and 500 meter depth)

Source:
Ocean Data Systems, Inc. for the U.S. Dept. of Energy, Division of Solar Energy, 1978.

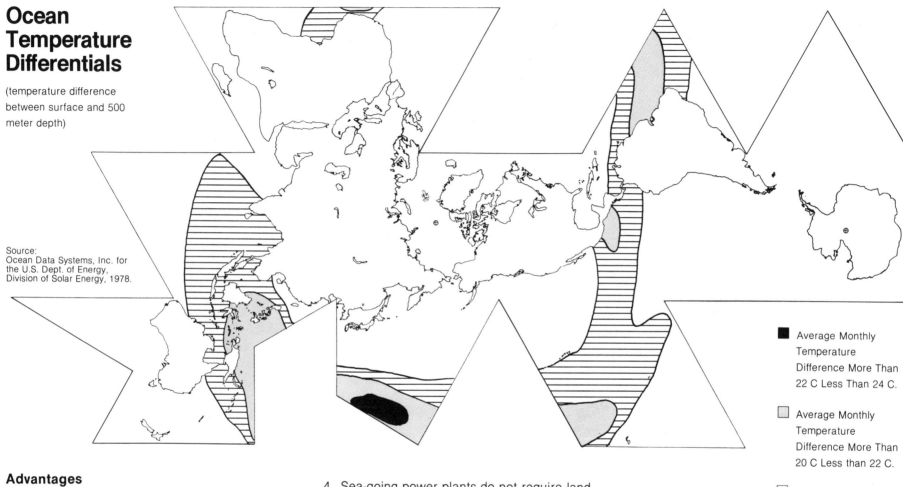

■ Average Monthly Temperature Difference More Than 22 C Less Than 24 C.

▨ Average Monthly Temperature Difference More Than 20 C Less than 22 C.

☰ Average Monthly Temperature Difference More Than 18 C Less Than 20 C.

Advantages

1. There is a continuous supply of this income energy source; it is one of the very few regenerative energy systems having base load capability.
2. Maximum power is available in hottest seasons when the demand is greatest.
3. There is no atmospheric, thermal, or water pollution.
4. Sea-going power plants do not require land area for plant site.
5. Low temperature materials can be used for construction.
6. Fresh water production could be a by-product.
7. Mariculture operations can utilize nutrient-rich deep ocean water for seafood production.

Suitable Coastal Water Locations for Temperature Differential Power Plants

SOURCE:
A. Lavi, C. Zener, "Plumbing the Ocean Depths: A New Source of Power."

▨ suitable coastal water locations for temperature differential power plants

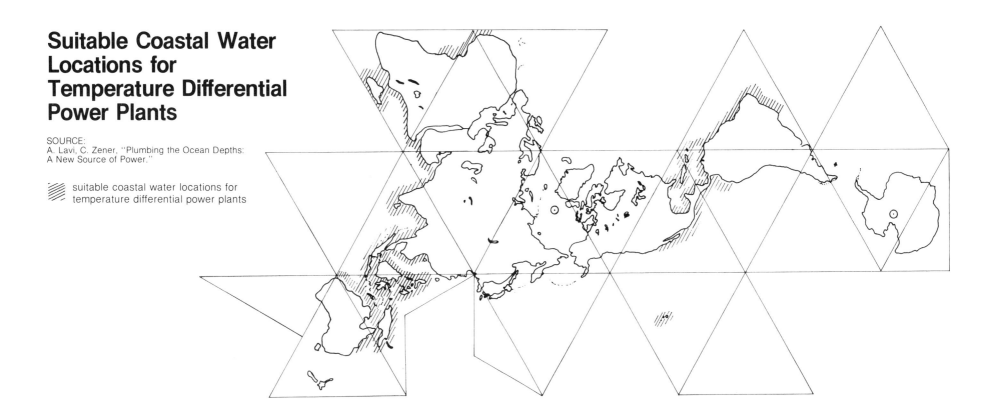

8. Machinery operates at constant load in a benign environment, never seeing more than 85° F, much more pleasant conditions than most land-based utility plants.

Disadvantages

1. Its impact on global weather and climatic patterns is unknown.
2. Its impact on marine environment is unknown.
3. Because of the relatively small temperature differences in deep and surface waters, ocean thermal power plant's potential operating efficiency is low—2–3%. Huge amounts of water would have to be moved, consuming about 30% of the gross power output of the plant.
4. Corrosion could be a severe problem if heat exchangers are built of cheap aluminum; if they are built of expensive titanium, one ocean thermal power plant would use an entire year's production.[6]
5. Mooring problems and stresses on the huge cold water intake pipe have yet to be adequately dealt with.
6. Upwelling of carbon-rich water from the ocean bottom could cause atmospheric CO_2 to increase substantially.[7]

... a 100 MW ocean thermal energy conversion (OTEC) demonstration plant can be operating by about 1985 and ... OTEC can supply a significant part of U.S. electric power at affordable rates by the year 2000.[5]

18 Hydrogen

Hydrogen is the simplest, lightest, and most abundant of the 92 regenerative elements in the universe, and the ninth most abundant on Earth. It is an "income" energy source that we will never exhaust since it is a material that recycles in a relatively short time. It can be made from water, so the potential supply is vast. Hydrogen has indirectly and invisibly functioned as humanity's energy source throughout history through hydrogen fusion reactions in the sun and in carbon-hydrogen combinations of petroleum and natural gas. Hydrogen can be used as a combustible fuel for tansportation, space heating, electric power generation, or industrial processes. It can be used as a material in industry for the hydrogenation of fats, oils, margarine, and soap; in the production of ammonia for fertilizers, and metal powders for annealing stainless steel; for inflating weather and other types of balloons; for cooling electric generators; and for the synthesis of chemicals for nylon, polyurethane, and glass. (It can also be converted directly to electricity in a fuel cell.) It is an exceptional way of storing intermittent energy sourcces, e.g., solar radiation and wind.

Hydrogen enjoys other advantages in efficiency and application over its fossil fuel counterparts. Not only are conventionally vented, flametype appliances readily adaptable to hydrogen, but ventless, flametype combusters are also possible. As much as 40% of the combustion energy is vented to the exhaust vent in conventional burners. On the other hand, nonvented hydrogen furnaces can deliver all the combusted energy to the heated space. (Hydrogen combustion yields no toxic substances, only heat and water.) Condensation and collection of exhaust water can provide

World Hydrogen Production

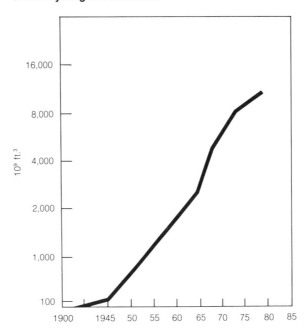

humidity control.[1] Even with present-day accounting perspectives, it may soon prove more economical for industry in high-density urban areas to convert to hydrogen than to install and operate air pollution control equipment mandated for fossil fuel use.

Production

Currently, hydrogen can be produced by four methods: electrolysis, thermochemical water-splitting, photolysis, and algae photosynthesis.

Electrolysis is the most widely known and at present the most convenient technique of hydrogen production. By passing a current through an electrolytic solution such as water, hydrogen and oxygen ions gather at opposite electrodes. Both the hydrogen and the oxygen can be tapped and stored. Theoretical efficiency of electrolysis is 2.79 kwh of electricity per m^3 of hydrogen in gaseous form. Presently, conversion efficiencies of electrical energy/hydrogen production are 60% to 75%. In the near future, electrolytic production efficiencies of 95% are expected.[1]

The electrolytic production of hydrogen from sea water holds potential for the recovery of economically significant quantities of the metals in sea water that are less active electro-chemically than hydrogen—silver, gold, mercury, and copper.[2]

Thermochemical splitting of water is a multistep process in which chemical disassociations and associations under high temperatures (700° to 1,000° C) change the valences of the closed-system recycling chemistries and decompose water into hydrogen and oxygen. Thermochemical water-splitting is now an experimental process, but with new technological developments in primary energy source conversion to heat it appears to have great competitive potential with the conventional electrolytic (electrical energy to hydrogen production) conversion. The advantage of thermochemical water-splitting is the direct utilization of heat to produce hydrogen, thereby avoiding the intermediate steps of electrical energy generation. (See geothermal section.)

Photolysis is a means of breaking molecular bonds with incident light photons. Water can be decomposed directly by light with a photocatalyst which absorbs the visible light to break water bonds. This process occurs naturally in the upper atmosphere, yet its direct applications by man at this point, under

Industrial Processes and Products Dependent on Hydrogen

Hydrogen					
TDI Explosives					
Nitroaromatics NO₃H					
Aniline and Dye Stuffs					
Fertilizer	Benzene				
Caprolactoamcyclohexane	Hydro Dealkylations				
Resins	Organic Chemistry				
Amines	Mineral Chemistry	Soap			
Urea	(HCl HBr H₂S)	Oils and Greases			
Acrylonitril	Pharmaceutical	Lacquers	Silicon Quartz		
Fibers	Industries	Lubricants	Floatglass	**Refrigeration**	
Ammonia	**Hydrogenations**	**Basic Industry**	**Glass Industry**	**Welding**	
				Fuels	
Hydrotreatments	**Methanol**	**Oxo Alcohols**	**Metallurgy**	**Electronics**	
Hydro Desulferization	DMT	Solvents	Pure Metals	**Artificial Stones**	
Hydro Cracking	Esters	Lubricants	Heat Treatments		
(Lubricants	Isoprene Rubber	Weed Killer			
Gasoline)	Formaldehyde	Detergents			
Hydro Treatments	Methyl Methacrylate	PVC-Plasticizer			
(Middle Light	Methyl Bromide	Paint and Lacquers			
Distillates)	Methyl Chloride				
Hydro Refining	Methylamine				
Hydro Refinishing	Acetic Acid				
(Lubricants)					

controlled conditions, do not demonstrate any promising conversion efficiencies.[2]

One system currently under investigation is an oxide of a rare earth element, rhodate, that, like a photovoltaic cell, generates electricity in direct sunlight but also splits water into hydrogen and oxygen.

Algae photosynthesis—the photosynthetic production of hydrogen using algae and bacteria to decompose water—is now being studied. The photoproduction of hydrogen was first observed in algae in 1942 and in bacteria in 1949.[3] Currently one algae in particular is receiving a great deal of attention because it not only produces hydrogen but fixes nitrogen as well, thereby producing fertilizer as well as energy. Recent research promises that this technique may be able to produce sufficient hydrogen to meet the energy needs of the entire planet.[4] Depending on our ability to control these photosynthetic processes systematically, algae could become a major source of hydrogen production.

Storage and Transport

Hydrogen can be stored in three physical states: gas, liquid, and solid (absorbed in metals such as lithium, palladium, and magnesium) and is readily accessible for use in all three states.

Gaseous hydrogen. By most techniques, hydrogen is produced as a gas. This gas can be compressed and stored, yet the volumes and pressures make it feasible only as a process in a centralized system. Hydrogen can be stored for up to several years in aquifiers and for longer periods (decades or centuries) in natural formations such as depleted oil and gas fields.[5] Hydrogen gas can be released from a compressed state to low pressures and be utilized in ways similar to that of natural gas by means of pipelines. It can be transported via pipelines for distances of 5,000 kilometers or more with present-day technologies.[5]

Liquid Hydrogen. Cryogenic (low temperature) technology, a spin-off of the space program, has developed sufficiently to show that liquid hydrogen can be seriously considered as a form of hydrogen storage. The liquid hydrogen rocket turbines that successfully propelled astronauts into lunar orbit were supported by a cryogenic technology able to produce, transport, and store liquid hydrogen at temperatures of 423° below zero. Pipelines, tanks, and transport trucks have all been developed for liquid hydrogen. Because of liquid hydrogen's extremely low temperature, a cryogenic pipeline might also be used for a superconducting transmission line for electricity.[6] The use of superconductors for electricity transmission results in almost no loss; thus, such a system would not only transport fuel for industrial, commercial, residential, and transportation uses but would greatly increase the efficiency of electricity distribution.

Hydrogen in a solid state. Storage of hydrogen in metal and intermetal hydrides promises to be an effective alternative to gas and liquid storage, avoiding the problems of gas compression and low temperatures. "Because of its small molecular size and high diffusivity, gaseous hydrogen is able to penetrate the lattice structure of solid metals or alloys and bind at various sites in the unit cell of the crystal. For many metals, such as titanium, the penetration is so great that

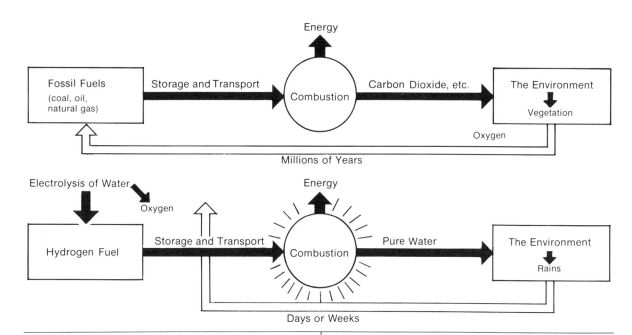

Energy Density Characteristics of Fuels

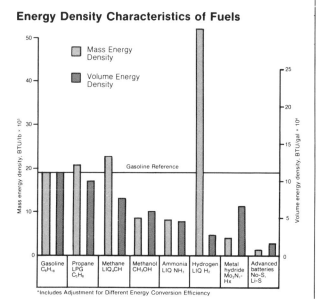

*Includes Adjustment for Different Energy Conversion Efficiency

Combustion Characteristics

The Flammability limits of a fuel are the rich and lean mixture limits outside of which combustion cannot be sustained. A fuel which possesses wide flammability limits allows the designer considerable latitude to optimize other factors such as waste or performance.[10]

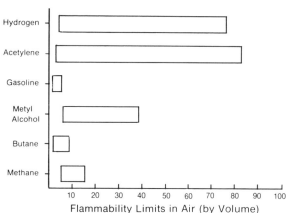

the concentration of hydrogen per unit volume is actually greater than in liquid hydrogen. The hydrides are formed by simply exposing the metal to pressurized hydrogen. Hydride formation is exothermic and can be reversed by the application of heat; waste heat from the combustion process can thus be used to free the hydrogen."[7]

One use of hydrogen storage in solid and liquid form would be in a land surface transport vehicle (automobile, train, or truck); hydrogen could be stored in liquid form in a small cryogenic tank and the boil-off gas (a problem with small cryogenic tanks is the inefficient surface-to-volume ratio and consequent boil-off) could be absorbed in a small metal hydride tank and released for combustion in the engine.

Studies by NASA and others indicate that hydrogen-fueled 747-size aircraft compare favorably with fossil-fueled aircraft, both in range and in projected economy.[10] Liquid hydrogen-fueled aircraft could be lighter, quieter, have smaller wing area, require shorter runways, and minimize pollution. Other studies indicate the economic viability of mass transit bus fleets powered by hydrogen. Because of savings associated with an estimated 50% increase in engine efficiency, reduced fuel costs, and engine maintenance, it is claimed that the costs for conversion to hydrogen could be paid back during the first year of operation.[11] An added, unexpected bonus of hydrogen use for intercity bus fleets and passenger cars is that, when the hydrogen-fueled bus or car is refueled, heat is generated. When the metal hydride storage medium absorbs hydrogen, significant quantities of heat are released. A hydrogen-fueled fleet of three-hundred buses would annually produce as much heat as a half-

Boeing 747 modified to carry all liquid hydrogen fuel in the expanded upper lobe.

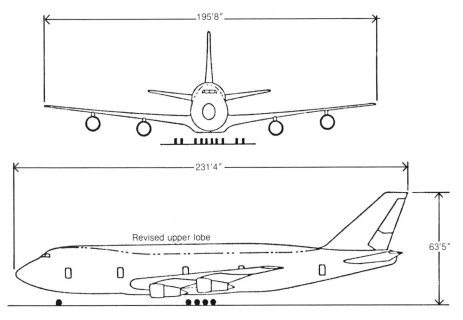

Subsonic cargo airplanes designed to carry payloads of 265,000 pounds for a range of 5,070 nautical miles.

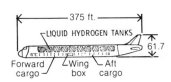

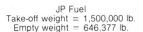

JP Fuel
Take-off weight = 1,500,000 lb.
Empty weight = 646,377 lb.

Liquid hydrogen fuel
Take-off weight = 915,000 lb.
Empty weight = 521,963 lb.

A closed-cycle hydrogen home could recycle waste heat generated by charging hydride tanks. Hydrogen piped to house from a central source is injected into storage tank under pressure. This causes a chemical bonding action with the hydride (a granulated metal alloy) that releases heat for hot water and space heating. Added heat comes from the family car as its onboard hydride tank is charged overnight from a separate tap. With an annual auto mileage of 9,000, total heat generated could be the thermal equivalent of 100 gallons of oil a year. The hydrogen gas itself is used for cooking, and also for heating the hydride as the gas pressure drops. Since discharging the gas has a cooling effect, a composite tank can provide air conditioning in summer. A refrigerator and freezer can also hook into the system.[12]

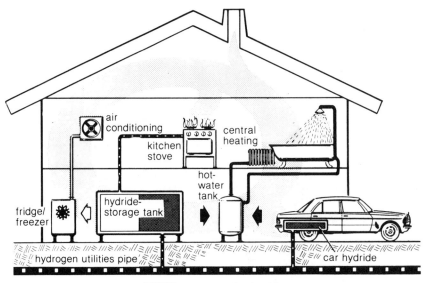

million gallons of oil (22 × 10^6 kwh). A passenger car with 9,000 annual miles would generate the thermal equivalent of 100 gallons of oil per year (4,400 kwh).[12]

Profile

- ☐ Kilowatt energy equivalent: 79 kwh for 1,000 cu. ft. hydrogen.[8]
- ☐ BTU energy equivalent: 50,000 BTU/lb (5 lb H_2 per 1,000 cu. ft.)[8]
- ☐ Methods of storing and transporting: in gas and liquid form, same as natural gas; as a solid, in metal hydrides.
- ☐ Methods of hydrogen production: electrolysis of water; heat/chemical cracking (thermochemical); photolysis (photochemical); algae photosynthesis (biological).

History

- 1500's Paracelsus dissolves a metal in an acid and uses the flammable gas (hydrogen) that results.
- 1766 England: Cavendish distinguishes hydrogen from other flammable gases.
- 1781 France: LaVoisier coins the name "hydrogen" (Greek for "made of water").
- 1783 France: Charles uses hydrogen for balloon inflation.
- 1890's U.S.: First commercial water electrolysis cell.
- 1898 England: Dewar produces liquid hydrogen.
- 1902 England: Verne prophesies economy based on hydrogen.
- 1927 Norway: Norsk-Hydro produces synthetic fertilizers from hydrogen produced electrolytically.
- 1931 Deuterium discovered.
- 1937 U.S.: German zeppelin Hindenberg burns.
- 1940 Canada: First large electrolyser in North America installed in Trail, B.C.
- 1969 U.S.: Hydrogen used to fuel rocket to the moon.

Advantages

1. It is the most abundant fuel source.
2. There is little or no air pollution; H_2 combustion yields H_2O and heat; exhaust particulate from a hydrogen engine is 1,000 times less than levels associated with gasoline engine exhaust.
3. Existing energy converters can readily switch to hydrogen; often with increased efficiencies.
4. Hydrogen has the greatest energy per unit mass of any chemical fuel; 2.5 times the energy per unit weight of gasoline, propane, and methane.
5. Hydrogen is cheaper to transport per unit of energy via pipeline than is electricity.
6. Hydrogen, unlike electricity, can be stored.

Disadvantages

1. Production of hydrogen is presently not as inexpensive as petroleum and coal mining.
2. Hydrogen has a low energy density on a volume basis; this prevents storage of sufficient quantities on-board some land vehicles to give a long range of operation. As heating value of H_2 (2,892 kcal/ m^3 of gas) is three times lower than natural gas, three times more hydrogen has to be transported to provide the same amount of energy.
3. The easiest way of producing hydrogen—by electrolysis—is inefficient.
4. Hydrogen requires special handling techniques.
5. There are some potential structural material problems associated with the use of hydrogen, such as embrittlement, corrosion, oxidation, and erosion.

Animate Energy

Human and animal muscle is a regenerative, nondepletable energy source, with low output relative to other inanimate energy sources. The total amount of power available from an idealized world population of four billion healthy individuals working 12 hours per day, 365 days per year would produce half a trillion kilowatt hours. Although an individual human being's energy output is small, it is ideally suited for low power and decentralized activities such as personal hygiene. (The energy necessary to produce, distribute, and operate electric toothbrushes is hardly warranted by their questionable hygienic, cultural, or aesthetic values.) In the past, muscle has produced many impressive achievements—the Great Wall of China, the pyramids of Egypt, Mexico, and South America, the great roadways of the Roman Empire, etc.—but only at the cost of immense human suffering. Obviously, humanity does not survive on planet Earth through muscle power—any jackass is stronger—but because of mental ability. Nevertheless, there is vast potential for the productive, noncoercive use of human (and animal) muscle power in all parts of the world, particularly in developing regions where unemployment is highest and inanimate power is in low supply.

Advantages

1. There is a plentiful, regenerative, and readily available supply of this energy source.
2. It is the most adaptable, nonspecialized energy source known to humanity.
3. Human energy expenditure is clean and has the least ecological impact of any energy source.
4. Reasonable expenditure of human energy is healthful to the individual.
5. Little or no additional expenditure of inanimate energy is needed to "harness" human energy.

Disadvantages

1. Low power outputs of human muscle—.04 to .08 horsepower for short periods of time. The work of ten people for one hour is about the equivalent of one horsepower hour, i.e., one person working a ten-hour day can produce about one horsepower-hour or about one-tenth the work of a horse.[1] One liter of gasoline fuel does about thirty man-hours of work.[1]
2. Human labor costs are high; cost of food necessary to produce a given amount of human muscle energy is 25–30 times more expensive than the cost of producing the same amount of energy from electric motors or gas engines.
3. Excessive human labor tends to tire human beings, leaving them little energy for education, recreation, etc.

Cogeneration

Cogeneration, in the broadest sense, refers to any multiple use of a particular energy flow. The underlying principle of cogeneration, the use of the waste energy from one energy using process to drive another, is prevalent throughout history and Nature; it is only in our day and age of specialization that cogeneration is thought of as something new. Heating water over a wood stove is a cogeneration unit, as is the heating in a car that comes from the waste engine heat. In Nature, "cogeneration" might better be referred to as "multigeneration" because not only does every flow of energy in the biosphere drive more than one process, but the waste from each process then drives another. Ultimately all energy flow ends up as low grade heat which warms the atmosphere and is then radiated out to space.

Large-scale cogeneration of process heat and electricity has long been technically and economically feasible. Generating steam and electricity together in the same power plant is more efficient than generating them both separately in different powerplants. Cogenerators were used extensively in the early years of American industry, accounting for about 22% of

---

An average paper plant could produce three to four times as much electrical power as it would consume.[2]

Cogeneration Plants

Separate Steam and Electric Power Plants[3]

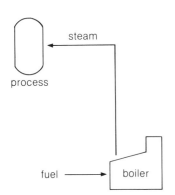

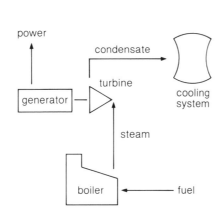

Back-Pressure Steam Turbine

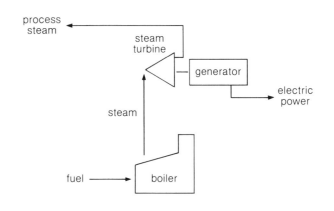

U.S. electric power supply in 1920. This percentage decreased to 18% in 1940 and to about 4% by 1976.[1] Much of the decline in cogeneration was a result of decreased fuel and capital costs for utilities in the 1950's and 1960's with a resulting low price for bulk power sold to industry. In addition, utilities discouraged industrial cogeneration by declining to buy surplus cogenerated power at a fair price while, at the same time, charging exceptionally high fees for backup power. Also, some potential cogeneration industries were not inclined to go into the power business because of historically low return on investment—they preferred to expand their primary market instead—and because of the fear of coming under the jurisdiction of utility regulatory agencies. (Many large paper companies were profitably cogenerating in the 1920's and 1930's but were forced out of the business when the Justice Department required them to decide whether they were in the paper or the electrical power business.) In Germany, because the above problems either do not exist or exist in a minor way, cogeneration still accounts for 30% of the electric power supply.

Most steam used for industrial processes or heating is at relatively low temperatures and pressures—usually less than 200 psi and 300 to 400 degrees F. In practice, however, the steam is generated from high grade fuels yielding temperatures over 3,000 degrees. "Topping" engines reduce this inefficiency by first producing electricity and them capturing waste heat for the production of process steam. This process has an overall efficiency of fuel use that is, typically, about twice that of conventional central power stations.

Gas Turbine—Waste Heat Boiler

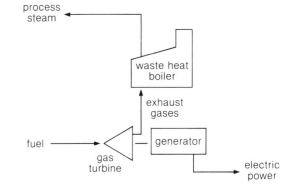

Alternatively, the temperature of exhaust products leaving many high temperature processes is sufficiently high to warrant the use of a "bottoming" cycle in which exhaust heat is used for the production of electricity.

Presently, steam turbines, gas turbines with heat recovery from turbine exhaust, and diesel engines with heat recovery from exhaust and/or cooling, are among the proven topping engines. Advanced Rankine and Brayton cycle equipment has been developed which operates efficiently in a bottoming cycle.

Most of the cogeneration units currently available are diesel or gas turbine engines with heat exchangers in the 150 to 10,000 kw size range.[4] These are, by and large, suitable for industrial needs and the heating requirements for large installations such as schools, businesses, hotels, etc. Small scale—5 to 20 kw cogeneration units that produce electricity and heat that could be suitable for use in the home or apartment complex have not been commercially developed yet. The only small scale cogeneration unit under development is the Fiat Motor Corporation's electricity production and heating unit. This ingenious device consists of a small Fiat engine with heat exchangers coupled to a 15 kw electrical generator. The heat produced by the engine is captured by the heat exchangers and then used to heat the building it is placed in. But in addition to just producing heat as a normal oil or gas fired furnace would do from the burning of their fuels, the Fiat engine drives the electrical generator that produces electricity. In Fiat's scheme, this electricity would be sold to the utility and the money received for this would pay for the fuel that powered the engine. According to Fiat's calculations, the building would essentially be heated for free. Electricity for the building would be purchased from the utility instead of getting it direct from the Fiat unit because the unit would only be running (and thereby producing electricity) when heat were needed. It would operate as a conventional furnace does in this respect. If electricity was to be obtained directly from the cogeneration unit, a backup battery storage system would be needed to store electricity for use when the heater/generator was not running.

Advantages

1. Cogeneration saves fuel, capital, and waste.
2. Cogeneration requires little or no cooling water relative to conventional steam electric power generating systems.
3. Substantial amounts of power generating capacity can be added to the local grid thereby reducing utility peak capacity requirements and helping to lower rate increases designed to finance electrical capacity additions.
4. The overall grid systems reliability can be increased through the decentralization of power generating sources.
5. Distribution and transmission costs will be lower.
6. The need for back-up power will decrease.
7. In light of uncertain and evolving electricity demand, cogeneration permits greater flexibility in planning future electrical generating capacity.
8. Cogeneration can come "on-line" much quicker than conventional power plants (1–4 years rather than 8–10).
9. Numerous cogeneration units provide more jobs than one centralized unit.

Disadvantages

1. Cogeneration needs a "friendly utility" to make it economically attractive (a utility is needed to purchase the surplus electricity that is produced).
2. Diesel cogeneration units have some air emission problems.

Exotic Energy Sources

Nature is constantly transforming. A seed transforms into a tree, ocean water transforms into water vapor, then into clouds, then into rain, creek, river, and ocean again. All biological, chemical, and physical processes in nature use energy in their transformations. Whenever there is an energy flow from one concentration of energy to another spot of lesser concentration, the second law of thermodynamics makes it clear that, in principle, power can be extracted. In some of these energy transformations in nature, humanity has been able to step in and tap the energy flow for its own uses. Solar energy, for instance, is one such naturally-occurring energy flow that humanity has been able to step in between it and the processes it "normally" drives to valve and shunt the solar energy flow in preferred directions to power human-engineered systems for heating, cooling, electricity, or food production. The energy flows associated with the winds, tides, and rivers are other places humans have inserted their propellers, dams, and turbines to obtain energy.

Energy flows in the Universe can be categorized into two fundamental types: those energy events or flows that happen quite regularly and are of relatively low intensity (such as sunlight), and those events or flows that happen rarely but are of great magnitude (such as hurricanes, tornadoes, or novas). Energy comes either regularly, diffusely, and with low intensity, or concentratedly, with high intensity, and infrequently. So far, humanity has focused most of its attention on harnessing those energy flows somewhere in the middle of this spectrum. This is probably because our senses are most in tune with this band-width of phenomena. We have thought of only the most obvious (to us) energy flows as energy sources or potential energy sources. We need to be aware of and to develop as many of our energy sources as possible. It is imperative that we break the mass-media/oil-conglomerate fix that only acknowledges oil, gas, coal, and nuclear (with a begrudging nod to "infeasible" solar) as players in the great energy game.

The following pages describe potential energy sources and means for tapping them that are of a more "exotic" nature than those we usually encounter, but nevertheless deal with the same all-embracing "nature" that our more conventional fossil, nuclear, sun, and wind harnessing devices do. (Describing how a nuclear powerplant works to the proverbial man from Outer Space would be far more exotic and complicated than many of the following.)

Water Salination

Water salination is a potential income energy source; it is a naturally occurring geo-chemical flux from which energy can be extracted by the mixing of fresh water with sea water through osmosis. When fresh and saline water mix, energy is released; when the opposite occurs, that is, when fresh water and salt water are unmixed (desalination) energy is required. Energy will also be released through the mixing of sea water with highly concentrated salt water bodies such as industrial waste water, the Dead Sea, the Great Salt Lake, or similar bodies of water in Russia, China, and elsewhere.

Total power available in the world from water salination is comparable to world hydroelectric power in magnitude. The global energy flux available from natural salination is equivalent to each river in the world ending in a waterfall 225 meters (738 feet) high.[1] This would produce more than three times the average rate of electric power generation in the world. A fresh water flow of one cubic meter per second into a salt water body could provide 2.24 MW of power; a salination power plant using only 10% of the flow of the Mississippi at an overall efficiency of 25% would deliver 1,000 MW of power.[2]

Salination energy is readily converted to mechanical or electrical energy. Two schemes currently exist for tapping this energy. One, depicted in Illustration 1, takes advantage of the high osmotic pressure that exists across a membrane when fresh and salt water come together. The two will naturally mix until the salt concentration is the same for both bodies of water. When separated by a semi-permeable membrane that allows only fresh water to pass, this mixing process can be harnessed to produce power. Water will actually flow uphill through the membrane and will continue to flow until the salt concentration is the same on both sides of the membrane. For a fresh water flow into the ocean, the pressure of the flow at the membrane will be 24 times the atmospheric pressure. (For a fresh water flow into a hypersaline lake such as the Dead Sea or Great Salt Lake this osmotic pressure difference is as high as 500 atmospheres. This translates into power outputs greater than 30 MW for each cubic meter per second of fresh water flow into the hypersaline lake.[3]) The water is raised in a pressure chamber and then dropped to turn a water wheel or turbine (see Illustration 1).

Another means of tapping the energy released from the mixing of fresh water with sea water is electrochemical cells or "salt batteries." An electrical potential is generated when fresh water is in contact with salt water. In a salt battery, an electrochemical effect (the movement of ions across a membrane) is harnessed instead of a mechanical effect (the movement of water across a membrane).

The theoretical potential for energy from salinity gradients is almost as great as the potential for ocean thermal energy conversion.[4]

If salinity power becomes economical, it may lend itself to decentralized power generation, utilizing many available small water supplies.[5]

Illustration 2 shows one design for such a device. Two types of membranes are used: one that passes only positive ions and another that only passes negative ions.

Besides the natural flow of fresh waters into a salty sea there are two other salinity gradients that could be developed for outputs of energy. The controlled input of sea water into dried lagoons or salt pans along arid and semi-arid coasts would create concentrated salt brines that could then be mixed with the more dilute ocean.[3] Another salinity gradient that could be developed is the subterranean formations of brine or solid salt located adjacent to or under the sea. These salt domes sometimes contain oil or gas. Recent analysis suggests that there is more energy available from the salt in these salt domes than could be obtained from the oil and gas.[3] (see Table)

Advantages

1. It is a nondepletable and clean energy source.
2. Large magnitudes of energy are available.

Disadvantages

1. It is now only at the theoretical stage.
2. It is potentially as ecologically disruptive as a hydroelectric power plant.
3. Where fresh water is scarce, power generation via water salination will compete with agriculture, industry, and personal use.

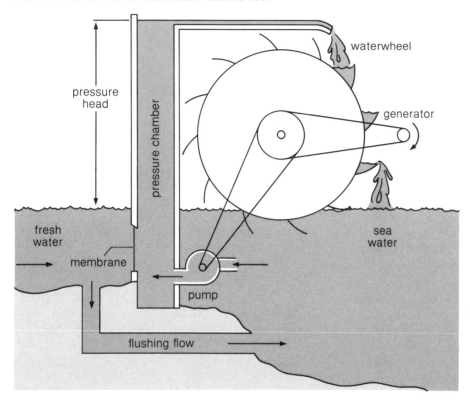

Illustration 1. Diagram of an Osmotic Salination Energy Converter, to extract power from the natural flow of freshwater into the sea.[2]

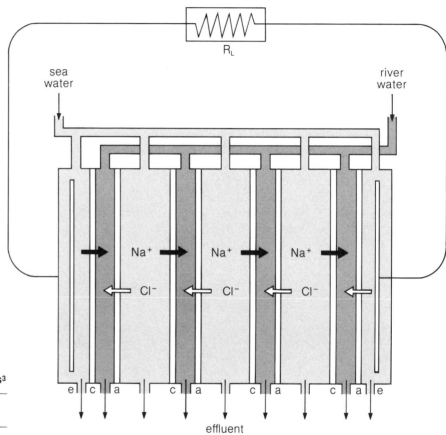

Illustration 2. A dialytic battery. The letters e, c, and a denote electrode, cation exchange membrane, and anion exchange membrane, respectively: R_L is the load resistance. In practice a stack would incorporate a large number of membranes, and the electrode compartments might be perfused separately.

Comparison of the Energy Available from the Salt and the Oil in Selected Salt Domes[3]

Dome	Salt volume (cubic miles)	Oil production (10^3 barrels)	Salt energy (MW-years)	Oil energy (MW-years)
High yield				
Thompson (Ft. Bend, Texas)	0.4	259,623	14,000	44,000
Hull (Liberty, Texas)	2.6	156,830	93,000	27,000
Humble (Harris, Texas)	9.8	138,639	350,000	24,000
Medium yield				
Avery Island (Iberia, La.)	4.0	53,054	140,000	9,000
Bayou Blue (Iberville, La.)	4.6	20,806	161,000	3,500
Belle Isle (St. Mary, La.)	1.9	10,316	68,000	1,700
Low yield				
Lake Hermitage (Plaquemines, La.)	0.9	2,475	32,000	420
Bethel (Anderson, Texas)	8.0	1,017	280,000	172
East Tyler (Smith, Texas)	4.3	55	150,000	9

Nuclear Fusion

Nuclear fusion is a capital energy source, but one of such vast potential as to render the distinction between income and capital energy sources nonfunctional. The source of power is the energy released when two light atomic nuclei fuse to form a single heavier nucleus. Fusion reactions can only be sustained at extremely high temperatures and pressures. For example, the deuterium-tritium reaction will only produce more energy than is required to initiate it if temperatures of 50,000,000° K. can be achieved. The principal problem encountered in fusion reactions is that of containment, since matter cannot exist as a solid at such temperatures. Two approaches are being examined. The older approach utilizes intense magnetic field configurations to form a magnetic "bottle" in which the reaction is contained. A more recent approach has been termed inertial containment. In this scheme, a pellet of fuel is heated by a pulse of intense laser light which vaporizes the pellet. A plasma (an ion gas) is formed. The rate of expansion of the plasma is limited by Newton's law of inertia (about 10^6 meters per second). If the plasma can be made hot enough by the laser beam before it becomes too diffuse, the reaction will yield significant amounts of energy.

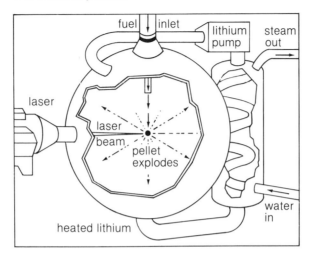

Laser fusion process

Experiments with fusion have so far not reached the break-even point (the point at which the reaction produces more energy than is required to initiate it). However, recent work indicates that the break-even point may be attainable with magnetic containment systems or laser fusion, possible in the 1980's.[6,7,8] The break-even point is then light-years away from being a viable, safe, commercial power source. The break-even point merely confirms the theoretical and technical possibilities, not the economical or ecological viableness. Most recent analysis indicates that no fusion option will "probably make a substantial contribution to the world energy supply until at least two decades into the next century."[9] All these assessments address themselves only to fusion's theoretical and technological possibilities. As nuclear fission has more than adequately demonstrated, many other factors need to be considered. Perhaps the most important being the answer to the question: Is what is appropriate for the Sun, 93 million miles away from the Earth, also appropriate within our delicate biosphere?

Several fusion reactions are known. The deuterium-tritium fuel cycle has been considered particularly attractive because it has the lowest ignition temperature known.[10,11] Deuterium is a common isotope of hydrogen present in seawater. Tritium is a rare radioactive isotope of hydrogen. However, it can be bred inside a fusion reactor from lithium. Estimated reserves of lithium are sufficient to supply energy needs for millions of years.[10] A deuterium-deuterium reaction is also possible. Deuterium could fuel the Earth's energy needs for billions of years.

Fusion, like fission, produces significant amounts of radioactive wastes and by-products. For the fusion reactor, the primary problem will be the loss of tritium.[11] Tritium is not only volatile, but tends to diffuse through high-temperature metal walls.[12] When combined with oxygen as water, tritium readily enters living systems. Long-lived radioactive wastes associated with a fusion power economy would be significantly less than that associated with a fission power economy; the magnitude of the problem would depend heavily on the structural materials used in constructing the reactor.[11,12] In the event of an accident, the radioactive inventory of a fusion reactor represents a biological hazard potential that is at least three orders of magnitude lower than in a fission reactor.[12] Fusion reactors are inherently incapable of a "runaway" accident. No critical mass is required for fusion. The fusioning plasma is so tenuous that there is never enough fuel present at any one time to support a nuclear excursion.[10]

There is at present only one known and tested way of producing energy from fusion reactors—the hydrogen bomb. Weaponry designers point out that the bomb is the only fusion concept for which the technology has been proven for over twenty years. Unbelievable as it may sound, there is in fact a "serious" proposal to harness fusion bombs, or "devices" as they are colloquially called. Scientists from Los Alamos National Laboratory and a research and development company in California have proposed a plan whereby 700 hydrogen bombs would be exploded every year (two per day) in a mile-deep cavity. The cavity would be filled with one million tons of water which would vaporize into high pressure steam at about 200 times atmospheric pressure by the twice daily thermonuclear explosions. The steam would continually circulate through surface power production facilities which would draw out heat to power turbine generators. New detonations would maintain the cavity temperature at about 500° C. The people planning this project talk of the mass production of 100,000 hydrogen bombs each year, more than the number of nuclear warheads in the global arms stockpile (the U.S. has an estimated 30,000 nuclear warheads). Although this is a "better" use of thermonuclear explosions, such a Strangelovian scheme is fraught with difficulties. Investing man-years of engineering time to develop the economical mass production of hydrogen bombs seems insane, to say nothing of the inherent problems of sabotage, theft, safety, or ecology.

History

1932 England: Cockroft and Walton, fusion first experimentally reproduced in particle acceleration experiments.
1951 U.S./U.S.S.R.: Development of magnetic containment systems.
1952 U.S.: Explosion of first fusion bomb.
1961 U.S.: Laser action first demonstrated.
1968 U.S.: Fusion reactions first initiated by interaction of laser light and a plasma.
1978 Princeton, U.S.: Large Torus Fusion device produces hottest temperature on Earth—60 million degrees Kelvin.

Uses

1. Generation of electricity.
2. Hydrogen generation.
3. As a fusion torch for the refinement and/or recycling of raw materials.

Advantages

1. There is a large fuel supply.
2. Long-lived radioactive wastes are significantly less than with fission reactors.
3. Fuel reprocessing is accomplished at reactor site.
4. Fusion fuel requires no combustion of the Earth's oxygen.
5. There are no air emissions such as carbon dioxide or other combustion by-products.
6. The ultra-high-density plasma from the exhaust of a fusion reactor can be used to disassociate and ionize any solid or liquid—a fusion torch—for recycling resources.

Disadvantages

1. It is not technically feasible at present to commercially produce energy from fusion.
2. There are potentially significant amounts of tritium leakage.
3. There are potentially significant amounts of thermal pollution.
4. A large fusion reactor could produce as much as 250 tons of radioactive wastes per year.[13]
5. Intense radioactivity of equipment would make maintenance almost impossible.[13]
6. The deuterium-tritium fusion reaction is based on lithium, which is not much more abundant than uranium.
7. Fusion reactors could breed fissionable bomb materials.[14]
8. Reactor vessels would have a short life span due to the deleterious effects of the radiation to which it is subjected. Periodic replacement will negatively affect plant availability and operating costs.[15]

Electrostatic Energy

Electrostatic energy is an income energy source that taps the Earth's electric field. Electrostatic induction is a means of charging certain properties with an electric force. The natural repulsion of similar charges (negative/negative or positive/positive) produces forces of considerable magnitude in which the electrons become concentrated and are actually discharged from the conductor. There are three types of electrostatic motors currently being studied: spark, corona discharge, and electret. Large electrostatic motors can be operated from the Earth's electric field, provided that appropriate aerials are used. Power for these motors can also be transmitted through the air direct without wires. One advantage of electrostatic motors is the high speeds that they can attain. One operating motor rotates at about 10,000 rpm.

On a clear day, the air above 1 square mile of the Earth's surfacs contains about 3 kilowatts of electric energy. During electric storms, however, the air above 1 square mile can contain upt to 10^9 kilowatts of electric energy.[16] At the present time this energy, in the form of electric currents, flows from the air into the Earth and from cloud to cloud. The power dissipated by these currents is estimated to be between 1 million to 1 billion kilowatts. The average thundershower releases about 100 million kwh of energy. It is not yet known what percentage of this energy can be converted into useful work and how fast the Earth's field would replenish itself once part of the energy has been extracted from it.[16]

There is no question that the Earth's electric field can be used to generate power.[16]

Electrostatic and Wind Power: Electrofluid Dynamics

A new energy harnessing technology has arisen that consists of an intriguing combination of wind and electrostatic energy. Called electrofluid dynamics, the process has no moving parts; it uses the wind to blow fine water droplets through a highly charged grid where the droplets are electrically charged. The wind continues to carry the now charged water droplets toward another charged grid where electrical current is extracted. Electrical current is produced because the electrostatic repulsion of similarly charged particles is overcome through the mechanical work done by the wind. This energy is transformed into electrical current.[8]

So far, there are wind-tunnel models of these units. One produces 1/10th of a watt. Scaled up to a full square meter, each unit would produce about 450 watts in a 22 mph wind.

Tesla Electric Transmitter

Another rather scientifically obscure technology for harnessing electrostatic energy could be what Nikola Tesla, the brilliant contemporary of Edison, was working on in his later years. Tesla was convinced that he could transmit electric power without wires, and in fact did so. His laboratory was lighted in this way, and a large transmitter was even constructed on Long Island (financed by J. P. Morgan) to test his ideas.[19]

Tesla said the atmosphere at ordinary pressure was a superior electrical conductor and one could therefore transmit large amounts of energy great distances without wires. His experiments "showed conclusively that, with two terminals maintained at an elevation of not more than 30,000 to 35,000 feet above sea level, and with an electrical pressure of 15 to 20 million volts, the energy of thousands of horsepower can be transmitted over distances which may be hundreds and, if necessary, thousands of miles."[19]

Geomagnetic Storms

Most electrostatic energy is generated by properties indigenous to the Earth, but another source—that associated with geomagnetic storms—comes from outside the Earth. Geomagnetic storms (and their related phenomena—the aurora borealis) are caused by the interaction of plasma clouds ejected from the sun and the Earth's magnetic fields. The most intense geomagnetic storms occur during the peak of the sunspot cycle and usually last from 12 to 24 hours. A concentrated current of about one million amperes flows along the aurora borealis and has produced large magnetic field fluctuations and the induction of electric fields in the earth or any other conductor—such as powerlines or pipelines— on the Earth's surface. Such a phenomenon may be tappable as an energy source.

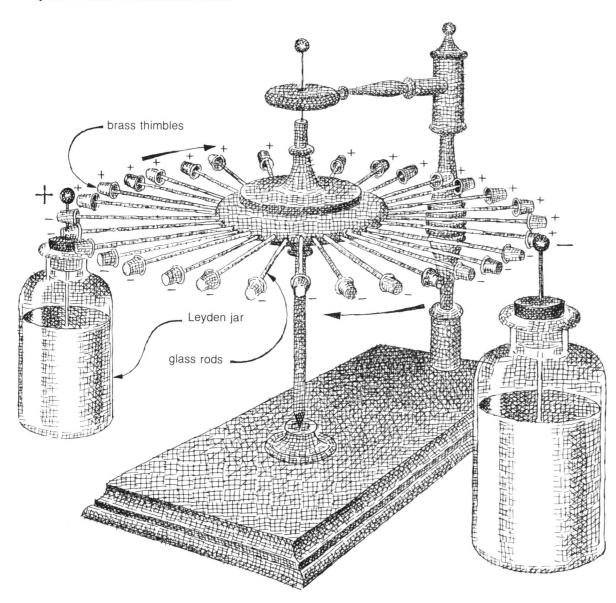

Benjamin Franklin's electrostatic motor[20]

- brass thimbles
- Leyden jar
- glass rods

About 50,000 thunderstorms occur each day, 2,000 at any given moment. They hurl thunderbolts at a rate of 100 per second. Each stroke discharges about 3 million kilowatts of energy, for a total of 8×10^{15} kwh per year.[17] *This is 100 times as much energy as the entire world uses.*

History

700 B.C.	Greece: Thales shows that amber, when rubbed, attracts light objects.. (The word "electron" comes from the Greek "elektron" meaning amber).
1663 A.D.	Germany: Von Guericke produces the first generator of electricity by holding his hand against a rotating sphere of sulfur.
1675	England: Newton suggests glass spheres for electric generator.
1746	Holland: Musschenbrock invents Leyden jar.
1746	U.S.: Franklin identifies lightning with electricity and develops electrostatic engine.
1782	Volta invents electrophore.
1789	France: Coulomb's work in electrostatics.
1860	Germany: Poggendorff invents corona electrostatic motor.
1882	Germany: Wimshurst develops an electrostatic generator.
1929	Germany: Van deGraff develops an electrostatic generator.
1970's	U.S.: Jefimenko demonstrates that large electrostatic motors can be operated from the Earth's electric field.

Deep Ocean Pressure

Water pressure at the bottom of the ocean is a potential income energy source. Water weighs about 62.4 pounds per cubic foot. A cubic mile weighs about 920 billion pounds. The deeper you go the more weight and pressure there is. At 25,000 feet the pressure is 12,500 pounds per square inch. These enormous pressures at the bottom of the sea could be used for power production. Two techniques currently exist for tapping this energy source. One involves the use of a semi-permeable membrane on the end of a pipe inserted vertically deep into the ocean. The membrane lets water pass but excludes salts. At depths below 8,750 meters, fresh water will rise (because fresh water is lighter than salt water and because of the intense ocean pressure at these depths) through the entire pipe to the ocean surface, until its weight equals the osmotic force across the membrane. This water can then be used as a fresh water source. Because the fresh water is forced up the pipe with considerable force it can be made to drive turbines within the pipe to generate power.[21]

Another scheme for tapping deep ocean pressure is a recently patented device that rests on the sea floor and uses the intense water pressure to drive a hydraulic turbine (Illustration 1). The device creates a space of normal pressure in deep seawater and then utilizes the resulting pressure difference for power production. Because of the unique design of the apparatus it is claimed that only a fraction of the power that is produced is used for the operation of the apparatus; the rest is available for use elsewhere.[22]

Advantage

It is a clean and vast source of energy and fresh water (in first technique discussed).

Disadvantages

1. It is only at the theoretical and preliminary drawing board stage.
2. Its impact on the ecology is unknown.

Illustration 1.[23] **Device to Utilize Deep Ocean Pressure to Drive a Hydraulic Turbine**

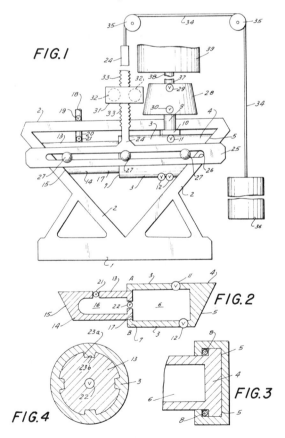

Potential Sites for Deep Ocean Pressure

Source:
National Geographic Oceans Map

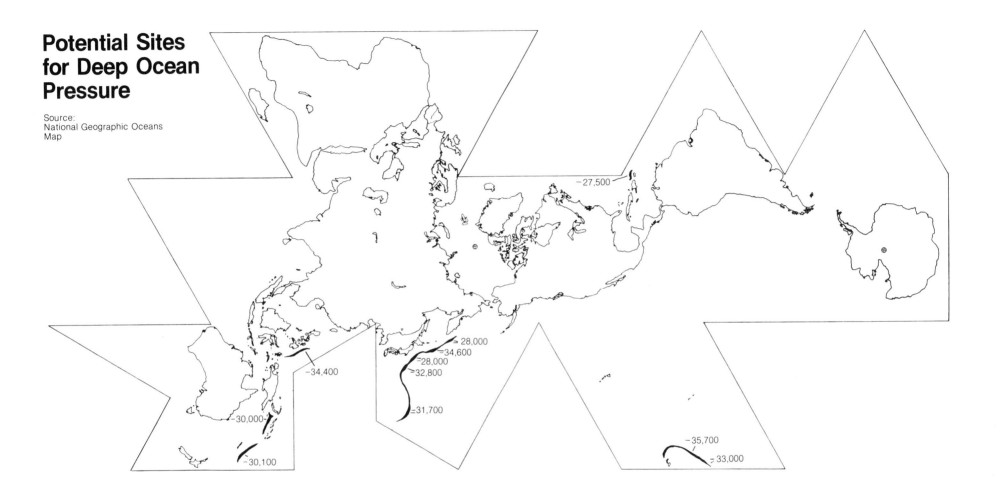

193

Phase Transformations

25

Nitinol Engine Schematic[24]

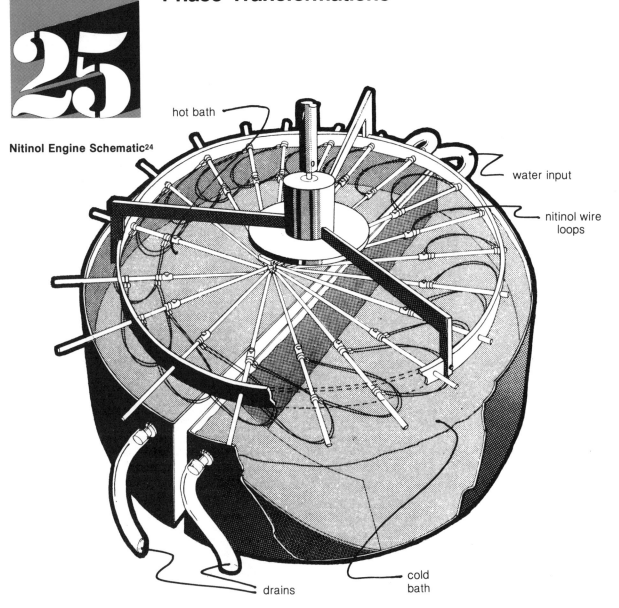

hot bath
water input
nitinol wire loops
drains
cold bath

Phase transformation is the process of a material changing state from a solid to a liquid or gas, or vice versa.

Transformations require or release energy, and some are theoretically harnessable for power production. One power-harnessable transformation which has gone beyond the theoretical stage is the solid state phase transformation of a nickel and titanium alloy called nitinol. Nitinol has a seemingly unique property for a solid metal: it bends easily when cooled and springs back to its original shape when warmed. This "shape memory," or ability to resume an original shape after being deformed, can potentially be harnessed for power. One scheme for such a purpose developed by Ridgeway Banks at Lawrence Radiation Laboratory in Berkeley, uses the unique physical characteristics of nitinol in an engine. The engine is a spoked wheel with 20 u-shaped stripes of nitinol hanging from individual spokes. When the wheel is lowered into a round pan divided into semi-circles of hot and cold water, it begins rotating because one set of wires are cold and go limp, while the others—those passing through the hot water—"fire" with a piston-like motion, thereby driving the wheel around. Power output of this first prototype has been measured at .23 watts.

Seebeck Effect Power

This type of engine could make use of numerous sources of waste heat, as in nearly all current power production plants and other natural sources that previously have been unsuitable for tapping.[24]

Advantages

1. There is no direct fuel consumption with a nitinol engine (though heat is "consumed").
2. There are no by-products from the power production—such as air or water emissions.

Disadvantages

1. Current nitinol engines have low power output.
2. Engines are only at prototype stage; there is as yet no information on size limitations and capabilities.

Another potential source of energy is the flow of electric current that results when two different metals (such as copper and bismuth) are joined to make a circuit and the two junctions are then maintained at different temperatures. This thermoelectric phenomenon, known as the Seebeck effect after the man who discovered it in 1812, is closely related to other effects that involve the conversion of heat into electricity and vice versa. One well-known example of these relationships is the production of heat whenever an electrical current flows through a conductor. The Seebeck effect has been used as the basis for a number of thermoelectric devices such as certain types of refrigerators, thermometers, and small power generators. In the 1930's and 1940's, the Russians used a simple thermoelectric device that ran off the heat produced by an oil lamp to power a radio.[25] The efficiency of such devices is about the same as solar cells but have the added advantage of being able to run when the sun isn't shining because they can use any heat source to generate electricity.

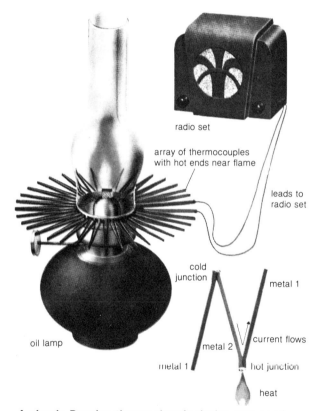

A simple Russian thermoelectric device was used during the 1930's and 1940's to power radios. It was a thermopile—thermocouples connected in series to give the necessary voltage—with the hot junctions heated by an ordinary oil lamp.[25]

27 Humid Air Power

About one-third of the solar energy that reaches the Earth's surface is used in evaporating water. As a result, an extremely large quantity of energy is contained in humid air in the form of latent heat.[26] If the water vapor in humid air is made to condense, its latent heat is released into the surrounding air. Humid air can be said to be a major energy source of thundershowers, hurricanes, and typhoons. Energy is released when water vapor turns into rain drops; one inch of rain falling over a 100 square mile area releases 10 trillion BTU's (2.9×10^9 kwh).[27] There is often over 1,000 times the amount of energy contained in humid air as latent heat than is contained in the same air as wind.[28]

There are currently three different theoretical ways of harnessing humid air for power production.[26] The first technique would utilize a tower similar to the cooling towers of large power plants (Fig. 1). Humid air would be processed through a vertical natural-draft condensation tower wherein it would become the working fluid for a heat engine that would convert the latent heat energy in the humid air to mechanical energy.

Another process for harnessing humid air is an expansion-compression mechanism (Fig. 2). Using this technique, work is extracted from the process of humid air expanding into a low pressure region. During expansion, moisture condenses at its dewpoint and is removed from the system. If the air and remaining water vapor is compressed and cooled enough, more work is produced from the expansion cycle than is put in from the compression cycle, resulting in a net output of energy.

The third technique for harnessing humid air would involve an absorption cycle wherein an absorbing solution such as calcium chloride or lithium bromide would be used to remove the moisture from the air.[26]

Capital costs for a 10.5 MW power plant of the expansion-compression type has been estimated at $1,500 per kw, with electricity cost being 4¢ per kw.[26]

Tropical storms release an estimated 550 to 1,000 $\times 10^9$ kwh of latent heat per day.

An average thundershower releases about 500 $\times 10^6$ kwh of energy through condensation of water.

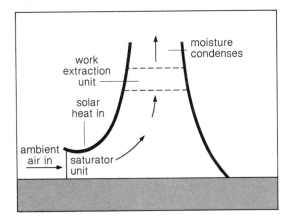

Figure 1.[26] Natural draft tower

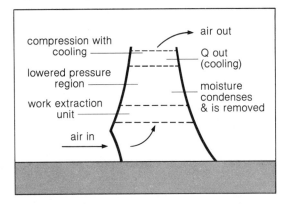

Figure 2.[26] Expansion-compression tower

The Purple Membrane

The Purple Membrane refers to a saltwater-tolerant bacteria whose purple-colored pigment can generate electricity directly in the bacteria cell. In "conventional" photosynthesis a complex series of over thirty enzymes release a stream of negatively charged electrons that make their way into fats, sugars, and proteins. The Purple Membrane does all this with just one enzyme, making both much more efficient and easier to tap the electron flow for electric current. In a sense, the bacteria is a living solar cell. Ten such cells can create an electrical potential of 3 volts; because of their microscopic size, it takes about 5,000 cells to make up one inch. Present proposals call for the generation of electricity or the desalination of water by large bins of these saltwater bacteria.

Advantages

1. The Purple Membrane would be an income-energy source of large potential.
2. Ecological effects should be low.

Disadvantages

1. Only small power outputs have been achieved so far.
2. Power production from the Purple Membrane is still at the theoretical and prototype state; there is as yet no information about size limitations, longevity, cost, etc.

Energy Fluctuation

Another electrical phenomenon that could be a source of energy, and which has received support from the U.S. Department of Energy, is what is called "power conversion from energy fluctuations."[29] In this system a series of small electrical devices convert thermal energy into electricity, the thermal energy being supplied from a number of sources—the Sun, waste heat, etc.

Fluctuation energy results from the thermal motions of electrons, and occurs for resistors (devices which resist the flow of electricity) and all other electrical circuit components. The higher the temperature, the higher is the fluctuation energy. This energy is detectable in a radio receiver as noise voltage or amplifier noise and the only way to reduce it is to lower the temperature of the electrical component which in turn lowers the velocity of the electrons. By increasing the temperature of a resistor, the voltage output is thereby increased. The potential power output from the voltage source is large if it can be converted to direct current. By miniaturizing such devices—as is already being done with electronic components—significant power could be produced at high efficiency.[29]

Gravity

In addition to the above potential energy sources there are other naturally-occurring energy flows that humanity has the theoretical possibility of tapping. Some of these leave even the realm of "exotic" and lean heavily to the pie-in-the-sky, back-of-the-envelope speculation or humorous categories, but are included for the sake of completeness as well as for their value as possible triggers for other, perhaps more "practical," energy harnessing schemes.

Gravity compares in magnitude with the Sun as a potential power source. In fact, gravity is the most abundant source of energy in the universe.[30] Gravitational waves (gravitational fields propagating at the speed of light) resemble electromagnetic waves in that they carry energy, momentum, and information. Electromagnetic waves interact only with electric charges and currents; gravitational waves interact with all forms of matter/energy. Recent, but unconfirmed discoveries by Weber seem to indicate the galactic center radiates an amount of energy which corresponds to the energy of 1,000 suns per year or more. This is about 10,000 times greater than all the light and radio waves emitted from the direction of the galactic center.

History

200 A.D.	Ptolemy formulates elaborate scheme of cycles and epicycles to explain planetary and solar orbit.
1530	Copernicus.
1589	Italy: Galileo performs series of experiments dealing with gravity.
1729	England: Newton formulates gravity laws and estimates gravitational force.
1740	France: Bouguer measures gravitational force.
1798	England: Cavendish measures gravitational constant using tension balance.
1911	Einstein's general theory of relativity offers explanation of gravity and prediction of gravity waves.
1960	U.S.: Dirac quantizes gravitational field.
1969	U.S.: Weber's preliminary finding concerning the possible detection of gravity waves.

Advantages

1. Gravity is the greatest source of energy known to humanity.
2. It is an income energy source.

Disadvantages

It is presently unharnessable.

31 Glacier Power

Glaciers are naturally-occurring phenomena that are relatively rare but are of very large magnitude. A glacier is essentially a moving river of ice that was formed from the gradual accumulation of snow and the metamorphosis of that snow into flowing ice. Glaciers only develop where summers are not warm. The movement of billions of tons of glacier takes an enormous amount of energy. It has been proposed that this glacial movement—velocities range up to 7 or 8 m (23–26 ft) per *day* in some Greenland glaciers—be tapped as an energy source. One way that this might possibly be done is to harness the movement of the glacier through enormous stepup gearings.

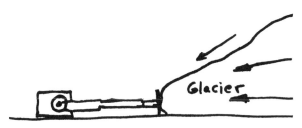

Another way of tapping glaciers is to melt a portion of them and use the fresh water runoff to power conventional hydroelectric facilities. One such scheme has been proposed for Greenland, whose 7.7 million square km ice cap could provide enough water to make Greenland's hydropower potential equal to the hydropower potential of all Europe (excluding the U.S.S.R.).[31] This proposal calls for the melting of parts of Greenland's glacier with solar radiation by putting soot or other dark material on the ice and snow and then collecting the resulting water behind conventional hydroelectric dams. Preliminary cost estimates indicate that total construction would run about $275 to $320 per kilowatt, and power generation run about 4.5 to 6 mills per kwh, figures that are more than competitive with all other presently utilized energy sources. This same plan calls for either the transmission of the generated electricity eastward to Iceland and then to Europe; westward to Canada and then to the U.S.; and/or the use of the electricity for the production of hydrogen through electrolysis and the export of the liquified hydrogen.

32 Other Pies in the Sky

Tectonic Plate Power

The surface of the Earth is comprised of a series of overlapping slowly-moving tectonic plates. The movement of these plates are similar to that of glaciers in that it is massive and for all practical purposes, probably untappable. The amount of energy involved in the movement of tectonic plates is very large (3×10^{12} kwh per year)[32] but as yet there are no proposals for harnessing it.

Tree Power

Another slow-moving natural process that expends energy is photosynthetic growth. One proposal is to harness the lateral or outward growth of trees through collars that would translate this outward growth into harnessable power.

Hydraulic jacket. Outward expansion of tree forces water into pipe at enormous pressure.

Freezing Water Power

Another similar potential power source is freezing water. Anyone who has seen a burst pipe is familiar with the incredible power of freezing water. The above illustration could conceivably work for the outward expansion of freezing water as well as for that of growing tree. The "frozen water drive" would only work in the sub-0° C temperatures in temperate or arctic climates, but could theoretically produce power.

Blackhole Power

A blackhole is a theoretical entity with a vast gravitational field. It is formed when a star explodes and leaves behind a condensed remains that is so dense that its gravitational field is inescapable for everything, including light waves, hence the term "blackhole." Two ways have been proposed to harness the immense gravitational pull of blackholes (assuming of course one can be found conveniently near the Earth). The first involves the sending of a thermonuclear fuel, such as heavy hydrogen, into the gravitational field. As the fuel is pulled into the blackhole it would be heated and compressed until ionization and nuclear fusion occurred, exploding the fuel and producing helium and energy. The force of the blast would blow the energy (as heat) out of the gravitational field where it could be collected and converted to electricity and then beamed to Earth via microwave.

The other technique is more imaginative still. In this scheme a body of unwanted waste (perhaps spoiled pies) would be sent into the gravitational field of the blackhole. Attached to the waste would be a *strong* wire that would do work (i.e., wind-up a spring or some other mechanical device) as the waste was pulled into the blackhole.

Matter/Antimatter Power

Recent theoretical work with quarks and their possible structures suggest that there may be unordinary kinds of matter and antimatter that will annihilate each other but not ordinary matter (or antimatter).[33] In other words, a bit of unordinary matter could be held in a container of ordinary matter, and then brought together with a bit of antimatter contained by unordinary matter. The resulting annihilation would yield about ten billion electron-volts per reaction. One of the major problems with this, the favorite power source of science fiction writers, is that it is predicated upon the existence of at least four sub-subatomic entities for which there is no experimental evidence.

Even More Sky in the Pie

Other rather romantic or otherwise schemes have been concocted for harnessing such things as noise, walking, automobile traffic, storm sewer flow, the Earth's (and the Moon's) rotation (the kinetic energy of the rotation of the Earth is about 100×10^{22} kwh or 1,000 billion kwh), the ICBM inventory of the U.S. and the U.S.S.R., the gravitational pull towards the center of the Earth, the nuclear and thermonuclear bomb inventories of the world, dynamite and all other explosives, the swaying of trees in the wind, wild animals, air pressure differentials, negative pressure in liquid, limestone (Nikola Tesla actually constructed a working engine that ran off of energy liberated from limestone through the action of sulfuric acid[19]), and the proverbial perpetual motion machine. In addition, it has been proposed by an anonymous energy wit that if the U.S. would postpone going metric until a 3-foot meter is developed, we would thereby, nation-wide, reduce commuting distances (and hence fuel consumption) by 10%. The same source also suggested lowering the boiling point of water so that it would take less energy to run our power plant turbines. Yet another proposal in a similar vein, is the deregulation of time. In this scheme, clocks would run faster during peak energy use and slower during low energy demand periods, thereby saving an undisclosed vast amount of energy. A graduated tax credit for sleep would also contribute to conserving energy. Every hour asleep is another hour with the lights off. An international strategy gaining an undisclosed but increasingly wide repute in some circles is the admittance of all OPEC countries into U.S. statehood. By this one stroke of statesmanship finesse, the balance of payments imbalance would be cleared up, dependence upon foreign oil would disappear and the U.S. could become energy independent almost overnight. Similarly by kicking the Northeast states out of the Union, the U.S. would dramatically reduce its energy imports. Finally, a crematorium cogeneration unit that would simultaneously produce heat and electricity from the combustion of its inputs would allow the deceased to make one last contribution to the planet's energy needs. With about 55 million such inputs per year and assuming an energy content about equal to biomass, the departed represent about 44 trillion BTUs (12.8×10^9 kwh) of energy per year.

33 Energy Conversion Techniques and Engines

The previous pages have primarily provided descriptions of energy sources; the following pages list techniques and engines for converting energy from energy sources to useful power.

Heat engines/converters

1. Steam engines
   (a) Savey
   (b) Newcomer
   (c) Watt
   (d) Cornish
   (e) Triple expansion
   (f) Parsons
   (g) Steam turbine
   (h) Hot air turbine
   (i) Stirling
   (j) Hinkley
2. Blast furnaces
3. Gas trubines
4. Heat pumps
   (a) Ocean-based temperature differential
   (b) Heating/cooling system, i.e., air conditioner, refrigerator
5. Heating systems
   (a) Hot water
   (b) Steam
   (c) Gas
   (d) Stove
   (e) Electrical resistance
6. Domestic appliances
   (a) Hot plate
   (b) Immersion heater
   (c) Toaster
7. Internal combustion engines
   (a) Four cycle
   (b) Two cycle
   (c) Diesel
   (d) Rotary
8. Jet engines
   (a) Turbojet
   (b) Turboprop
   (c) Ramjet
   (d) Turbofan
9. Rockets
   (a) Liquid
   (b) Solid (unusual application—rocket engines for electric power)
10. Nuclear reactors
    (a) Light water fission reactor (LWR)
    (b) High temperature gas-cooled fission reactor (HTGR)
    (c) Sodium fission reactor
    (d) Breeder reactor
    (e) Fusion reactor

Direct conversion

1. Magnetohydrodynamics (MHD)
2. Thermionics
3. Fuel cell
4. Battery
5. Photovolatic (solar cells)

Electric conversion

1. Generators (dynamo)
2. Electric motors
   (a) AC
   (b) DC
   (c) Electrostatic motor
3. Lights
   (a) Incandescent
   (b) Fluorescent
   (c) High intensity
   (d) Neon
   (e) Mercury vapor
4. Electric heating

Mechanical Conversion

1. Water wheels
   (a) Undershot
   (b) Overshot
2. Water turbines
   (a) Pelton
   (b) Francis
   (c) Kaplan
3. Sail
4. Helium or hydrogen airship (lift)
5. Contraction turbine
   (a) Banks engine
   (b) Sussman engine
6. Wind power
   (a) Horizontal axis
   (b) Vertical axis

Methods of Energy Transport and Storage

Energy Transport	Coal	Oil	Natural Gas	Uranium	Falling Water	Geothermal	Methane	Alcohol	Wind	Waves	Tidal	Ocean Currents	Solar	Temperature Diff.	Hydrogen	Algae/Bacteria	Electrostatic	Gravity	Electricity
Ship	■	■	■			■	■								■				
Rail	■	■	■	■		■	■								■				
Truck	■	■	■	■		■	■								■				
Pipeline		■	■			■	■								■				
Conveyor Belt	■																		
A.C. Transmiss.																			■
D.C. Transmiss.																			■
Cryogenic Cable															■				■

Energy Storage	Coal	Oil	Natural Gas	Uranium	Falling Water	Geothermal	Methane	Alcohol	Wind	Waves	Tidal	Ocean Currents	Solar	Temperature Diff.	Hydrogen	Algae/Bacteria	Electrostatic	Gravity	Electricity
Bulk Storage (Tanks)		■	■				■	■							■				
Underground Mines	■	■	■	■			■								■				
Storage Battery									■				■						■
Heat Storage	■	■	■	■		■							■						■
Pumped Water					■				■		■								■
Compressed Air									■	■		■							■
Hydrogen via Electrolysis	■	■	■	■	■	■	■		■	■	■	■	■	■					■
Vegetation													■			■			
Flywheels																			■

Relative Costs per Distance of Energy Transmission and Transportation

(Source: From "An Agenda for Energy," by Hoyt C. Hottel and J. B. Howard in *Technology Review*.)

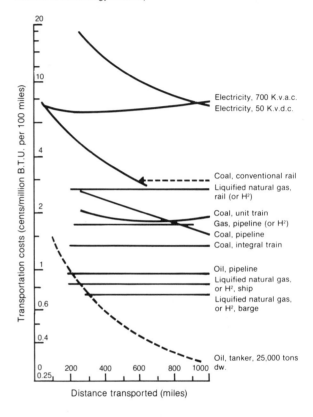

Energy can be transported in fluid form for about one-tenth the cost of transmitting the energy equivalent in electricity.

Two percent of the total electric power is lost in the first 50 miles of transmission due to both transformer and line losses. Conversely, a 48-inch oil pipeline can transport petroleum about 3,600 miles for the same two percent energy loss. The coal train can go for 900–1,200 miles before losing two percent of its energy.[1]

Comprehensive Energy System Outline

Energy Sources
1. Coal
2. Oil
3. Gas
4. Fission
5. Falling Water
6. Geothermal
7. Solar
8. Bioconversion
9. Wind
10. Tides
11. Waves
12. Ocean Currents
13. Temperature Differential
14. Hydrogen
15. Conservation
16. Animate
17. Fusion
18. Exotic

Energy Conversion Techniques and Engines
1. Heat engines/converters
   (a) Steam engines
   (b) Gas turbines
   (c) Heat pumps
   (d) Internal combustion engines
   (e) Nuclear reactors
   (f) Rockets
   (g) Jets
   (h) Heating systems
   (i) Blast furnaces
   (j) Domestic appliances
2. Direct conversion
   (a) MHD
   (b) Thermionics
   (c) Fuel cell
   (d) Battery
   (e) Photovoltaic
3. Electric conversion
   (a) Generators
   (b) Electric motors
   (c) Lights
   (d) Heating
4. Mechanical conversion
   (a) Water wheels
   (b) Water turbines
   (c) Contraction engines
   (d) Helium/hydrogen balloons
   (e) Sails/wind

Energy Storage
1. Bulk (tanks)
2. Underground
3. Mines/caverns
4. Electrochemical battery
5. Pumped water reservoir
6. Compressed air
7. Hydrogen
8. Flywheel
9. Heat in water
10. Sodium etc.
11. Vegetation

Energy Transport
1. Ship
2. Rail
3. Truck
4. Pipe
5. AC transmission
6. DC transmission
7. Cryogenic cable/pipeline

Energy Utilization
1. Industry
2. Transportation
3. Residential/commercial
4. Agriculture
5. Utilities

Power Output of Basic Machines 1700–2000 A.D.

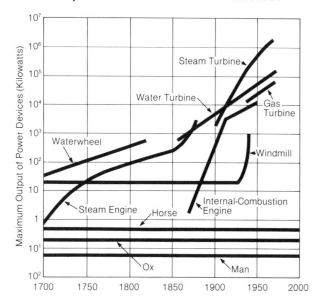

Power output of basic machines has climbed more than four orders of magnitude since the start of the Industrial Revolution (ca. 1750). For the steam engine and its successor, the steam turbine, the total improvement has been more than six orders, from less than a kilowatt to more than a million. All are surpassed by the largest liquid-fuel rockets (not shown), which for brief periods can deliver more than 16 million kilowatts.[2]

Energy Conversion Efficiency.

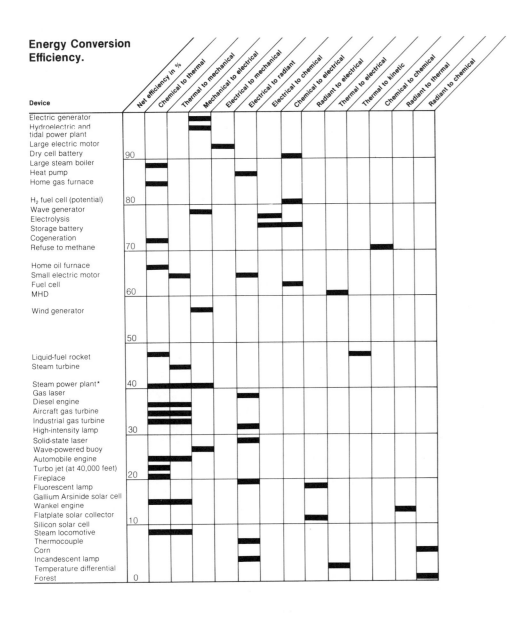

*Natural gas, coal, oil, nuclear, geothermal.

World Energy Data Chart

The World Energy Data Chart which follows integrates much of the quantifiable data of Part II on energy sources; as such, it forms a basis for the next section on strategies.

Column 1 lists known energy sources; Column 2 their current yearly use or consumption rates by humanity; Column 3 the total ultimately recoverable reserves, that is, the total amount of the given resource estimated to exist on Earth; Column 4 the total *known* or *proven* reserves (different to those *estimated* to exist in Column 3); Column 5 the economically recoverable reserves (a yet smaller fraction of proven and ultimately recoverable reserves); Column 6 the energy content of ultimately recoverable reserves (changing technology, economics, and exploration will change the quantities of proven and economically recoverable reserves); Column 7 the present life expectancy of a given energy reserve at projected use rates; Column 8 the presently known ultimate potential available from the nondepletable energy sources that is available per year; Column 9 the yearly amount of the energy that could be harnessed with *present day* know-how in 10 years of intense development, given that the massive committments outlined in Part III are carried out. Columns 10 and 11 extend the non-depletable energy source development of column 9 to 25 years and 50 years respectively; Columns 12 and 13 list the current smallest and largest size power plants that humanity's know-how has been able to produce and operate successfully. They help to illustrate the limit cases of current technology for each energy source. Column 14 lists the current planning and construction time of power plants; Column 15 the present average life-span of a given power facility; Column 16 the capital costs in terms of dollars per installed kilowatt; Column 17 the costs of the energy delivered in terms of mills per kilowatt-hour; and Column 18 the cost per million BTU. Column 19 lists the cost in dollars per unit of power production—that is, cost of nuclear power plant, wind mill, etc.

These data help to illustrate the amounts of energy available from each source as well as the time frame in which new sources of energy could be phased in and obsolete facilities phased out. This chart and the ones on the previous pages dealing with energy efficiency, output, transport, advantages, and disadvantages help in comparing all known energy sources and their widely differing potentials, capabilities, and actualities, as well as helping to set the boundaries for what is possible in the global energy development strategies which follow.

	1	2		3	
		Current Annual Production/Consumption, 1977	References	**World Ultimately Recoverable Reserves**	References
Energy Source		In given units	In 10^{12} kwh equivalents	In given units	In 10^{12} kwh equivalents
Coal		$2,774 \times 10^6$ mt	23.9	1	7.6×10^{12} mt	60,800	10
Petroleum		$2,959 \times 10^6$ mt	36	1	2×10^{12} bbl	3,200	10
		21.8×10^9 bbl	36	1			
Oil shale/tar sands/ heavy oil		47.4×10^6 bbl	.08	2	1.5×10^{12} bbl	2,520	2,10
Natural gas		49×10^{12} ft^3	14.7	1	12.8×10^{15} ft^3	3,870	
Nuclear electric		398×10^9 kwh	.40	1			
Uranium ore		22.3×10^3 mt		1	3.41×10^6 mt	174	11
Subtotal			75			70,564	
Hydro-large		1.46×10^{12} kwh	1.46	1		40/yr	
Hydro-small						25/yr	
Geothermal		9×10^9 kwh	.009	2		500/yr	
Solar: terrestrial heat		9.33×10^9 kwh	.0093	3		70,956/yr	
Solar: terrestrial photovoltaics		5.25×10^6 kwh	.000005	4		70,956/yr	
Solar: extraterrestrial						173,000/yr	
Bioconversion		4.97×10^{12} kwh	4.97	5	146×10^9 mt/yr		12
Fuelwood		1.2×10^9 m^3	1.93	1,6			
Wind		6.6×10^9 kwh	.0066	7		30,000/yr	
Tides		2.1×10^9 kwh	.0021			363/yr	8
Waves		26×10^3 kwh	$.26 \times 10^{-9}$			262/yr	
Ocean currents						1.75/yr	
Temperature differential						22,000/yr	
Hydrogen		20×10^6 mt	.64	8,9	90% of the Universe is H$_2$		8
Animate							
Salinity							
Electrostatic						9,000	13
Fusion						2.2×10^{12}	
Exotic							
Conservation							
Sub-Total			8.36				
Total			83.3				

Energy Source	4 Total Known or Proven Reserves (In given units)	4 Total Known or Proven Reserves (In 10^{12} kwh equivalents)	References	5 Economically Recoverable Reserves (In given units)	5 Economically Recoverable Reserves (In 10^{12} kwh equivalents)	References	6 Energy Content of Ultimately Recoverable Reserves (In 10^{21} thermal joules)	6 Energy Content of Ultimately Recoverable Reserves (In 10^{15} thermal kwh)	References	7 Present Life Expectancy at Projected Use Rates (In years)	8 Present Ultimate Potential per Year (In 10^{12} kwh)
Coal	1.327×10^{12} mt	10,125	2	737×10^9 mt	5,623	2	188	52.2	10	73	
Petroleum	658×10^9 bbl	1,079	2	230×10^9 bbl	377	17	12.2	3.4	10	20	
Oil shale/tar sands/ heavy oil	270×10^9 bbl	443	14	230×10^9 bbl	370	18	4.1	1.1	10		
Natural gas	2.23×10^{15} ft^3	699	2				14	3.9	10	22	
Nuclear electric Uranium ore	1.73×10^6 mt	88	11				28.2×10^6	78,400	10	17	
Subtotal		12,434			6,370			78,460			
Hydro-large		40/yr			30/yr		14	.04/yr		10^3's	50
Hydro-small		25/yr			15/yr		9	.025/yr		10^3's	50
Geothermal		800×10^6	15		100/yr		28.8×10^6	80,000		10^2–10^6's	500
Solar: terrestrial heat		70,956/yr			500/yr		255	70.9/yr		10^9's	70,956
Solar: terrestrial photovoltaics		70,956/yr			500/yr		255	70.9/yr		10^9's	70,956
Solar: extraterrestrial		173,000/yr					623	173/yr		10^9's	173,000
Bioconversion					300/yr					10^9's	
Fuelwood					100/yr					10^9's	
Wind		175/yr	16		100/yr	19	2.5/yr	.7/yr		10^9's	700 (44)
Tides		26.3/yr			15/yr		.094/yr	.026/yr		10^9's	26.3
Waves		262/yr			50/yr		.94	.26/yr		10^9's	262
Ocean currents											1.75 (21)
Temperature differential		22,000/yr			100/yr		79.2/yr	22/yr		10^9's	22,000
Hydrogen	13.5% of earth's crust		8							10^9's	
Animate											
Salinity											
Electrostatic		9,000			unknown			8/yr		10^9	8,000
Fusion							8×10^9	2.2×10^9		10^6–10^9's	
Exotic											
Conservation				1%/yr	.83/yr	20		.083			83

Energy Source	9 Potential Development in 10 years (In 10^{12} kwh)	10 Potential Development in 25 years (In 10^{12} kwh)	11 Potential Development in 50 years (In 10^{12} kwh)	12 Largest Size Power Plant (In MW)	13 Smallest Size Power Plant (In MW)	14 Planning Construction Implementation Time (In years)	15 Present Lifespan of Power Plant (Average in years)	16 Capital Cost (1976 $) ($/kw installed)	17 Delivered Energy Cost (Mills/kwh)	18 $ per 10^6 BTU ($)	19 $ per Unit (10^9 $)
Coal				2,500 (USSR)	25	5–6 elect. (26) 3–7 mine 6–10 gas plant (2)	20–40 (28)	600 (30)	24 (30)	1.20 (37)	.05 (41)
Petroleum				3,200 (US)	25	5 elect. 3–12 offshore (27) 1–3 on shore	20–40	350 (30)	28 (30)	1.87 (37)	
Oil shale/tar sands/ heavy oil						5–8 prod plant (26)	20–40			2.60–10.30 (38)	4 (42)
Natural gas				2,000	25	5 elect. 5–10 pipeline	20–40			2.00 (39)	
Nuclear electric Uranium ore				2,500 (U.K.)	3	9 (26) 6–7 (2)	30 (29)	650 (30)	15–21 (30)		1–3
Subtotal											
Hydro-large	7	9	12	12,600 (Brazil)		8–12 (26)	50–100				
Hydro-small	3	6	10		5 kw	1	50–100				
Geothermal	10	25	50	400	12.5	4–6 (26)	25–40	230–470	2.5–8		
Solar: terrestrial heat	100	200	300	10 actual 1,000 est.	1 watt	1–.5 heating 3 tower	25+ 25–40	13–200 (31)		3.61 (28)	10^2–10^3 10^8
Solar: terrestrial photovoltaics	100	200	300	.25	1 watt	1	20–40				
Solar: extraterrestrial		25	50	100 w actual 1–10,000 est.	10 watt	8–25					
Bioconversion	25	75	200			3	50+			2.10–4.20 (40)	
Fuelwood	10	25	50			5–10	50+				
Wind	15.6 (22)	39 (23)	78 (24)	3	24 watt	.2–2	50+ (29)	350–400 (32)	13 (29)	29	10^6 (43)
Tides	2	5	10	240	.3	6	60–70				
Waves	5	15	30	50 est.	10 watt	3–10	50+	1,000–2,000 (33)	27 (35)		
Ocean currents		.5	5								
Temperature differential	1	10	25	400 est.	70 est.	3–10		1,100–1,500 (34)	9 (34)		
Hydrogen											
Animate									4,000 (36)		
Salinity											
Electrostatic				20 w (est.)	.02 w						
Fusion	negligible	unknown	unknown	50–2,000 (25)		10–40 for development					10^{11}
Exotic											
Conservation	8.3	20.75	41.5			.5–5					
Sub-Total	284.5	634.5	1,161.5								

References

The Role Design Science

1. For more details see: Gabel, M., *Ho-ping: Food for Everyone*, Anchor Books, Garden City, NY, 1979; and Brown, H., Cook, R., and Gabel, M., *Environmental Design Science Primer*, Earth Metabolic Design, 2016 Yale Station, New Haven CT 06510.

2. Adapted from: Ben-Eli, M., "Steps to World Game," Design Science Institute, 3500 Market St., Philadelphia PA 19104.

Coal

1. Zabetkis, M. E., and Phillips, L. D., "Coal Mining," U.S. Dept. of the Interior, Supt. of Documents, Stock No. 024-00011-00, Washington DC.

2. Osborn, E. F., "Coal and the Present Energy Crisis," *Science*, Vol. 183, 2-8-74, pp. 477–481.

3. ERDA, Weekly Announcements, Vol 3, No. 18, 5-3-77.

4. Bassham, J. A., "Increasing Crop Production through More Control of Photosynthesis," *Science*, Vol. 197, 8-12-77.

5. U.S. Steel, 71 Broadway, New York.

6. Zuilich, F., letters, *Scientific American*, June 1976, p. 8.

7. National Coal Association, "Bituminous Coal Facts," 1972.

8. *World Energy Supplies 1972–1976*, Series J, No. 21, U.N., New York, 1978.

9. Enzer, H., Dupree, W., and Miller, S., "Energy Perspectives," a U.S. Dept. of the Interior presentation of major energy and energy-related data, U.S. Government Printing Office, Washington DC, February 1975.

10. Wilson, C. L., *Energy: Global Prospects 1985–2000*, McGraw-Hill, New York, 1977.

11. *Environmental Aspects of Energy Production and Use with Particular Reference to New Technologies*, U.N. Economic and Social Council, Economic Commission for Europe, Senior Advisors to ECE Governments on Environmental Problems, Geneva, Feb. 1976.

12. Meyer, R. A., et. al. "Desulfurization of Coal," *Science*, Vol. 177, 9-29-72, p. 1187.

13. National Academy of Sciences, *Mineral Resources and the Environment*, National Academy of Sciences, Washington DC, 1975.

14. Ramsay, W., "Unpaid Costs of Electrical Energy—Health and Environmental Impacts from Coal and Nuclear Power," Resources for the Future, reported in *Science News*, February 17, 1979, p. 105.

15. Hayes, D., *Ray of Hope: The Transition to a Post-Petroleum World*, W. W. Norton, New York, 1977, pp. 40–41.

16. Hammond, A. L., "Coal Research IV: Direct Combustion Lags Its Potential," *Science*, Vol. 194, 10-8-76, p. 172.

17. Pimentel, D., et. al., "Land Degradation: Effects on Food and Energy Resources," *Science*, Vol. 194, 11-8-76, pp. 149–155.

18. "Ecology: Prize for a Stripper," *Newsweek*, 12-1-75, p. 88.

19. Sheridan, D., "A Second Coal Age Promises to Slow Our Dependence on Imported Oil," *Smithsonian*, August 1977.

20. "Energy Technology to the Year 2000," a special symposium, *Technology Review*, M.I.T., Cambridge MA 02139, October–November 1971.

Gasification

1. Hammond, A. L., "Coal Research (II): Gasification Has an Uncertain Future," *Science*, Vol. 193, 8-27-76, p. 750.

2. O'Hara, et. al., "Material Considerations in Coal Liquification," *Metal Progress*, November 1977, pp. 33–38.

3. Hammond, A. L., "Coal Research (III): Liquification Has Far to Go," *Science*, Vol. 193, 8-3-76.

4. Osborn, E. F., "Coal and the Present Energy Situation," *Science*, Vol. 183, 2-8-74, pp. 477–81.

5. Wilson, C. L., *Energy: Global Prospects 1985–2000*, McGraw-Hill, New York, 1977.

6. Hammond, A. L., "Questioning the Synthetic Fuels Option," *Science*, Vol. 193, 7-27-76, p. 752.

7. "Federal Coal Research: Status and Problems to Be Resolved," Report to Congress by the Comptroller-General of the U.S., U.S. General Accounting Office, 2-18-75.

Peat

1. Boffy, P. M., "Energy: Plan to Use Peat as Fuel Stirs Concern in Minnesota," *Science*, Vol. 190, 12-12-75, p. 1066.

2. "Energy Short Finland Gets Help from Peat," *Christian Science Monitor*, October 1977.

3. Carter, L. J., "Peat for Fuel: Development Pushed by Big Corporate Farm in Carolina," *Science*, Vol. 199, 1-6-78, p. 33.

4. Based on No. 1 above; U.S. has 5% of world total peat; Minnesota's peat bogs produce 15×10^6 tons/yr. Assuming Minnesota's renewal rates for the rest of the world yields 300×10^6 tons/yr.

5. "Peat for Energy Use," in *Important for the Future*, ed. by J. Barnea, UNITAR, New York, Vol. III, No. 4, September 1978.

Petroleum

1. Hayes, D., *Rays of Hope: The Transition to a Post-Petroleum World*, W. W. Norton, New York, 1977.

2. *National Geographic*, June 1974, pp. 792–825.

3. Wilson, C. L., *Energy: Global Prospects 1985–2000*, McGraw-Hill, New York, 1977.

4. *World Energy Supplies 1972–1976*, U.N., New York, Series J, No. 21, 1978.

5. Hubbert, M. K., "Outlook for Fuel Reserves," in *McGraw-Hill Encyclopedia of Energy*, Lapedes, D. N., editor-in-chief, McGraw-Hill, New York, 1976.

6. *McGraw-Hill Encyclopedia of Energy*, Lapedes, D. N., editor-in-chief, McGraw-Hill, New York, 1978, p. 559.

7. "Pumping Oil the Hard Way," *Newsweek*, 4-30-79.

8. U.S. Energy Research and Development Administration, "Enhancement of Recovery of Oil and Gas," *Progress Review No. 1*, January 1975.

9. Commoner, B., "Reflections," *The New Yorker*, 4-23-79.

10. Enzer, H., Dupree, W., and Miller, S., "Energy Perspectives," a U.S. Dept. of the Interior presentation of major energy and energy-related data, U.S. Government Printing Office, Washington DC, February 1975.

11. *Environmental Aspects of Energy Production and Use with Particular Reference to New Technologies*, U.N. Economic and Social Council, Economic Commission for Europe, Senior Advisors to ECE Governments on Environmental Problems, Geneva, February 1976.

12. Adapted from "Energy Technology to the Year 2000," a special symposium, *Technology Review*, October–November 1971, M.I.T., Cambridge MA 02139.

Gas

1. Wilson, C. L., *Energy: Global Prospects 1985–2000*, McGraw-Hill, New York, 1977.

2. Barnea, J., "Energy Sources for the Future," a paper presented at UNITAR Conference, UNITAR, New York.

3. Rosenblum, D., U.S. G.A.O. report on safety of LNG, as reported in *New York Times*, 1-26-78, p. A18.

4. *World Energy Supplies: 1972–1976*, U.N., New York, Series J, No. 21, 1978.

5. Enzer, H., Dupree, W., and Miller, S., "Energy Perspectives," a U.S. Dept. of the Interior presentation of major energy and energy-related data, U.S. Government Printing Office, Washington, DC, February 1975.

6. "Energy Technology to the Year 2000," a special symposium, *Technology Review*, M.I.T., Cambridge MA 02139, October–November 1971.

7. Hammond, A. L., "Questioning the Synthetic Fuels Option," *Science*, Vol. 193, p. 752.

8. *Environmental Aspects of Energy Production and Use with Particular Reference to New Technologies*, U.N. Economic and Social Council, Economic Commission for Europe, Senior Advisors to ECE Governments on Environmental Problems, Geneva, February 1976.

Nuclear Power

1. Douglas, J. H., "The Great Power Debate (2) Breeder Reactors," *Science News*, Vol. 109, 1-24-76.

2. Sullivan, W., "U.S. Space Mishaps in '64 and '70 Noted," *New York Times*, 1-25-78.

3. Polman, N., Paolucci, D. A., "What Killed the Scorpion?" *Sea Power*, May 1978, pp. 13–19.

4. Wilson, C. L., *Energy: Global Prospects 1985–2000*, McGraw-Hill, New York, 1977.

5. Wicker, T., "Paying the Nuclear Piper—II," *New York Times*, 9-30-77.

6. Based on the estimated $3.5 billion for decommissioning 400 facilities that will be obsolete by 1981; G.A.O. thinks the estimate is too low.

7. Based on U.S. costs of $15 billion and U.S. as having 35% of the world's nuclear facilities.

In fact, U.S. has 30% of the commercial reactors and 32% of the research reactors.

8. "Radioactive Waste: Some Urgent Unfinished Business," *Science*, Vol. 195, 2-18-77, p. 661.

9. Hayes, D., *Rays of Hope: The Transition to a Post-Petroleum World*, W. W. Norton, New York, 1977.

10. "The Uranium Mill Tailings Cleanup: Federal Leadership at Last?" G.A.O. EMD-78-90, June 1978, p. 39.

11. These are: N.R.C., F.T.C., B.O.E., F.E.A., T.V.A., E.P.A., C.E.O., N.A.S.A., N.A.S., N.S.F.; Departments of Labor, Commerce, Interior, Agriculture, State, Defense, Transportation; and H.E.W. G.A.O., O.T.A., I.A.E.A., U.N., Export-Import Bank, D.E.P., F.B.I., C.I.A., N.S.A., state regulatory agencies, I.E.A., state environmental agencies; police, fire, safety, and insurance agencies.

12. *World Energy Supplies: 1972–1976*, U.N., New York, Series J, No. 21, 1978.

13. "Why Atomic Power Dims Today," *Business Week*, 11-17-75, p. 98.

14. Hammond, A., "Fusion: The Pro's and Con's of Nuclear Power," *Science*, Vol. 178, 10-13-72.

15. Weinberg, H. M., "Social Institutions and Nuclear Energy," *Science*. Vol. 177, 7-7-72, p. 32. "...Plutonium 239 with a half-life of 24,000 years will be dangerous for perhaps 200,000 years."

16. *New Scientist*, 4-3-75, p. 30.

17. Sullivan, W., "Scientists Believe Cosmic Rays Generate Electricity that Causes a Stroke of Lightning," *New York Times*, 12-8-77, p. 38.

18. Ramsay, W., "Unpaid Costs of Electrical Energy—Health and Environmental Impacts from Coal and Nuclear Power," Resources for the Future, reported in *Science News*, February 17, 1979, p. 105.

19. Henderson, H., "The Revolution from Hardware to Software," *Technological Forecasting and Social Change*, December 1978, pp. 317–24.

20. "Energy Technology to the Year 2000," a special symposium, *Technology Review*, M.I.T., Cambridge MA 02139, October–November 1971.

Income Energy Sources Introduction

1. Lee, S. M., "Insulation and Wind: A Natural Combination for Self-sufficient Power Systems," 12th IEEE Photovoltaic Conference November 1976, Naval Weapons Center, China Lake, CA.

2. Lovins, A. B., *Soft Energy Paths: Towards a Durable Peace*, Ballinger, Boston MA, 1977.

Hydroelectric

1. *World Energy Supplies: 1972–1976*, U.N., New York, 1978, Series J, No. 21, 1978.

2. Waring, J. A., "Solar Energy Conversion," paper presented to the Forum on Energy of the World Future Society, April 1974, 4916 St. Elmo Ave., Washington D.C.

3. Hubbert, M. K., "Energy Resources," in *Resources & Man: A Study and Recommendations by the Committee on Resources & Man*, National Research Council, W. H. Freeman, San Francisco CA.

4. "Power Production at Federal Dams Could Be Increased by Modernizing Turbines and Generators," U.S. General Accounting Office Audit Report, 3-16-77, EMD-77-22.

5. Smil, V., "China's Energetics: A Systems Analysis," in *Chinese Economy Post-Mao*, A compendium of papers submitted to the Joint Economic Committee, Congress of the U.S. 11-9-78, U.S. Government Printing Office.

6. Jones, C., "Small Dams—Enough Power for New York City," *Christian Science Monitor*, 6-29-77.

7. U.S. Dept. of Energy Information, "Weekly Announcements," Vol. I, No. 11, 12-23-77, p. 3.

8. Kocivar, B., "Lifting Foils," *Popular Science*, February 1978, pp. 71–73.

9. Hasson, E. M., *Windpower Conference*, Vol. II, pp. 4294–95, Melbourne, Australia, 1962.

10. Edison, W., "New Energy Sources for the Future: A Report to Congressman Edgar," *Congressional Record*, 8-1-77, p. E 4998.

11. *Technical Progress in Israel*, No. 142, October 1975, published by Association of Engineers and Architects in Israel, Engineers' House, 200 Dizengoff St., Tel Aviv, Israel.

12. McNichols, J. L., et al., "Thermoclines: A Solar Thermal Energy Resource for Enhanced Hydroelectric Power Producton," *Science*, Vol. 203, 1-12-79, p. 167.

13. Craig, P., et al., editors, *Distributed Energy Systems in California's Future*, Interim Report, Vol. 1 & 2, U.S.D.O.E. document HCP/P7405-03, May 1978.

Geothermal

1. "The Earth and Its Hot Unknown Interior," in *Important for the Future*, ed. by J. Barnea, UNITAR, New York, February 1976.

2. White, D. E., Williams, D. L., editors, *Assessment of Geothermal Resources of the U.S. 1975*, Geological Survey Circular 726, Washington D.C., 1975, p. 148.

3. Maugh, T. H., "Natural Gas: U.S. Has It and the Price Is Right," *Science*, Vol. 191, 2-13-76, p. 550.

4. Robson, G. R., "Geothermal Electricity Production," *Science*, Vol. 184.

5. Dr. C. Otte, Geothermal Division, Union Oil Co.

6. Yamamura, A., "Outline of a Report of Comprehensive Investigation of the New Modes of Power Generation," October 1970, Japan Electric Machine Industry Association.

7. Based on correspondence from Donald Finn, California Energie Inc.; 165×10^6 feet drilled annually (about 32,000 wells); 55% = 90×10^6 feet drilled.

8. Northrup, C. J. M., et. al., "Magma: A Potential Source of Fuels," *International Journal of Hydrogen Energy*, Vol. 3, No. 1, 1978, p. 1.

9. Duffield, C., "Geothermal Technosystems and Water Cycles in Arid Lands," Univ. of Arizona, Office of Arid Lands Studies, Tucson AZ, 1976.

10. Klass, D. L., Ghosh, S., *World Magazine*, 4-23-74.

11. Eaton, W. W., "Geothermal Energy," ERDA, 1975, p. 8.

12. Penner, S. S., Iserman, L., *Energy: Vol. II Non-Nuclear Technologies*, Addison-Wesley Publishing Co., Reading MA, p. 575–76.

13. *Alternative Energy Sources for Hawaii 1975*, Report of the Committee on Alternative Energy Sources for Hawaii of the State Task Force on Energy Policy, published by Hawaii Natural Energy Institute, University of Hawaii, February 1975.

Wind Energy

1. "Wind Energy: Large and Small Systems Competing," *Science*, 9-2-77, p. 971.

2. Gustavson, M. R., "Limits to Wind Power Utilization," *Science*, Vol. 206, 4-6-79, pp. 13–17.

3. Villeco, M., "Wind Power," *Architecture Plus*, May–June 1974, pp. 64–78.

4. Eldridge, F. R., *Windmachines*, The Mitre Corp., October 1975.

5. Green, W., "A Novel Windmill Challenges U.S. Policy on Wind Energy," *New York Times*, 2-13-79.

6. Stabb, D., "Wind," *Architectural Design*, April, 1972, pp. 253–54.

7. *Edison Electric Institute Statistical Yearbook for 1976*, Edison Electric Institute, 90 Park Ave., New York NY 10016, 1977.

8. Total U.S. electric generation is about 2.0×10^{12} kwh; the amount theoretically available from the more than 900,000 newly-outfitted transmission towers operating at .50 loadfactor is 11.5×10^{12} kwh.

9. Merriam, M. F., "Wind Generators and the Oil Reserve," letters, *Science*, 3-9-79, pp. 954–55.

10. Bergy, K. W., "Wind Power Potential for the U.S.," *Aware Magazine*, October 1974, Vol. 49, pp. 2–5.

11. Metz, W. D., Hammond, A. L., *Solar Energy in America*, AAAS, Washington, D.C., 1978.

12. Thomas, R. L., Shoales, J. E., "Preliminary Results of the Large Experimental Wind Turbine Phase of the National Wind Energy Program," NASA Lewis Research Center, Cleveland OH, NTIS PC A02/MF A01.

13. Bereson, L., "Sail Power for the World's Cargo Ships," *Technology Review*, M.I.T., Cambridge MA 02139, March–April 1979.

14. Dornis, L., *Catch the Wind*, Four Winds Press, New York, 1976.

15. Zelby, L. W., "Hydrogen as an Energy Storage Element," The Hydrogen Economy Miami Energy (THEME) Conference March 18–20, 1974, Miami Beach Florida Conference Proceedings, the School of Engineering and Environmental Design, Univ. of Miami, Coral Gables FL, edited by T. Nejat Veziroglu.

16. Todd, C. J., et al. "Cost-effective Electric Power Generation from the Wind," U.S. Dept. of the Interior, Bureau of Reclamation, Denver CO, 1977.

Solar Energy

1. Georgescu-Roegen, W., *The Entropy Law and Economic Process*, Harvard University Press, Cambridge MA, 1971.

2. Rice, R. A., "120 Million MW for Nothing," *Technology Review*, M.I.T, Cambridge MA 02139, June 1972, p. 59.

3. Metz, W. D., "Solar Thermal Energy: Bringing the Pieces Together," *Science*, Vol. 197, 8-12-77.

4. Westman, W., "How Much Are Nature's Services Worth?" *Science*, Vol. 197, 9-2-77, p. 960.

5. Hayes, D., *Rays of Hope: The Transition to a Post-Petroleum World,* W. W. Norton, New York, 1977.

6. Metz, W. D., Hammond, A. L., *Solar Energy in America,* AAAS, 1978.

7. George Porter, Nobel Prize in Chemistry.

8. Fuller, R. B., "Man with a Chronofile," *Saturday Review,* 4-1-67; and "Buckminster Fuller Retrospective," *Architectural Design,* December 1972.

9. Okress, E. C., "The Franklin Institute Has High Hopes for Its Big Balloon," *IEEE Spectrum,* December 1978, pp. 41–46.

10. Clarke, Jr., V. C., "Solar Power Mountain Concept," NASA Technical Briefs, Summer 1977, p. 212.

11. Lindmayer, J., et al., "Energy Requirement for the Production of Silicon Solar Arrays," Final Report, Sept. 1977, Solarex Corp., Rockville MD, NTIS PC A07/MF A01.

12. Edwards, H. B., "Optics for Natural Lighting," NASA Technical Briefs, Summer 1978, p. 209.

13. Lovins, A. B., "Energy Resources," in *World Energy Strategies: Facts, Issues, and Options,* Ballinger, Boston MA, 1975.

14. "Wind Energy: Large and Small Systems Compete," *Science,* Vol. 197, 9-2-77.

15. *Technical Progress in Israel,* No. 142, October 1975, published by Association of Engineers and Architects in Israel, Engineers' House, 200 Dizengoff St., Tel Aviv, Israel.

16. Stobaugh, R. and Yergin, D., *Energy Future,* Random House, New York, 1979.

17. O'Neill, G. K., "Space Colonies and Energy Supply to the Earth," *Science,* Vol. 190, 12-5-75.

Bioconversion

1. "Fuel from Organic Wastes," *Chemical Technology,* Nov. 1973, pp. 689–98.

2. "India Puts Its Waste to Work," *Development Forum,* August 1978, p. 9.

3. Reed, T. B., Lerner, R. M., "Methanol: A Versatile Fuel for Immediate Use," *Science,* Vol. 182, December 1973, pp. 1299–1304.

4. Garner, W., Smith, L., "Disposal of Cattle Feedlot Wastes by Pyrolysis," EPA-R273-096, January 1973.

5. Appell, H. R., et al., "Converting Organic Wastes to Fuel," U.S. Government Pub. No. 1574, 1971.

6. Fry, L. J., Merrill, R., "Methane Digestors," *The New Alchemy Institute Newsletter No. 3,* 1973.

7. Gest, H., Kamen, M. D., "Photoproduction of Molecular Hydrogen by Rhodospirillum Rubrum," *Science,* Vol. 109, June 1949, pp. 558–59.

8. Robson, G. R., "Geothermal Energy Production," *Science,* Vol. 184, 4-19-74, pp. 371–72.

9. Hayes, D., *Rays of Hope: The Transition to a Post-Petroleum World,* W. W. Norton, New York, 1977, p. 195; and Metz, W. D., Hammond, A. L., *Solar Energy in America,* AAAS, 1978, p. 106.

10. Smil, V., personal communication 10-79; China is committed to having 20 million biogas facilities by 1985. Also see: Smil, V., "Intermediate Technology in China," *Bulletin of the Atomic Scientists,* February 1977; and "China's Energetics: A Systems Analysis" in *Chinese Economy Post-Mao,* A compendium of papers submitted to the Joint Economic Committee, Congress of the U.S., 11-9-78, U.S. Government Printing Office.

11. Penner, S. S., Iserman, L., *Energy: Vol. II Non-Nuclear Technologies,* Addison-Wesley Publishing Co., Reading MA, 1975.

12. Blankenship, D. T., Winget, G. D., "Hydrogen Fuel: Production by Bioconversion," Proceedings of the Eighth Intersociety Energy Conversion Engineering Conference, held August 13–16, 1973, pp. 580–82.

13. Calvin, M., "Solar Energy by Photosynthesis," *Science,* Vol. 184, 4-19-74, pp. 375–81.

14. Lovins, A. B., *Soft Energy Paths: Towards a Durable Peace,* Ballinger, Boston MA, 1977.

15. Hammond, A. L., "Photosynthetic Solar Energy: Rediscovering Biomass Fuels," *Science,* Vol. 197, 8-19-77, pp. 745–46.

16. Crabbe, D., McBride, R., *The World Energy Book,* Nichols Publishing Co., New York, 1978, p. 194.

17. See World Energy Data Chart at end of this section.

18. *World Energy Supplies: 1972–1976,* Series J, No. 21, U.N., New York, 1978.

Tidal Power

1. *New Sources of Energy,* Proceedings of the United Nations Conference on New Sources of Energy, held August 1961 in Rome, Vol. 1, U.N., New York, 1964, p. 127.

2. Scarlett, C. A., "Tidal Power," in *McGraw Hill Encyclopedia of Energy,* Lapedes, D. N., editor-in-chief, McGraw-Hill, New York, 1976.

3. Duff, G., "Tidal Power in the Bay of Fundy," *Technology Review,* M.I.T., Cambridge MA 02139, November 1978, p. 34.

4. Penner, S. S., Iserman, L., *Energy: Vol. II Non-Nuclear Technologies,* Addison-Wesley Publishing Co., Reading MA, 1975.

5. Charlier, R. H., "Tidal Power," *Oceanology International,* September/October 1968, p. 33.

6. Hayes, D., *Rays of Hope: The Transition to a Post-Petroleum World,* W. W. Norton, New York, 1977.

7. *The Renewable Energy Handbook,* Energy Probe, Univ. of Toronto, Toronto, Ontario, Canada.

8. *The Illustrated Science and Invention Encyclopedia,* H. S. Stuttman Co., Inc., New York, 1978.

Wave Power

1. Bott, A. N. W., "Power Plus Proteins from the Sea," *Journal of the Royal Society of Arts,* July 1975, p. 499.

2. *The Renewable Energy Handbook,* Energy Probe, Univ. of Toronto, Toronto, Ontario, Canada.

3. Jones, E. B., "Wave Power," *Alternative Energy Sources Magazine,* No. 12, October/November 1973, Rt. 1, Box 36B, Minong WI 54859.

4. "Waving Japan Onward," *Science News,* 6-24-78, p. 402.

5. Kamogawa, H., " A Review of Japanese Ocean-Based Solar Energy Conversion Systems," Summaries of Technical Presentations, Vol. II, pp. 5–10. Japanese-U.S. Symposium on Solar Energy Systems. NSF/RANN.

6. "Waves a Million," *New Scientist,* 5-6-76, pp. 309–10.

7. Crabbe, D., McBride, R., *The World Energy Book,* Nichols Publishing Co., New York, 1978, p. 194.

8. Peterson, C. A., U.S. Patent No. 4,086,775, May 1978.

9. Semo, M. S., U.S. Patent No. 3,353,787, November 1967.

10. "Energy from the Ocean," *Energy Perspectives: Information from the Battelle Energy Program,* No. 13, Battelle Memorial Institute, 505 King Ave., Columbus OH 43201, August 1974.

11. Fuller, R. B., "Floating Breakwater," U.S. Patent No. 3,863,455, issued 2-4-75.

12. Middleton, P., et al., "Canada's Renewable Energy Resources: An Assessment of Potential," A Study Prepared for the Office of Energy Research and Development; Dept. of Energy, Mines, and Resources, 2-20-76.

Ocean Currents

1. Kamogawa, H., "A Review of Japanese Ocean-Based Solar Energy Conversion Systems," Summaries of Technical Presentations, Vol. II, pp. 5–10. Japanese-U.S. Symposium on Solar Energy Systems. NSF/RANN.

2. *Large-Scale Current Measurement in Lake Superior,* 1976, U.S. Government Printing Office: C55.13: ERL 363-GLERL8.

Temperature Differential

1. Lavi, A., Zener, C., "Plumbing the Ocean Depths: A New Source of Power," *IEEE Spectrum,* October 1973, pp. 22–27.

2. Isaac, and Schmitt, "Resources from the Sea," *International Science an Technology,* June 1963.

3. Anderson, J. H., Anderson, Jr., J. H., "Power from the Sun by Way of the Sea?" *Power,* Feb. 1965, pp. 63–66.

4. Anderson, J. H., Anderson, Jr., J. H., "Thermal Power from Seawater," *Mechanical Engineer,* April 1966, pp. 42–46.

5. Whitmore, W. F., "Solar Energy: The Prospect for OTEC," letters, *Science,* Vol. 198, 12-9-77.

6. Metz, W. D., Hammond, A. L., *Solar Energy in America,* AAAS, 1978.

7. Williams, R. H., "The Greenhouse Effect for Ocean-Based Solar Energy Systems," Working Paper No. 21, Center for Environmental Studies, Princeton Univ., Princeton NJ, October 1975.

8. *The McGraw-Hill Encyclopedia of Energy,* Lapedes, D. N., editor-in-chief, McGraw-Hill, New York, 1976.

Hydrogen

1. Sharer, J. C., Pangborn, J. B., "Utilization of Hydrogen as an Appliance Fuel," Institute of Gas Technology, ITT Center, Chicago IL, THEME, Univ. of Miami, Coral Gables FL 33124, 3-20-74.

2. Paleocrassas, S. N., "Photolysis of Water as Solar Energy Conversion Process," THEME, Univ. of Miami, Coral Gables FL 33124, 3-20-74.

3. Mitsui, A., "The Utilization of Solar Energy for Hydrogen Production by Cell-Free System of Photosynthetic Organisms," THEME, Univ. of Miami, Coral Gables FL 33124, 3-20-74.

4. Blankenship, D. T., Winget,

G. D., "Hydrogen Fuel: Production by Bioconversion," Proceedings of the Eighth Intersociety Energy Conversion Engineering Conference, 8-13 to 8-16-73, pp. 580–82.

5. Weingart, J., "The Helios Strategy," Mitchell Prize Winner, 1977.

6. Whitelaw, R. L., "Electric Power and Fuel Transmission by Liquid Hydrogen Superconductive Pipeline," Mechanical and Nuclear Engineering, Virginia Polytechnic Institute and State Univ., Blacksbury VA.

7. "Hydrogen, Synthetic Fuel of the Future," Science, Vol. 178, 11-24-72.

8. Reed, T. B., Lerner, R. M., "Methanol: A Versatile Fuel for Immediate Use," Science, Vol. 182, December 1973, pp. 1299–1304.

9. Environmental Alert Group, Public Interest Report, "Solutions to the Energy Crisis," 1973, 1543 N. Martel Ave., Los Angeles, CA 90046.

10. Korycinski, P. F., Snow, D. B., "Hydrogen for the Subsonic Transport," NASA Langley Research Center, Hampton VA, The Hydrogen Energy Economy Meeting, THEME, Univ. of Miami, Coral Gables FL 33124, 3-20-74.

11. Billings, R. E., "A Hydrogen-powered Mass Transit System," International Journal of Hydrogen Energy, Vol. 3, No. 1, Pergamon Press, 1978, pp. 49–59.

12. Scott, D., "Hydrogen Bus," Popular Science, December 1978, pp. 72–73.

Animate Energy

1. Pimentel, D., Pimentel, M., "Counting the Kilocalories," Ceres: FAO Review on Agriculture and Development, FAO, U.N., Via delle Terme di Caracalla, 00100, Rome, Italy, September–October 1977.

Cogeneration

1. Williams, R., "Cogeneration," The Center for Environmental Studies, Princeton Univ, Princeton NJ, 1978, p. 84.

2. Lovins, A. B., "Energy Resources," A Paper Presented to the U.N. Symposium on Population, Resources and Environment, Stockholm, Sweden, September–October 1973.

3. Bos, P., The Potential for Cogeneration Development in Six Major Industries by 1985, A Report to the Federal Energy Administration by Resource Planning Associates, Inc., Cambridge MA, 1977.

4. Hagler, H., A Technical Overview of Cogeneration: The Hardware, the Industries, the Potential Development, A Report to the Division of Industrial Energy Conservation, Dept. of Energy, by Resource Planning Associates, Inc., Washington DC, 1977.

Exotic Energy Sources

Salination

1. Loeb, S., "Osmotic Power Plants," Science, Vol. 189, 8-22-75, p. 654.

2. Norman, R. S., "Water Salination: A Source of Energy," Science, Vol. 186, 10-25-74, pp. 350–52.

3. Wick, G. L., Asaacs, J. D., "Salt Domes: Is There More Energy Available from Their Salt than Their Oil?" Science, Vol. 199, 3-31-78, p. 1436.

4. McCormick, Michael, ERDA Ocean Systems Branch; as reported in ERDA Weekly Announcements, Vol. 3, No. 30, 7-29-77.

5. Technical Progress In Israel, No. 142, Oct. 1975, published by Association of Engineers and Architects in Israel, Engineers' House, 200 Dizengoff St., Tel Aviv, Israel.

Nuclear Fusion

6. Metz, W. D., "Magnetic Containment Fusion: What Are the Prospects?" Science, Vol. 177, 10-72, pp. 291–93.

7. "Soviet Reversing Energy Policy Puts Stress on Coal," New York Times, 1-7-75, pp. 43–48.

8. Metz, W. D., "Laser Fusion: A New Approach to Thermo-nuclear Power," Science, Vol. 177, 9-72, pp. 1180–82.

9. Holbien, J. P., "Fusion Energy in Context: Its Fitness for the Long Term," Science, Vol. 200, 4-14-78, p. 168.

10. Gough, W. C., Eastland, B. J., "The Prospects of Fusion Power," Scientific American, Vol. 224, 2-71, pp. 50–64.

11. Parker, F. L., "Radioactive Wastes from Fusion Reactors," Science, Vol. 159, 1-5-68, pp. 83–84.

12. Steiner, D., "The Radiological Impact of Fusion," New Scientist, 12-16-71, pp. 168–71.

13. Hayes, D., Rays of Hope: The Transition to a Post-Petroleum World, W. W. Norton, New York, 1977, p. 27.

14. Lovins, A. B., Soft Energy Paths: Towards a Durable Peace, Ballinger, Boston MA, 1977, p. 188.

15. Parkins, W. E., "Engineering Limitations of Fusion Power Plants," Science, Vol. 199, 3-31-78.

16. "Electric Motors Powered from Thin Air," New Scientist and Science Journal, 3-18-71, p. 612.

17. Steinhart, C., Steinhart, J., Energy Sources: Use and Role in Human Affairs, Duxbury Press, 1974, p. 24.

18. "13 Wind Machines," Popular Science, September 1978, p. 74. Also see: Gourdine, M. C., "Electrogas-dynamics," Science & Technology, July 1968, pp. 50–55.

19. Nikola Tesla: Lectures, Patents, Articles, published by Nikola Tesla Museum, Belgrade, Yugoslavia, 1956, Appendix, p. 137.

20. Strong, C. L., "Electrostatic Motors Are Powered by the Electric Field of the Earth," Scientific American, 9-74.

21. Levenspiel, O., de Nevers, N., "The Osmotic Pump," Science, 1-18-74.

22. Molinar, S. J., "Apparatus for Power Generation in Deep Seawater," U.S. Patent No. 3,994,134, issued 11-30-76.

23. U.S. Patent No. 3,994,134, issued 11-30-76.

24. Gardner, F., "Nitinol: Torque of the Town," CoEvolution Quarterly, Spring 1975, p. 68.

25. The Illustrated Science and Invention Encyclopedia, H.S. Stuttman Co., Inc., New York, 1978.

26. Oliver, T. K., et al., South Dakota School of Mines and Technology, Rapid City SD, "Clean Energy from Humid Air," paper presented at the AIAA 12th Thermophysics Conference, Albuquerque NM, 6-77.

27. Hidore, J., Physical Geography and Earth Systems, Scott Foresman and Co., 1974.

28. Oliver, T. K., et al., "Energy from Humid Air," Journal of Energy, Vol. 2, No. 1, January/February 1978. Assuming wind speed at 15.4 mph (6.89 m/s), average kinetic energy per kh of air processed is 23.7 joules; a kilogram of humid air with a dewpoint of 23.9° C. (75° F) contains 46,642 joules.

29. "Converting Solar Energy into Electricity: A Major Breakthrough?" Hearing before a Subcommittee of the Committee on Government Operations, U.S. House of Representatives, 6-11-76, U.S. Government Printing Office.

30. Dyson, F., "Energy in the Universe," in Energy and Power: A Scientific American Book, ed. by Scientific American, W. H. Freeman, San Francisco, 1971.

31. Barnea, J., editor, Important for the Future, UNITAR, Vol. III. No. 2, p. 4.

32. Steinhart, C., Steinhart, J., Energy Sources, Use and Role in Human Affairs, Duxbury Press, North Scituate MA, 1974, p. 17.

33. Freund, P. G. O., Hill, C. T., "A Possible Practical Application of Heavy Quark Physics," Nature, Vol. 276, No. 5685, 11-78; "Energy Storage—with Quarks," Science News, 12-2-78, p. 393.

Energy Conversion

1. Appell, H. R., et al., "Converting Organic Wastes to Fuel," U.S. Government Printing Office, Publication No. 1574, 1971.

2. Starr, C., "Energy and Power," Scientific American, 9-71, p. 38.

World Energy Data Chart

1. World Energy Supplies 1972–1976, U.N., New York, Statistical Papers, Series J, No. 21, 1978.

2. Energy: Global Prospects 1985–2000, Report of the Workshop on Alternative Energy Strategies, M.I.T., McGraw-Hill, New York, 1977.

3. This number is strictly a "ball park" estimate; it assumes a world total of 15,000 solar-heated houses (12,000 in U.S.), 2.5×10^6 solar hot water heaters and 200 industrial solar energy applications. Hot water heaters are assumed to produce 10 kwh per day of energy; solar heated houses and industrial solar applications 38 kwh per day. These figures do not take into account any of the work done for humanity by the sun in growing our food, heating our atmosphere, drying our clothes on the clothesline, etc. It only attempts to count the solar energy that is used consciously or volitionally by humanity. For a more thorough discussion on the limitations of this number see Baer, S., "The Clothesline Paradox," CoEvolution Quarterly, Winter 1975.

4. 1977 worldwide sales of photovoltaic cells, 750 kw; cumulative sales are about 2 megawatts. 2 megawatts installed capacity operating 30% of the time = 5.25×10^6 kwh.

5. Based on 7×10^6 biogas facilities in China (Reference No. 10, Bioconversion section) each producing $2,400 \times 10^6$ BTU/yr. and 50×10^3 biogas facilities in India (Reference No. 9, Bioconversion section) each producing $3,600 \times 10^6$ BTU/yr. (from Makhijani, A. and Poole, A., Energy and

Agricultures in the Third World, Ballinger, 1975, Charts 4–6, p. 109). The total figure is speculative because there is a large difference of reported BTU output of China's biogas facilities (two orders of magnitude difference). In Makhijani et al. reported output for Chinese biogas facilities is $8,400 \times 10^6$ BTU/yr.; Smil (in personal communication), put the figure closer to 11×10^6 BTU/yr. The figure used here ($2,400 \times 10^6$ BTU/yr.) is less than a third of the larger figure. Clouding the issue still further is the fact that the energy produced by the decentralized biogas facilities does not have to be delivered, thus saving energy, and that the leftover sludge is a valuable fertilizer, thus saving further energy.

6. Based on 20×10^6 BTU per cord of wood and 128 ft^3 per cord (m^3 = 35.31 ft^3); production/consumption figure from *World Energy Supplies 1972–1976*, Series J, No. 21, U.N., New York, 1978; BTU and ft^3 figures from Shelton, J., *The Woodburners Encyclopedia*, Crossroads Press, Vermont, 1977.

7. Estimate, based on 500,000 windmills (350,000 in U.S.; 300,000 pumping, 50,000 electric, Dorf, R.C., *Energy, Resources and Policy*, p. 22), average size 3 kw, operating 50% of the time, and three 2 megawatt power plants operating 60% of the time.

8. *McGraw-Hill Encyclopedia of Energy*, Lapedes, D. N., editor-in-chief, McGraw-Hill, New York, 1976.

9. This is not all used as an energy source; it includes hydrogen used in petrochemical processes, ammonia production, methanol production, and specialty gas uses, such as in steel fabrication, etc.

10. Hubbert, M. K., in *A National Fuels and Energy Policy Study*, U.S. 93rd Congress, 2nd Session, Senate Committee on Interior and Insular Affairs, Sec. No. 93-40 (95-75), 1974.

11. This is how much electricity is available by converting the uranium in light water reactors; not the total energy content of the uranium. This figure would be near $78,000 \times 10^{12}$ kwh for economically ($30/lb) recoverable uranium.

12. Total non-fossil carbon production per year for the world.

13. 50,000 thunderstorms/day; 100 thunderbolts/second; 3×10^6 kw/thunderbolt = $9,459 \times 10^{12}$ kwh/year.

14. Linden H. R., Parent, J. D., *Analysis of World Energy Supplies*, Institute of Gas Technology, Chicago, IL, 1975.

15. This is just heat content down to six miles beneath the Earth's surface.

16. This figure does not include the amount of energy available over the Antarctic or the oceans.

17. Assuming 35% recovery from proven world reserves of 658×10^9 bbl.

18. Hayes, D., *Rays of Hope: The Transition to a Post-Petroleum World*, W. W. Norton, New York, 1977. p. 44.

19. Figure from *McGraw-Hill Encyclopedia of Energy*, D. N. Lapedes, editor-in-chief, McGraw-Hill, New York, 1976, but only includes land-based generators.

20. One percent per year for 40 years; or 40% of the total 1976 energy consumption in 2016.

21. For twelve currents only.

22. Based on 100,000 25-megawatt wind turbines per year for 10 years operating 70% of the time and 1×10^6 6-kw wind turbines per year for 10 years operating 50% of the time.

23. Based on 100,000 25-megawatt wind turbines per year for 25 years operating 70% of the time and 1×10^6 6-kw wind turbines per year for 25 years operating 50% of the time.

24. Based on 100,000 25-megawatt wind turbines per year for 50 years operating 70% of the time and 1×10^6 6-kw wind turbines per year for 50 years operating 50% of the time (in "Energy Ideas," No. 11, 3-78, National Recreation and Park Association).

25. Fifty to two hundred megawatts estimated for laser fusion; 500 to 2,000 MW estimated for tokamak.

26. U.S. Department of the Interior, "Energy Perspectives," a presentation of major energy and energy-related data by Enzer, Dupree, and Miller, U. S. Government Printing Office, Washington, D.C., February 1975.

27. Ibid., three years minimum for Gulf of Mexico wells, twelve years maximum for Gulf of Alaska wells.

28. Lovins, A. B., *Soft Energy Paths: Towards a Durable Peace*, Ballinger, Boston MA, 1977, p. 141.

29. "Tidal Power May Now Make Sense," *Business Week*, 11-9-74, p. 115.

30. Bethe, H. A., "The Necessity for Fission Power," *Scientific American*, Vol. 234, No. 1, 1-76.

31. Metz, W. D., "Solar Thermal Energy: Bringing the Pieces Together," *Science*, Vol. 197, 8-12-77, p. 650; this figure assumes mass production; at $20 per kilowatt it is about the cost of an automobile engine.

32. Bergey, K. H., "Wind Power Potential for the U.S.," *Aware Magazine*, 10-71, Frank Eldridge of MITRE Corp. estimates that mass production of costs of wind energy systems will be $300 to $1,000 per installed kilowatt for small systems (1 to 10 kilowatt) and $600 to $1,000 for large systems.

33. Central Electricity Generating Board in Great Britain.

34. *Science*, Vol. 188, 12-5-77.

35. Japanese estimate for their 2-megawatt wave power facility under development. *The Renewable Energy Handbook*, Energy Probe, Univ. of Toronto, Toronto, Ontario, Canada.

36. Poole, A. D., "Questions on Long-Term Energy Trajectories," quoted in Lovins, A. B., *Soft Energy Paths: Towards a More Durable Peace*, Ballinger, Boston MA, 1977. This figure refers to human labor in a "dirt-poor pre-industrial society."

37. Wade, N., "Windmills: The Resurrection of an Ancient Energy Technology," *Science*, 6-7-74, pp. 1055–58.

38. Based on oil shale cost per bbl of $15–$50.

39. Hammond, A. L., "Photosynthetic Solar Energy: Rediscovering Biomass Fuels," *Science*, Vol. 197, 8-19-77.

40. Ibid., Energy Plantations $2.10/10^6 BTU, Agriculture Residues: $3.00/10^6 BTU; Marine Energy Plantation: $4.20/10^6 BTU.

41. For new 2+ million meter deep mine.

42. From Reference No. 1; for 200,000 bbl/day plant.

43. For largest size wind turbine mass-produced; Hewson, E. W., "Wind Power Potential in Selected Areas in Oregon," Technical Report, Department of Atmospheric Sciences, Oregon State Univ., Cowallis, 1973.

44. Based on Lorentz, E., *The Nature and Theory of the General Circulation of the Atmosphere*, World Meteorological Organization, Geneva, 9-67; if 175 is just land area and land area of world is one-fourth of world, total world including oceans is 700.

General References

Clark, W., *Energy for Survival: The Alternative to Extinction*, Doubleday, Garden City NY, 1974.

Daniels, F., *The Direct Use of the Sun's Energy*, Yale Press, New Haven CT, 1964.

Dittrich, R. F., Allon, K. D., *PSE&G Cogeneration Evaluation*, Report No. 36.76.12, Public Service Electric and Gas Co., Newark NJ, 1977.

Federal Power Commission, *National Power Survey*, U.S. Government Printing Office, Washington D.C., December 1971.

Hagler, H., *A Technical Overview of Cogeneration: The Hardware, the Industries, the Potential Development*, Report to the Division of Industrial Energy Conservation, Dept. of Energy, by Resource Planning Associates, Inc., Washington D.C., 1977.

Hammond, A. L., Metz, W. D., Maugh, T. H., II, *Energy and the Future*, AAAS, Washington D.C., 1973.

Hickel, W. J., "Geothermal Energy," NSF RA/N 73 003, NSF GI 34313, 93P, Univ. of Alaska, Anchorage, September 1972.

McGraw-Hill Encyclopedia of Energy, Lapedes, D. N., editor, McGraw-Hill, New York, 1976.

Nydick, S. E., Davis, J. P., Dunlay, J., Fam, S., Sakuja, R., *A Study of Inplant Electric Power Generation in the Chemical Petroleum Refining, and Paper and Pulp Industries*, Report to the Federal Energy Administration by Thermo Electron Corp., Waltham MA, 1976.

Proceedings, UN Conference on New Sources of Energy, Vols. 1–7. UN ST/ECA/49, 1970.

Science, AAAS, 4-19-74 and 2-10-78.

Scientific American, "Energy and Power," September 1971.

Thirring, H., *Energy for Man: Windmills to Nuclear Power*, (Repr. of 1958 ed.), Greenwood Press, Westport CT, 1968.

General Hydrogen References

Adt, R. R., Jr., Greenwell, H., and Swain, M. R., "The Hydrogen/Methanol-Air Breathing Automobile Engine," School of Engineering and Environmental Design, THEME, Univ. of Miami, April 1974.

Billings, R. E., "Hydrogen Storage in Automobiles Using Cryogenics and Metal Hydrides," Billings Energy Research Corp., Provo, Utah, THEME, University of Miami, April 1974.

Bockris, J. O'M., *Energy: The Solar-Hydrogen Alternative*, Adelaide, Australia, 1975.

Escher, W. J. D., and Hanson, J. A., "Ocean Based Solar-to-Hydrogen Energy Conversion Macro System," Escher Technology Associates, St. Johns, M.I., and Oceanic Inst., Walmanalo, Hawaii, THEME, University of Miami, April 1974.

Finegold, J. G., and Van Vorst, W. D., "Engine Performance with Gasoline and Hydrogen Fuels: A Comparative Study," UCLA, THEME, University of Miami, April 1974.

Graves, R. L., Hodgson, J. W., and Tennant, J. S., "Ammonia as a Hydrogen Carrier and Its Application in a Vehicle," University of Tennessee, Knoxville, Department of Mechanical and Aerospace Engineering, THEME, Univ. of Miami, April 1974.

Gregory, D. P., and Wurm, J., "Production and Distribution of Hydrogen as a Universal Fuel," *Proceedings, Seventh Intersociety Energy Conversion Engineering Conference*, San Diego, September 1972.

Heronemus, W. E., "Only Solar Energy Processes Will Bring Us to the Hydrogen Economy," University of Massachusetts, Amherst, THEME, Univ. of Miami, April 1974.

International Journal of Hydrogen Energy, Pergamon Press.

Jones, L. W., "Liquid Hydrogen as a Fuel for Motor Vehicles: A Comparison with Other Systems," *Proceedings, Seventh Intersociety Energy Conversion Engineering Conference*, San Diego, September 1972.

Kippenhan, C. J., and Corlett, R. C., "Hydrogen-Energy Storage for Electrical Utility Systems," *Proceedings, Seventh Intersociety Energy Conversion Engineering Conference*, San Diego, September 1972.

Korycinski, P. F., and Snow, D. B., "Hydrogen for the Subsonic Transport," NASA Langley Research Center, Hampton, VA, THEME, University of Miami, April 1974.

Laskin, J. B., "Electrolytic Hydrogen Generators," Teledyne Isotopes, Electrochemical Department, Timonium, MD, THEME, University of Miami, April 1974.

Martin, F. A., "The Safe Distribution and Handling of Hydrogen for Commercial Application," Union Carbide Corp., Linde Division, *Proceedings, Seventh Intersociety Energy Conversion Engineering Conference*, San Diego, September 1972.

Mingle, J. O., Eckhoff, N. D., and Rash, L. A., "An Engineering Assessment of the Hydrogen Economy," Kansas State University, Manhattan (Mingle and Eckhoff), and Beech Aircraft Corp., Wichita, THEME, University of Miami, April 1974.

Murray, R. G., Schoeppel, R. J., and Gray, C. L., "The Hydrogen Engine in Perspective," *Proceedings, Seventh Intersociety Energy Conversion Engineering Conference*, San Diego, September 1972.

Ohta, T., "Hydrogen Energy Systems as to Be Applied in Japan and the Key Technologies," Yokohama National University, Department of Electrical Engineering, THEME, University of Miami, April 1974.

Salzano, F. J., Cherniavsky, E. A., Isler, R. J., and Hoffman, K. C., "On the Role of Hydrogen in Electric Energy Storage," Brookhaven Nat'l Lab., Upton, NY, THEME, University of Miami, April 1974.

Sharer, J. C., and Pangborn, J. B., "Utilization of Hydrogen as an Appliance Fuel," Institute of Gas Technology, ITT Center, Chicago, THEME, University of Miami, March 20, 1974.

Veziroglu, T. N., and Basar, O., "Dynamics of a Universal Fuel System," University of Miami at Coral Gables, Department of Mechanical and Industrial Engineering, THEME, University of Miami, April 1974.

Weil, K. H., "The Hydrogen I. C. Engine—Its Origins and Future in the Emerging Energy-Transportation-Environment System," *Seventh Intersociety Energy Conversion Engineering Conference*, San Deigo, September 1972.

Winsche, W. E., Hoffman, K. C., and Salzano, F. J., "Economics of Hydrogen Fuel for Transportation and Other Residential Applications," Brookhaven Nat'l Lab, Upton, NY, *Proceedings, Seventh Intersociety Energy Conversion Engineering Conference*, San Diego, September 1972.

Wiswall, R. H., and Reilly, J. J., "Metal Hydrides for Energy Storage," *Proceedings, Seventh Intersociety Conversion Engineering Conference*, San Diego, September 1972.

Making the World Work

Global Strategies for a Regenerative Energy System

1 Introduction

Present-day energy planning can be categorized as the attempt to solve fifty-year global problems with four-year local solutions staffed with two-year personnel funded with one-year allocations that have been budgeted by bureaucrats who can not see more than six months, the next election, or vacation (whichever comes first) in advance and who know next to nothing about energy other than it does not, like other problems, seem to go away if ignored. It is the attempt to solve vast problems with half-vast solutions. The best that comes out of this process seems to be that which has the greatest probability to fail slowly. Clearly, something better is needed.

Global energy problems, because they are not going to go away, are sooner or later going to be dealt with, or already are being dealt with in part, on a national, regional, and local scale by present-day energy planners. Unfortunately, present-day energy planning is often crippled by a number of serious impediments. The usual frame of reference of energy planners (and by this term is meant anyone who attempts to solve energy related problems on any level of social organization) only allows for the recognition of a select few of the problems or factors impinging upon the overall energy situation. One of the more disabling of the impediments is a tragically short-sighted parochial focus. This narrowing of vision brought about by the cataract of overspecialization limits the energy planner to a bandwidth of sight that not only makes it exceedingly hard to find solutions but makes it even harder to see the actual scope of the problem situation. Another infirmity is the use of scarcity models. This methodological apoplexy assumes a fundamental inadequacy of life-support resources. It's either "us" or "them" that will get the last pieces of the last energy pies. The real tragedy of this morbid paranoia is that it leads to further complications that manifest themselves as arms buildup, brain-drain weapons research, limited skirmishes, resource waste, and a host of other disorders, which are, as is well documented, contagious and highly fatal. Needless to say, energy planning is made exceedingly more difficult in this neurotic context.

Another disabling disease is crisis-to-crisis management. In this ailment, the energy planner is comatose until awakened by the magic wand of catastrophic calamity. Upon awakening, a fantastic attempt is frantically made to deal with overwhelming emergencies with whatever is at hand or underhand in makeshift manner until the energy planner finds himself awakened once again by the magic wand of catastrophic calamity. In the scramble to avoid Armaggedon by way of cataclysm, "preventive medicine" is at best a vague dream of how things could be handled in a world of forty-hour days and no epidemics. Yet another crippler is linear thinking; that is, the more of more or the "if ten is good twenty is better" pathology. The energy planner sometimes catches this social disease through the use of a seriously flawed (to say the least; terminally incomplete to say a little more), method of attempting to *predict* the future by extrapolating from the past and then building energy facilities to meet the hypothetical demand postulated by the fabricated future trend. Major complications set in when the conditions that created the past—such as cheap oil, gas, electricity, etc.—are no longer present (and much less likely to be "future").

Another debilitation of present-day energy planning is the pervasive mind-parasite and its attendant side-effects that go under the name of reductionism. In this disease, the complex world system is reduced (usually by ignoring it) to the less complex energy system, which is further reduced to the major problem within the energy system, that is reduced yet further to the dominant symptom of that problem, which is reduced still further and further into the proposed solution or the study(s) needed to bring about the first stage of that "solution" or the action needed to bring about the actions needed to . . . ad infinitum, into the endless paper-shuffling horizon of energy bureaucracy. Complex systems are reduced to simple systems. Unfortunately, complex systems behave differently than simple systems, and solutions to the problems of simple systems do not solve the problems of complex systems; in fact, because they ignore the inter-relationships of systems, solutions to simple systems' problems tend to exacerbate the whole complex system's health or functioning. It is this type of thinking that leads utility company planners to reduce society's entire energy needs to electric power needs and to reduce that to increasing centralized electric power generating capacities and to reduce that to the building of nuclear power plants and that to the spending of millions of dollars in public relations and advertising to convince the public that nuclear power is safe; ignoring through such reductions society's need for liquid fuels (by far the largest energy need), conservation and the lowering of energy demand, decentralized energy production, environmental and social impacts of nuclear power, and the host of other complex interactions that the "simple" solution of increasing nuclear power has in the complex system of the world. In addition, such "solutions" are nearly always thought of as causal; that is, solution A will cause the elimination of problem B. There is a one-to-one correlation of solutions to problems and vice-versa. Problems are reduced until they fit "the" solution. And the effects of any solution are only looked at in regards to how they effect the isolated and reduced problem/symptom. How the solution affects the rest of the energy system and the rest of society and the rest of the world is not paid attention to. Just as modern medicine is criticized for not treating the whole person, energy planning should be criticized for not looking at and treating the whole system. Energy planners need to treat the whole system, not the disemboweled symptoms of their special interest versions of that system. "The" solution rarely exists to a problem of a complex system; a system of strategies within the constellation of problem areas of the whole needs to be developed.

One of the further complications that sets in as a consequence of the insidious mind parasites of reductionism is the cerebral palsy of the thinking process—mechanistic models. In this degenerative disease, the unfortunate energy planner sees the world (if he can see that far) and the energy system as a machine. He or she actually believes the energy system to be like a giant car engine, the smooth running of which depends upon the planner's screwdriver adjustment to the carburetor. If the energy planner could just *find* the carburetor, everything would be fine, he believes. Unfortunately, the screwdriver is also missing. Also unfortunately, no one has even seen the energy system's carburetor, though rumors have had it located in either Washington, D.C., Riyadh, or alive and well in Argentina. Instead of looking for a hypothetical carburetor's location in the back pages of the *National Enquirer,* the energy planner needs to develop non-mechanistic models of the world, the energy system, and his or her role in those living systems. One of the cures for reductionism is a shot of awe mingled with respect for the incredible biological complexity that greets the energy planner every time he opens his eyes and *sees* a gnat, bird, worm, tree, plant, Sun, and the interrelationships between these living systems. The energy system also deserves our respect as a complex living system, albeit almost infinitely less complex than our friend the gnat, but nevertheless livingly interconnected with everything else.

Last, "but not the least" is the most horribly disfiguring disease of them all, the cancer of ideology. This disease attacks without warning and mercilessly until it renders apparently normal brains into vegetable pulp that surrounds a tiny tape recorder with ten prerecorded stock answers to every and any question. One of the tell-tale signs of this disease is a prerecorded message which blames foreigners for all energy problems. Another message usually centers around "the good ol' days" and how good it used to be. Unfortunately, ideology rarely "solves" anything other than the reason for itself by furnishing the justifications for its own existence. Energy planners have to stop responding in terms of labels, pigeonholes, and ideologies but on the basis of what will work.

Any attempt to solve a problem at the expense of the planet will create a larger problem down the road.

All the above infirmities have helped the world limp into our present energy state. Now that we are an interdependent planetary society in the midst of the greatest transformation in our history in the midst of the most dangerous world human existence has yet witnessed, we need to do better.

To design a system that will produce the preferred energy state described at the end of the first section of this book demands a plan and planning process that is both comprehensive and anticipatory: that is global in perspective, local in the extent of its considerations, and with long and short range perspectives that are in tune with each other. Any design for solving the world's energy problems has to deal with the whole Earth, all the people on Earth and their needs, as well as the world's total resources and know-how. Without dealing with the larger, global system, the local systems' "solutions" degenerate into mere treatments of symptoms; "home remedies" for warning signs rather than cures or preventive medicine for systemic disorders. Dealing with parts will never solve a problem of the whole; the sum of local perspectives and solutions will never equal or come close to adequately dealing with a global problem. Conversely, the paradoxes, cul-de-sacs, and intractable situations of local problems and their proposed solutions can often be eliminated by the synergies of a global solution.

Ignoring the larger system is worse than sticking one's head in the sand; trying to treat the U.S.'s energy problems without treating the rest of the world's energy problems while the

Problems that do not fit neatly into categories tend not to be addressed.

U.S. is part of an interdependent global system is murderous at "best" and suicidal at "worst." In either case, the patient (and doctor) dies.

Any effective design needs to include three basic elements: 1) emergency measures to relieve the dangers of the immediate crisis; 2) conservation measures to reduce or stabilize demand on the energy system, to reduce energy loss, and to reduce environmental impact; and 3) long-term measures to make sure the problem does not recur. Additionally, any plan needs to allow for a maximum amount of participation in the fleshing out and evolution of the overall plan by the people who will be affected by the plan.

The fundamental energy problem is not what is popularly depicted in the newspapers or on television; the basic problem is not the price of gasoline or fuel oil or gas deregulation or insulation tax credits or oil imports or windfall profits or nuclear proliferation. It is also not nuclear vs. coal; or coal vs. solar; or "hard" vs. "soft" energy paths; or depletable vs. non-depletable energy sources. In some sense it is all of these, but most fundamentally it is quite simple: how do we get enough energy to *everyone* on Earth to meet all their life-support needs? How do we harness enough energy so that 100% of humanity—those alive tomorrow as well as today—have all the energy they need to have optimally functioning life-support facilities? How do we get the energy that is needed to feed, clothe, shelter, educate, furnish health care and recreational opportunities, and insure social well-being? And, how do we do this in the cleanest, safest, and quickest way? And, last but not least, how do we do all this, not at the expense, disadvantage, or coercion of anyone, but, ideally, through spontaneous cooperation?

The complexity of all the world's problems (not just the energy problem) demands a new level of awareness in our dealings with these problems. These days, nothing is solved by simply treating local symptoms. We are all part of a larger system, and it is the larger system that has the problems. The energy problem is inextricably tied to all aspects of development. It is important to realize that to tackle the energy problem is to deal with almost all the world's problems. There is no one answer because there is not one problem. We need to deal with all aspects of the global energy situation. This is not as impossible as it may first sound because we are *already* dealing with the entire energy system. What is needed is to emphasize certain things, to develop new structures to coordinate existing processes into synergetic totals, and to coordinate the whole system towards goals that are beneficial to all of humanity.

The overall guiding principles or values of development used here are: 1) to provide global humanity with a regenerative structure for its peaceful sustenance and growth; and 2) to maximize diversity, flexibility, and stability. The decision-making criteria listed in the first part of this book are the specific guidelines for development used in designing the strategies in this document. In designing anything it is best to start from scratch; to design the ideal system as if the existing system did not exist, and then work back from this preferred state to the present by identifying those things in the present system that could be changed or that need to be developed or phased out to bring about the preferred state. We need to start with what we want, not what we think the "invisible hand" is going to do next. The opposite approach, that of starting with the present-day, problematic situation, limits possibilities and prejudices the problem-solver to deal with the difficulties of the system instead of its possibilities. As was said earlier, dealing with the preferred state and working back is dealing with the behavior of the whole system; dealing strictly with the problems of the present system is dealing with parts.

Using the preferred state and the decision-making criteria of this book as guidelines for resolving the planet's energy problems leads to very different strategies than those currently being employed by the decison-makers of local, regional, national, and international governments and organizations of the world.

From the superficial look of things, the individual nation-states of the world did not learn very much in recent years about energy, its supply, its use, and the dangers of obtaining a still-growing quantity of that energy from one or two very finite sources in a world where more and more demands are being placed on fewer and fewer resources. The imminent crunch has been shouted from the proverbial rooftops as well as plastered on billboards, but seemingly nothing has changed. In spite of all the alarms and plans that were generated by the discovery that the United States' supply of oil could be drastically reduced by the other oil-producing nations, our dependence has not decreased. In fact, to the contrary, the U.S. in 1979 imports nearly 50% of its oil, up from 35% in 1973. Meanwhile, every major energy-producing country in the world is continuing to rely almost entirely on fossil and nuclear fuels as well as to consume more and more fuel. This does not say much for the intelligence or the vision of the political leaders of the world. Part of the reason for this terminal nonchalance or seeming madness is that a large segment of the public has not significantly experienced the deprivation that was experienced in other crisis situations—such as the 1930's depression, the dustbowl, or in wartime situations. Many people do not know what to believe about the nature of the problem, and so tend to continue their current habits. Except for the minor inconveniences caused by the 1973 oil embargo and the 1979 gasoline shortage, the energy crisis has been an "energy experts' crisis."

However, those inclined to take an optimistic view of the situation may point out that almost every major energy-using country has some research and development underway into alternative energy supplies and use, be it wave power in England, wind power in Denmark, methanol use in Brazil, or solar and wind power in the U.S.

The constellation of global energy policies and strategies that follow outline one vision of one possible future. They are paths that could be taken that would lead to a preferred state, or closer to a preferred state. *Every move or strategy is based on present day technology or organizational know-how.* Nothing new is needed to reach the preferred energy state other than our commitment. These paths deal primarily with large-scale strategic moves. It should be borne in mind that every strategy, no matter how grand, ends up in someone's backyard. Therefore, to implement any strategic plan it must be transformed (preferably by the "backyard" 's user) into specific, locally-

---

The short-term problem is not technology, but accelerating diffusion.
—M. Maidique, Harvard Business School

---

appropriate tactical plans that are considerate of all local social and cultural habits, beliefs, and patterns, as well as the meteorological, geological, hydrological, physiographical, limnological, soil, vegetation, wildlife, and land-use patterns. As the now familiar litany goes, "The technology should be made appropriate for the specific area, not vice-versa." The locality should not be forced into the mold of a technology that is not suitable for it. Short-range tactical local planning and problem-solving needs to become one, or in tune, with the long-range aspects of global-scale strategic planning. Future generations and/or the environment should not be traded off against the pressing needs of the moment; long-range and short-range sould be merged together instead of pitted against each other. In this context, education becomes the overall, ideal, comprehensive energy strategy and planning methodology. Decentralized decision-making in the context of planetery needs becomes the necessity of the time. "Locally appropriate planetary planning by the individual" is an apt slogan for the future.

The rapid changes called for and postulated in the following pages are not assumed to come about as a result of traditional top-down energy planning, although progressive leadership in this area would obviously be exceedingly helpful. Although it is always overlooked by top-down centralized planners, the rapid and extensive popular response to crisis as well as opportunity situations has been one of the strongest forces in the evolution of humanity. Populist, decentralized, almost spontaneous actions by large segments of populations have brought about rapid changes in society. The proliferation of citizen's band radios in 1975–78 is a recent case in point; an example of a decentralized response to a problem situation is the overwhelming public response to the water shortage in northern California in 1977, where voluntary and good-natured response went far beyond all government and industry expectations; an example of a similar populist response, not to a crisis or problem situation but to an opportunity can be found in the Homestead Act of 1862 and the resulting settlement and development of vast tracts of lands. The genius of the Homestead Act was that it set certain ground rules and parameters, and then let society go to it. The challenge to present-day energy planning is to devise energy Homestead Act analogues and variations and then stand back and let the people of the world do what they see is needed. Support for self-help initiative through loans and technical assistance would help, as well as the courage to try new approaches and new infrastructure arrangements not tied to the maintenance of what already exists. Of obvious crucial importance to this whole process is energy education. It can even be said that the most critical challenge facing the world energy system is how to get the most people, in the quickest amount of time, "in the know" or educated, about their energy life-support systems. Education, in a very real sense, is long and short range planning. The first steps in getting anything done rapidly on a societal level is to establish the credibility of the problem. Once that is done, "planners"—in the sense of the top-down, centralized, they-lead-you-follow notion—are not necessary; what is, though, is coordination and accurate information; lots of it.

Given these overall values and guidelines, the following outlines one design science plan that attempts to deal holistically with the world's energy problems.

2 Blueprints for Energy Survival

The following sections outline various individual policies, strategies, or tactics that comprise the overall global energy development plan. On the first page are four graphs which help illustrate the major strategic themes of the global energy strategies. The next section is simply a linear listing of major "moves," policies, or strategies, the third deals with energy uses and possible strategies and tactics for increasing the efficiency of our use of energy; the next section is a more detailed explication of specific tactics dealing primarily with the development of energy sources depicted graphically in terms of "first things first" and "what happens where"; and the last section is a global scenario that attempts to weave these various elements together and presents some of the possible long-range consequences of solving our regenerative energy needs.

All the themes, strategies, and tactics are to be viewed as *one* interdependent system of actions or developmental plan. Many of the separate strategies are known, some are new; some are bound to strike some people as radical, subversive, idealistic, impractical, or uneconomic. Taken alone, some are just that. Any simple action, by itself, could be very difficult to implement or accomplish without the other moves. The totality outlines or blueprints what needs to be done to reach the preferred global energy state that the book describes. The parts should be viewed as a synergetic sum; they have full meaning only in relation to the whole.

Strategic Themes

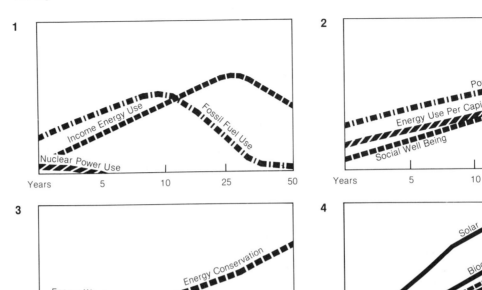

Strategic Moves

1. *Develop and Change Over to Income Energy Sources:* Extensively developing the Earth's non-depletable income energy sources would help eliminate the world's dependence on the dwindling supplies of fossil fuels, provide cleaner, more abundant, more equitably distributed, less inflationary, and, in the long-run, less expensive sources of energy. There should be an international agreement to develop *every* non-depletable energy source and technique for harnessing that source to the prototype stage.

2. *Diversify Energy Sources:* Diversifying the global energy system's primary and secondary energy sources would help stabilize the global system and insure a continuous energy supply. If one energy source went "out," others would be ready to pick up the increased load; adaptability and flexibility would increase.

3. *Decentralize Energy Sources:* By decentralizing energy sources wherever possible and appropriate, the control over those sources will increase for the user, fewer people will be dependent upon each power source, the increased economics of mass-production can be utilized, and the system itself will be more flexible, adaptable, and stable.

4. *Match Energy Needs with Energy Sources:* Matching the end-use energy need with the appropriate energy source will help eliminate the waste inherent in using high quality or high temperature energy sources—such as coal or nuclear reactors—to heat water to relatively low temperatures. Matching needs with sources also helps eliminate distribution losses, and helps diversify and decentralize much of our energy usage.

5. *Consciously Use Fossil Fuels for Transition to Non-depletable Energy Sources:* Phasing out fossil fuels in a planned way would allow them to be used in a regenerative way to power the transition to the Earth's "main engines"—the non-depletable energy sources—relatively quickly and without any major changes in lifestyle. Phasing out our use of the disappearing fossil fuels would allow humanity to be in control of the potentially volatile situation that will occur (if these fuels were not phased out) when everyone begins to scramble and fight over the last pieces of the fossil pie. Some of the remaining supply of the world's fossil fuels should be stored for future contingencies. Comprehensive plans for the use of the presently known remaining supply of the fossil fuel pie should be drawn up that will show how society can get from "here" to "there" without a decrease in quality of life.

6. *Institute Global Energy Utility:* Instituting a non-political, regionally-organized, non-profit, consumer-oriented Global Energy Utility that would have as its focus for existence meeting the regenerative energy needs of 100% of humanity, instead of a select nation-state minority or the highest paying customer, would greatly aid in stabilizing the global energy situation. In addition, the Global Energy Utility could be the agent for developing out-of-the-way energy sources in the world and would foster international cooperation in the process. It would monitor the global energy situation; disseminate educational materials; coordinate research, development, and training; compile, compare, and point out whole system contradictions of all the individual nation-state energy plans; and assist in the distribution of energy to areas that were energy short.

7. *Develop Income Energy Sources in Out-of-the-Way Areas:* Developing income energy sources in out-of-the-way areas such as the winds in Antarctica or ocean-based bioconversion plants, would increase the world's supplies of non-depletable energy, foster international cooperation, and also provide income to the Global Energy Utility to help it carry out its other activities.

8. *Phase Out Nuclear Power in Developed Regions:* Phasing out nuclear power in an orderly, controlled way would help stop nuclear weapons proliferation, stop the accumulation of nuclear fuel waste, decrease the risks of sabotage, decrease the demand on limited capital, decrease the burden we are placing on the future generations on Earth, and possibly decrease the increasing centralization of power sources.

9. *Undersell Nuclear Power in Developing Regions:* Purchasing all nuclear power plants, especially in developing countries, at profitable prices to the sellers and replacing the nuclear power capacity with non-depletable energy sources at little or no cost (preferably with locally-produced technologies) would help stop and retrench nuclear proliferation.

10. *Monitor the Global Energy Situation:* Monitoring the global energy situation on all levels of energy use as well as energy sources, distribution, potential needs, etc., would greatly aid in decision-making. Collecting and correlating the diverse sources of energy information, such as from industry, government, and research organizations, would help in obtaining reliable energy figures for planning.

11. *Develop Energy, Food, and Water Autonomous Houses and Communities:* Developing energy, food, and water-harvesting homes and communities that would be autonomous or even net producers of energy, food, and water would decrease the load these systems have on the existing energy system (29% of all energy consumed), increase the amount of available energy, help decentralize the energy system, help match energy sources to energy needs, and help eliminate energy loss due to distribution. Existing structures can be outfitted with heating/electricity production co-generation units that would also help accomplish the above.

12. *Conservation/Do More with Less:* Doing more with less throughout the energy system—developing new products that perform more per given energy input, and retrofitting the already existing energy-users, would save energy, reduce environmental impact, and

conserve energy and materials for other uses. A massive conservation effort—whereby every dwelling and industry is made as energy-efficient as possible—would not only save large amounts of energy but also provide jobs.

13. *Institute Two-Way Decentralized Utilities:* Developing utilities that would *buy* as well as sell electricity (and perhaps methane, alcohol, and hydrogen as well) to their customers, no matter how large or small the producer, would help reduce the need for more large, centralized power plants, provide a market and incentive for small-scale energy producers, provide a market for the homeowners' or communities' surplus energy, and, finally, increase the flexibility and adaptability of the utility, its customers, and the region.

14. *Decentralize Utilities/Rehabilitate Inner City:* By having the utility install small-scale heat and electricity producing cogeneration systems in dwellings that need new heating systems in the run-down sections of the urban environment, along with neighborhood biomass-processing methane-production facilities, and making loans available to the house owner to improve the dwelling and to establish biomass mini-farms on vacant lots, rooftops, and backyards, the products of which can be sold (to repay the loans) to those same methane production facilities, would generate employment, improve the general urban environment and living conditions of each rehabilitated house, provide inexpensive electricity to the utilities for distribution and sale, provide heat for the outfitted dwellings, and supply fuel for the operation of those electricity and heat-producing units.

15. *Centralize Energy Production and Use Wherever Appropriate:* Large-scale centralized and highly automated energy facilities in areas and for processes where this type of arrangement would be appropriate would increase energy supplies and take advantage of increased efficiencies of scale.

16. *Develop New Energy Carrier Mediums:* Developing new energy carrier mediums such as hydrogen, methane, or methanol to complement electricity and to store energy derived from the intermittent non-depletable sources would make the various income energy sources more viable than they already are. In addition, these new energy carriers would replace the existing liquid and gaseous hydrocarbons in existing power plants and make the entire energy system more flexible, adaptable, and stable.

17. *Develop Small-Scale Income Energy Sources:* Developing small-scale, non-depletable energy sources such as small-scale hydroelectric or tidal would increase the amount of energy available, increase the flexibility and adaptability of the global energy system, and help match appropriate energy sources to energy end-uses.

18. *Develop a Global Energy Game:* Developing computerized global energy "games" that would accurately model the world's energy situations and be widely accessible to policy maker, researcher, and student (even at the home computer and high school level) where alternative strategies for meeting the world's energy needs could be tested would help increase the efficiency of decisions by policy makers, the awareness of the energy problems by the public, and the amount of creative thinking that is brought to bear on the world's energy problems.

19. *Develop Urban Energy Systems:* Developing energy systems appropriately matched to urban energy needs and possibilities, such as district heating, power from urban waste, or biomass crops grown on urban vacant lots, or wind turbines powered by the aerodynamics of large buildings, would decrease energy waste and increase the availability of energy in the urban areas.

20. *Develop Rural Energy Systems:* Developing energy systems appropriately matched to rural energy needs and possibilities, such as feedlot methane production or wind-powered water pumps, would aid in making more energy available in rural areas and would decrease the amount of wasted energy.

21. *Evenly Divide the World's Energy Resources:* Collectivizing all the world's energy sources evenly, "globalizing" the world's energy resources—the coal, oil, gas, solar, wind, etc., energies of the *entire* planet, regardless of where they are located—would increase world interdependence, help decrease overconsumption, put a proper value on those resources in limited stock, increase world economic stability, and guarantee social well-being for even the poorest countries. In addition, within five years *every* country would have more energy per-capita after divison than before, because the solar energy potential of the oceans and other non-nation-state territories would be included in the global pie. Everyone would have more after division because the global pie being cut would be so much larger than the nation-state pie.

22. *Phase Out Inefficient Energy Consuming Artifacts:* Phasing out the most inefficient and wasteful of the world's energy consuming artifacts wherever appropriate to cultural and other dictates would be a constant ongoing activity that would continually save energy and decrease waste.

23. *Develop Safe, Durable, Low-Resource-Using, 100 m.p.g. Automobile:* By developing an automobile that was almost totally safe, that would last for at least 100,000 miles, used fewer resources in its manufacture, was not being constantly changed for cosmetic and sales purposes, was easily serviced by its owner, and which got *at least* 100 miles per gallon of gasoline is technologically possible, ecologically desirable, economically imperative, and would do more to wipe out the U.S.'s (and every other developed country's) balance of payments deficit than any other single move.

24. *Amass Energy Reserves:* Storing large quantities of different energy carriers, such as the fossil fuels, or hydrogen, methane, or methanol for an emergency back-up supply would increase the flexibility and stability of the world energy system.

25. *Develop Regenerative Energy Supplies and Tactics for Developing Regions:* Powering developing regions with fossil and nuclear fuels and then switching to income energy sources is not as economical or as efficient as starting off with regenerative energy supplies and tactics. Developing regenerative energy supplies and tactics right at the start will greatly benefit the regions under development in terms of employment, balance of payments, long-term expense, and pollution control, as well as regional stability and flexibility because the energy supply will not be subject to international economic and political vacillations.

26. *Stabilize Population:* Continuing to stabilize the world's population, both in the energy-starved and energy-obese areas of the planet, will relieve some of the pressures on the global energy system.

27. *Disseminate Energy Information:* Educating the people of a region about the planet's and their own unique energy resources, uses, and prospects will aid in bringing about the lifestyle changes that could reduce energy consumption that is caused by negligence and ignorance, and possibly increase energy availability through the decentralized production of energy. The more people who have more information the greater will be the stability and flexibility of the system.

28. *Provide Access to Energy Decisions:* With energy literacy comes response-ability. Mechanisms for decison-making and planning for the locality by the locality will produce an energy system more adapted to local requirements and possibilities, thereby reducing waste, providing more autonomy, and increasing the entire system's flexibility and stability.

29. *Regulate Multi-national Energy Corporations:* Regulating multi-national energy corporations through international guidelines, laws, incentives, and taxes would stop abuses of the present system and bring the vast resources and talents of these entities to bear on constructively meeting the energy needs of 100% of humanity. A 1% annual tax on profits would be used to finance the research and development of the Global Energy Utility.

30. *Regulate Exploitation of the Ocean's Energy Resources:* Regulating the exploitation of the world's ocean energy resources would decrease the chances of irreparably harming the oceans, and provide for the orderly development of these resources, thereby increasing energy availability.

31. *Develop New Energy Sources:* Developing new energy sources, both exotic and new variations of old sources, will help diversify the energy system as well as provide more energy and increase humanity's understanding of the Universe.

32. *Develop More Energy-Efficient Food Production, Industrial Processes, Transportation, and Utilities:* Developing more efficient food production, industrial processes, mass transportation, and utilities will decrease the amount of waste and increase the amount of energy available for other tasks as well as the stability of the global food, industrial, transportation, and utility systems.

33. *Provide Small-Scale Energy Development, Low or Interest-Free Loans:* Using the money that would have been spent on more large-scale power plants to provide low- or no-interest loans to homeowners and industry for energy efficiency improvement could save as much or more energy than the new power plant would provide.

34. *Establish Global Food Service:* Establishing a Global Food Service to deal with the food problems of the world—similar in intent and purpose to the Global Energy Utility—would greatly aid in solving the world's food problems and also assist in the solution of the world's energy problems. Once starvation and malnourishment are ended on Earth, food producers could continue to grow more and

more food or biomass because of their increasing skills, better food distribution, less food loss, new genetic improvements, etc. These surplus crops could be purchased (thereby helping to keep farmers solvent) for conversion to energy.

35. *Establish a World "Energy Fund for Food":* Instituting a special energy fund or reserve for use by regions that are too poor to pay for the energy needed to produce their own food would help enormously to end food shortages and starvation, as well as help stabilize population and aid overall development.

36. *Establish "Energy Extension Services":* Establishing "Energy Extension Service" educational and training outposts throughout the world, in a format similar to the U.S. Agricultural Extension Service, would greatly aid in the proliferation of new energy sources and energy conservation measures.

37. *Limit Military Expenditures:* Quantifying military energy expenditures and then not allowing them (by SALT agreement) to exceed that level for five years, thereafter reducing the total by 1% per year would help curtail the international war machine, help impose "sane" limits upon the development of ever more energy intensive ways of ending life, and foster basic research into "doing more with less" that could have far-reaching spin-off effects on the "home-front."

38. *Develop Extensive and Humane Life Support and Educational Systems throughout the World:* If the world's clean technological systems continue to produce more life support with less resources, if the world's population stabilizes as it is tending to, and if the world makes the transition to regenerative energy supplies, the world's *current* energy problems will be an item of history. The more options and choices that the billions of people, their energy, time, and intelligence have to grow into, the greater will be humanity's next step(s) in evolutionary development.

39. *Eliminate Human Drudgery/Increase Reinvestable Time:* When technological, social, and economic development can reduce mechanical, dull, or demeaning coerced human labor in all or part of the energy system, people will be free to pursue their own educational, cultural, or spiritual interests and this would lead to greater diversity, stability, and evolution.

Energy Uses and Abuses/ Conservation and Ephemeralization

The previous section lists major strategic moves or policies. The following section takes a little more detailed view of our energy situation by presenting possible strategies and tactics for saving energy in various categories of energy use.

There are numerous ways of looking at the energy problem and possible energy development strategies; just as you can slice an orange one way and see a cross-section, slicing it at a right angle to that will produce an entirely different view—a longitudinal one—that shows orange segments. It's the same orange, but offering entirely different views. The energy problem is not, unfortunately, as simple as an orange. There are more than two ways to slice it. The following section is another "slice" and the one after this is yet another. The latter deals with our energy problems primarily from the perspective of energy sources and their development while the former deals with energy from the viewpoint of energy uses and abuses; what is commonly referred to as "conservation." There are obvious overlaps, redundancies, and interrelationships between the sections. The tactics associated with each energy use and abuse category are presented as options for a wide variety of decision-makers: from farmer, home-owner, corporate and governmental policy-maker, to the general public.

As long as human beings continue to learn more, conservation and ephemeralization can be viewed as a non-depletable energy source. Conservation means, "to preserve from waste"; ephemeralization means, "doing more with less"; both terms are fundamental to design science. (See Decision-making Criteria/Economic Accounting Systems.) As know-how increases we will learn both what waste is and how it can be reduced. Today, we acknowledge certain types of waste because advances in technology have made us aware of more efficient ways of utilizing our resources. Advances in our understandings of who, what and where we are have also made us more aware of the limits we are operating within and what we want to do with the energy we have. As our awareness of what is waste and wasteful and our abilities to eliminate these progresses, conservation and ephemeralization can be viewed as an energy source.

There is enormous room for energy conservation and ephemeralization in all realms of energy use. As categorized in the charts on page 29, energy is used by industry, transportation, commerce, residences, and the utilities. Within each of these areas, there are many opportunities for reducing waste. In North America, there is such a multitude of opportunities for conservation that recent analysis[1,2,3] discloses the possibility of furnishing the same standard of living, or higher, with roughly one-third to one-half less energy than is currently squandered. In Sweden and West Germany, the standard of living is roughly equivalent to that found in North America, but energy consumption per person is about one-half. In New Zealand, it is half again what it is in Sweden and West Germany. Even the U.S. consumed half as much electrical energy in 1963 as it does today. Clearly, the U.S. standard of living has not doubled since 1963.

Advantages:
1. Resources can do more than they could before.
2. Less waste means less pollution; less environmental disruption.
3. The more functions performed per given unit of energy, the more energy that is available to perform other functions.
4. Conservation provides more jobs and is safer than any other source of energy.

Disadvantage:
Progressive change in technology can produce dislocation and necessitate re-education for those who are involved in the old technology.

Shelter
One major area for both progressive ephemeralization and conservation is in new shelter design and in retro-fitting existing shelters. Depending on the criteria for judging, between one and three billion people on Earth are without adequate shelter. In developed regions, shelter or the residential sector consumes 29% of all energy. All shelters can

---

Energy saved is more valuable than energy produced.

The most compelling factor for conservation is the cost of not conserving.

---

easily be built to use less than half the amount of energy they usually consume;[4] many shelters can be built to be energy autonomous and others, in ideal geophysical locations, could actually be net energy producers, that is, produce more than they use. These advances over present shelter systems can be accomplished through the use of all or some of the following: proper insulation; storm windows; weather stripping; prevention of excessive ventilation; size and placement of windows; use of natural illumination; roof overhangs; exterior finishes; building orientation to the Sun's angle and path across the sky and prevailing wind direction; landscaping; underground construction; reduction of non-essential lighting and energy-intensive building materials; establishing minimum energy and shelter performance standards and having all houses "labeled" with such (as well as energy requirements for heating, cooling, and major appliances) on the outside and inside; having "wattmeters" installed in every kitchen which would show instantaneously energy consumption throughout the house; developing zoning regulations to insure solar and wind rights; eliminating the more than three thousand obsolete and conflicting building codes with updated codes that allow and encourage energy efficiency; creating mortgage and tax benefits for energy efficiency; coordinating individual house, neighborhood, and community energy systems; adding on solar water and space heating and cooling equipment; adding on wind energy systems for electrical or pressurized-air needs; substituting composting toilets for the

> *More than 272 million barrels of oil a year could be saved by use of dual-purpose appliances.*[5]

five-gallon-flush sewer-connected toilet; adding on woodburning heating systems; lowering the water-heater temperature; utilizing dual purpose appliances such as a combination furnace and hot water heater; and using efficient and well maintained equipment.

Still other innovations hold the possibilities of not only reducing shelter energy needs, but of turning the residential sector of the economy into an energy producer. For example, the Fiat Unit mentioned in the Cogeneration Section, which consists of a heat exchanger, a 15-kw electric generator, and one of Fiat's standard small engines (which can be run on methane, hydrogen gasoline, natural gas, methanol, LPG, or biogas), is designed to heat a house with the heat from the running engine (just as the normal furnace will heat your house from the "running" furnace) but instead of just producing heat with the combustion of the fuel as does the furnace, the engine also produces electricity by powering an electric generator. Hundreds of thousands of these units throughout a region would decentralize the utility and reduce or eliminate the need for more large power plants. Because it can take advantage of the economics of mass production and is built from proven components taken from the automobile industry the unit price is such that 100,000 of the units would out perform one 1,500 megawatt nuclear power plant at less than one third the cost ($286 million vs. $1 billion).[6] In addition to being less expensive than a single large centralized facility, the approximately 100,000 decentralized electric and heat producing units of Fiat that would replace the centralized unit could all be installed in less than two years (compared with approximately ten years for a nuclear power plant), start recovering capital almost immediately after investment, and produce more jobs than a centralized facility. Because both heat and power are being utilized, the units are 90% efficient; because they are modular, efficiency could be improved still further. An estimated 40% savings in primary energy consumption can be achieved through the use of these systems,[6] that is, 40% less fuel would be consumed by the use of cogeneration systems to produce both heat and electricity than to have electricity produced by central generators and heat produced on site separately.

A related strategy would be to decentralize the electric utility and rehabilitate the urban environment simultaneously in one synergetic move. Instead of building a large central power plant for $1 billion the utility could install the same electrical generating capacity for less cost in urban dwellings that need new heating systems. In addition, the utility could subsidize the start up of urban farms for biomass production that would be turned into methane to run the small cogeneration units. By building methane production facilities and guaranteeing a market for biomass, large numbers of relatively unskilled people could be employed in growing crops in the urban environment. For example, the Bronx section of New York City has over 700 acres of vacant land that could be used to produce biomass. Additionally, there are backyards, rooftops, median strips, etc. that could be used. Such a strategy would aid in the redevelopment of the inner cities of the world and provide employment for large numbers of people, either in the production of biomass and methane or in the installation and servicing of

> *If all new houses in the next twelve years were built as efficiently as we know how to build them, we would save about as much energy as we expect to get from the Alaskan North Slope.*[2]

the cogeneration systems.

Similarly, a home- or neighborhood-size hydrogen fuel cell could reduce or eliminate the need for more power plants.

Large buildings could also be equipped with their own cogeneration or fuel cell systems. In addition, power could be produced from the sewage of a large building. Neighborhoods and communities can have their energy systems interlinked and be even more efficient than single-dwelling units. Neighborhood solar collectors and heat storage facilities and district heating are two examples of increased efficiencies obtained at this level. Community-sized solar systems would allow seasonal storage of solar energy—something that is presently impractical for individual dwellings.[7] By building new communities from the perspective of energy efficiency, towns could be made to function just as well or better while consuming only a fraction of the energy a traditional city consumes. A well-designed new town requires less than half as much gasoline to get around than does a sprawl community; in addition, it has been estimated that 16% of urban car trips in the U.S. could be replaced by telecommunications. A few energy-saving designs proposed through the years by Buckminster Fuller include large domes over cities and smaller domes ventilated in such a way that they are in effect "chilling machines." Such shelters would eliminate the enormous energy demand of air conditioners. One example of the energy savings that could be

Low pressure cone above dome draws down a central cold air core countering major rising thermal spiral column. Interior motion is an involuting torus (doughnut shape).

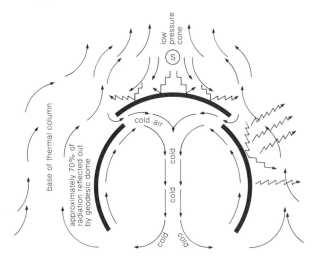

achieved by large domes over cities is a proposed dome over Manhattan. The skyscrapers beneath the dome are in fact high-rise "spikes" similar to the vents of a huge air-cooled engine; the total surface exposed to the weather by all the buildings under the dome is eighty-five times the surface of the dome covering all the structures. Heating and cooling the entire domed area would take less energy than heating and cooling each of the skyscrapers separately. Fuller estimated that saving the cost of snow removal alone for ten years would pay for the dome.

Innovative design for financing home energy improvements includes bank and savings and loan funds that would be loaned out and then repaid with the savings from the reduction or elimination of electricity and/or fuel bills or the recasting of the original home mortgage to a longer time period to keep the new monthly payments as near as possible to the original payments. Federal and state backed "solar banks" could also loan out funds in this manner.

Transporation

Transportation also offers a host of opportunities for conserving energy. In developed regions, the transportation system uses about 26% of the total energy supply. The efficiency of the transportation system can be increased through the proper matching of the task needed to be performed with the appropriate transport mode that will perform it most efficiently. For example, by diverting long-haul trucks to rail, goods can be moved long distances at lower energy costs (700 vs. 3,000 BTU/ton); by diverting short-haul passenger transport from air to rail (a 100-mile flight consumes 2.5 times as much fuel per passenger-mile as a 1,000-mile flight); and by forming car/van pools for urban commuters. These require less fuel (four people in one car/van require one-fourth as much fuel as four people driving separate cars), and cause less pollution. In addition, limiting engine size, car weight, fuel allotments, ultimate car speed; developing more efficient engines such as the Stirling cycle or gas turbine; establishing higher fuel efficiency performance standards; raising fuel prices to their replacement value; eliminating the stock of older, heavier cars that get less than ten mpg; eliminating taxes on mass transit; decreasing mass transit fares

---

There is no definable limit to conservation, at least not until we approach both thermodynamic limits and the exhaustion of our ingenuity to modify and refine tasks.[8]

---

through subsidization or providing free public transportation in cities; redesigning mass transit to provide personal freight accommodations and to improve reliability/safety, privacy, and attractiveness; developing personal rapid transit such as flexible bus lines with smaller buses; instituting exclusive bus/car pool lanes for commuting; electrifying short-distance cars and buses; developing pedestrian and bicycle transitways; using 1% of the Highway Trust Fund for bicycle paths; building more "park and ride" centers where commuters would park on outskirts of the city and ride mass transit into the city; taxing peak-hour traffic; reducing parking spaces or banning cars in congested areas; enforcing speed limits; re-establishing sailing clipperships and lighter-than-air ships for bulk transport (sailing ships consume only 5–10% as much fuel as conventional ships); reducing airplane cruise speeds, increasing cruise altitudes, and eliminating delays (which account for 4.2% of fuel consumption in airplanes[9]); using hydrogen, methane, or methanol as a fuel thereby reducing pollution; taxing pleasure craft and their fuels; pricing fuel according to vehicle miles per gallon; and reducing and limiting military transport would all help conserve energy. In addition, cities could be redesigned to eliminate the need and use of private cars and the urban car could be redesigned. For example, one design calls for a car for two people, with the two passengers back-to-back with the engine in the middle. This arrangement allows cars to be built so narrow that the normal highway lane can now become two lanes, thereby doubling highway capacity and halving traffic and parking congestion. Another problem of vehicles is their overall engine efficiency; for example, the automobile engine itself runs at 10–20% efficiency, but should it be running all the time? Buckminster Fuller's metaphor points out the problem very well: "In America, at all times, two million cars are halted at stoplights with their engines running. With each car engine producing about 10 horsepower while at the stoplight, this means that there is roughly the equivalent of 20 million horses jumping up and down and going nowhere."[10]

One solution to this wasted automotive energy that Fuller proposed in 1943 in his automobile design for Henry Kaiser was to have each wheel of his three-wheel car individually powered by separate air-cooled engines, each coupled to its own wheel by variable fluid drive. The engines were always run at optimum speed; the speed of the car was controlled by varying the quantity of fluid in the coupling. The engines were of low (15 to 20) horsepower with only one engine required to maintain cruising speed, so once starting inertia was overcome the car could average 60 to 80 miles per gallon of gas.

Most present United States automobiles are twice as big and half as efficient in energy use (miles per gallon) as some smaller Japanese and European cars. Because fully half of United States crude oil is consumed by motor vehicles, such a redesign of motor vehicles would result in huge energy savings. And this saving in energy consumption during actual use of the automobile is only part of the story. A smaller car would not only save energy in use because it weighed less, but also because it required less energy and other resources to be produced. Less steel used per car means less energy expended in the mining, refining, distribution, production, and recycling of the car, and the more steel available for other uses. Another energy-saving factor is having fewer model changes. This would tend to decrease energy use by discouraging rapid turnover of cars. Nearly 8 million BTU's (2,400 kwh) per car could be saved in production cost alone by reducing the average 3,600-pound American automobile to 2,700 pounds.[11] Nearly 5,000 kwh (17 million BTU's) could be saved by reducing auto weight to 1,800 pounds.

There is no technological impediment to mass producing by 1990 an automobile that gets 100 or more miles per gallon, will last for a minimum of 100,000 miles, is almost 100% safe for its occupants, can be slept in by two people, and is easily maintained by its owner. The idea of legislating that cars get 30 m.p.g. by 1990 is ludicrous in light of present-day capabilities. Through such an automotive redesign all transport needs could be met with domestic oil or biomass-derived alcohol. An option for such a car could be a still for producing alcohol from waste biomass for running the car.

Another energy conserving transportation strategy is the immediate implementation of gasoline rationing.[12] As one brilliant analysis points out, U.S. petroleum demand will continue to grow (and balance of payments, weakening of the dollar, inflation, etc. will continue to get worse) because there is no natural or structural self-preservation mechanism in the oil demand system similar to the structural limits or self-preservation mechanism inherent in the electric grid, telephone system, or in any living system.[12] For example, when the phone system is overloaded, you get a busy signal; when the electric grid is overloaded it sheds loads or shuts down to save the system; when any living system's survival is threatened, built-in self-preservation mechanisms take over, and depending on the nature of the threat, program behaviors such as fight, flight, immunilogical response, etc. When the oil demand system is overloaded it continues to meet demand, in fact gleefully does so, as long as oil customers continue to have money to spend, regardless of the impact such action has on the larger system of the national or international economic, environmental, or cultural well-being. "In effect, the petroleum subsystem will survive by parasitically debilitating the larger national (and international) system."[12]

Immediate gas and diesel fuel rationing would add a self-preserving limit to the oil demand system, thereby enabling the consumption of oil (and all its resulting problems) to be controlled. The most equitable rationing proposal yet devised would distribute equal gasoline entitlements to everyone over 18 (regardless of whether they owned a car or had an operators license). These individuals would then "exchange their gas rations for fuel or sell them on the open market to motorists who wanted to consume more than their fair share of the nation's gasoline supplies and were willing to pay for the privilege."[12] One rationale for this scheme is that it does not punish those people who are not consuming gasoline. By giving gasoline allotments to just licensed drivers or registered car owners, as past gasoline rationing schemes proposed, the rationing system inadvertently punished those who didn't drive—those who conserved energy by walking, bicycling, or taking mass transit—and rewarded those who consumed energy. By giving equal shares to everyone (two gallons per day has been proposed for the U.S.), everyone is rewarded for conserving energy. Rationing gas would act to bring about many desirable energy-conserving actions, such as car and van pooling, use of public transportation, cars in better mechanical adjustment, and bicycling and walking.

Industry

Industry consumes about 40% of the total amount of energy in an industrialized region. There are many well-documented ways of increasing energy efficiency in industry without decreasing output. Such tactics as insulation of skid rails in steel production, furnace insulation, combustion control, burner positioning, heat-recovery equipment to recapture some of the heat lost in stack gases, preheating combustion air (1,000° F preheating reduces the total fuel consumption 30%), or on-line computer controls (in the steel industry such controls executed a carefully devised operation, which resulted in a 25% reduction in fuel consumption per ton of production and a 12% increase in the plant's rate of production), and continuous stack-gas analysis (5–30% savings of fuel consumed) can all be adapted to present-day industrial installations with estimated savings of more than 500,000 barrels of oil per day (3 trillion kwh per year).

Besides improvements in existing plants or in new plants of conventional design, it is also possible and highly desirable to implement currently known new plant designs to improve industrial fuel efficiency in areas such as glass and cement production (a cement kiln of European design is 55% more efficient than kilns now used in the United States) and in combining the industrial production of process steam (17% of the total fuel consumption in the United States) with electric power generators (see Cogeneration Section). In addition, heat left over from the industrial process could be used for district heating or cooling or heating surrounding agricultural land to increase productivity.

Tangential to energy conservation and overall energy efficiency is the improvement of fossil fuel mining techniques. Current recovery of oil is about 30% efficient; for every 100 barrels of potentially available oil in the Earth, current practice leaves 70 barrels still in the Earth. Increasing this percentage to 60% would double our petroleum supply.

In the global energy development strategy all these methods would, of course, be used. It is assumed that the overall present-day energy-use efficiency rates will be greatly improved because many present-day energy converters are old non-state-of-the-art equipment. We are assuming that all new equipment, facilities, etc., that will have to be built to realize this energy strategy will be built utilizing the highest possible energy conversion efficiencies. What this means in terms of the energy development strategy is illustrated below.

In 1977, 83×10^{12} kwh produced all the world's goods and services at an overall energy conversion efficiency of about 5%. New equipment and facilities built after this are assumed to operate at an overall efficiency of 10%. This has the effect of producing much more goods and services than would merely doubling the energy consumption of our present-day equipment. It is also assumed that some of the equipment operating today will no longer be operative in ten years and that the replacements will be state-of-the-art equipment. The following illustration from the aluminum industry effectively makes this point.

One way to produce the income energy-harnessing artifacts that will be needed for the overall global energy development strategy in a relatively short period of time (again, testing technological limit conditions) is to tap the vast production capabilities of the North American and other developed regions automotive and aerospace industries.

If a moratorium were put on all new automobile production for a few years, as was done in World War II, all the production tools could be used for producing income energy-harnessing artifacts and spare parts for existing cars. The same amount of materials, energy, and manpower used to turn out nearly eleven million one-to-two-ton automobiles each year could instead produce thousands of millions of energy-harnessing artifacts. Instead of consuming vast quantities of energy (approximately 111 million barrels of petroleum

---

Cogeneration and more efficient use of electricity could together reduce U.S. use of electricity by a third and central station generation by 60%.[2]

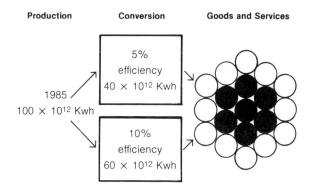

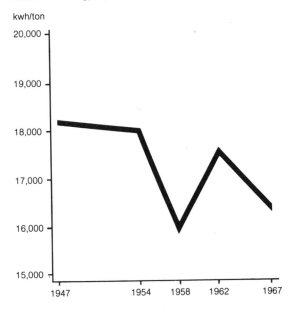

Unit Energy Consumption in Aluminum Reduction

Source: Rand Corporation, *Interim Report: The Growing Demand for Energy*, April 1971.

NOTE: 1958 was a low production year. Consequently, *only the most efficient plants were used*, which accounts for the low figure shown

Aggregate U.S. Data, *Census of Manufacturers*

per year /180 billion kwh per year) for the duration of thier lifespan as do automobiles, the energy-harnessing artifacts would be producers of energy. Besides this increase in energy availability from energy-harnessing artifacts and the potential decrease in energy consumption through fewer high-energy-consuming automobiles, there would be other benefits from such a switch-over of end-products. A moratorium would give automobile designers and manufacturers the time to come up with the energy-efficient (100 + m.p.g.), durable, safe, and ecologically acceptable alternative to the automobile mentioned in the previous section. Ten to thirty miles to a gallon of gasoline is virtually a crime in light of current engineering know-how. (The old record for most miles per gallon—376—set by a modified 1959 General Motors Opel in 1974 by Shell engineers at their Wood River, Illinois, laboratory was shattered in 1977 by some teenage apprentice workers at the Mercedes-Benz factory in Stuttgart, West Germany, whose diesel-powered vehicle gets 1,585 miles per gallon!)

It should be an industrial ethic that *all* industries strive to manufacture enough products that produce more energy over their entire life cycle than they consume in their production. The strategy for the automotive industry described above, for example, could possibly turn that industry into a net energy producer rather than consumer.

In addition to the already-mentioned tactics, energy savings could accrue through using solar, wind, and geothermal energies to provide power, heat, and mechanical requirements for operating machinery; instituting around-the-clock operations to utilize machinery continuously; eliminating idle time and power surges to start machinery as well as relieve commuter congestion; designing products to last and to be recyclable; setting plant and product energy performance standards; providing tax credits for any energy efficiency improvement; reducing energy-intensive advertising; using less packaging and even less energy-intensive packaging; reducing model changes and unnecessary redundancy of products; reducing unnecessary illumination of hallways, "dead space," and exteriors; using more natural ventilation, reducing air-conditioning, and using remote thermal scanning to spot heat losses.[13] This last item could be done from airplane or satellite and could disclose, through thermal infrared scans, heat losses through building roofs and heating system distribution lines.

Food

Food production is the planet's largest sector of energy use. Solar energy in particular is directly responsible for nearly all our food through the solar energy trap of photosynthesis. In addition, the global food system uses more non-solar energy than any other single activity. Meeting the energy needs of the global food system is of paramount importance because of the threats of starvation, malnourishment, sickness, and brain damage that could result if the world does not produce and distribute all the food that will insure that everyone on Earth has an adequate diet.[14] As with all energy users, food production could be much more efficient. There are numerous ways to produce more food using less energy. The possibility even exists for the world agricultural system to make a major contribution to solving the world's energy problems by being a large producer of energy through the growing of "energy crops."[14]

The overall efficiency of the global food system could be increased by using more efficient tractors; reducing the use of energy-intensive chemical fertilizers and pesticides; using more animal waste, compost, green manure, sewage, sludge, crop rotation, and biological control of pests; reducing unnecessary tillage (saving as much as 100×10^{12} BTU/29×10^9 kwh per year);

substituting grazing for animal feed (500×10^6 BTU/146×10^3 kwh could be saved from converting just 20% of present feed grain acreage to pastures for direct grazing by livestock); reducing energy-intensive food processing and packaging (e.g., returnable bottles, paper instead of aluminum foil, etc.); listing energy content (along with food value) of each food product; substituting hydrogen as a feedstock for nitrogen fertilizer manufacture; increasing local, decentralized food production, thereby reducing processing and transportation costs; integrating livestock feedlots with feed-producing areas so as to recycle wastes and save transportation; developing new machinery ideally suited for polyculture; developing new food sources that are more energy efficient in terms of photosynthesis, need less fertilizer, cultivation, processing, storing, and are easier to harvest; integrating food (or energy crop) production with power production to tap the CO_2 produced by combustion thereby improving crop productivity; reducing consumer insistance on "perfect" non-blemished fruit and vegetables thereby reducing waste; reducing meat, fat, and other energy-intensive food consumption; increasing consumption of energy-efficient foods such as raw fruits and vegetables; using solar energy or methane produced from animal or plant waste for drying crops (instead of propane or natural gas); and using wind or solar energy for electricity needs or for providing mechanical power for such things as pumping water.

From Sub-City to Electric City

Utilities consume about 26% of the total primary energy in a highly developed industrialized area. Until recently no one questioned the prevailing notion that "what was good for utilities was good for the country"; that an all-electric world was anything but desirable; that electricity was the end-all and be-all of the forms of energy. The reasons for this tacit assumption were related to the fact that electricity itself is clean and efficient when used for what it was originally used for (lighting), is the quickest way of getting energy from one place to another (186,000 miles per second), and could be used for a wide variety of tasks. Utilities grew because of electricity's advantages over other forms of energy and because of the decreased costs of large-scale electricity production, "increased reliability through interconnection, sharing of capacity among non-simultaneous users, centralized delivery of primary fuel, ease of substituting primary fuels without retrofitting many small conversion systems, localization and hence simplified management of residuals and other side effects, ability to use and finance the best high technologies available, ease of attracting and supporting the specialized maintenance cadre that such systems require, and convenience for the end user, who need merely pay for the delivered energy, purchased as a service without necessarily becoming involved in the details of its conversion."[2]

Utilities are being critically examined today because some of the above are either no longer true in all cases or they have been superseded by new understandings and values. The major deficiency of electricity and centralized utilities is the inefficiency associated with it. Fifty to seventy-five per cent of the primary energy is lost in the initial conversion from coal, oil, gas, or uranium to electricity. Of the electricity generated, another 10–30% is lost in distribution. For the small power user, transmission and distribution in fact accounts for nearly 70% of the cost of electricity that is generated at large power stations.[2] Other disadvantages include the huge capital requirements for large-scale power plants (1,000 MW nuclear reactors cost more than $1 billion); high incidence of "down time" of large power plants; high operation and maintenance costs; diseconomies of large-scale systems such as the increased need for adequate backup systems; lack of flexibility and diversity; long lead times associated with bringing any new system on-line; and the less tangible, but real, disadvantages to the society-at-large by having power sources centralized, remote, and un-understandable or controllable by the people they affect. In addition, such sources tend to "concentrate political and economic power, encourage urbanization, distort political structures and social priorities, increase bureaucratization and alienation, compromise professional ethics, [inhibit] greater distributional equity within and among nations, inequitably divorce costs from benefits, enhance vulnerability and the paramilitarization of civilian life, introduce major economic and social risks, reinforce current trends toward centrifugal politics . . . and nurture, even require, elitist technocracy whose exercise erodes the legitimacy of democratic government."[2]

> *All we want our customers to know is that when they press the switch they will have power. We'll worry about the rest.—Donald C. Cook, President, American Electrical Power*

> *... although a particular energy conservation measure may save only one unit of electricity, this indicates a savings of over three units of primary energy required to generate the electricity.[8]*

Some of the strategies that have been proposed for dealing with these issues and with increasing the overall efficiency of the utilities are: deregulating the price of primary fuels; making energy prices equal what it will cost to replace them with non-fossil or non-nuclear fuels; life cycle costing; cogeneration; two-way decentralized utilities (see Shelter section); public non-profit control of utilities; regulations to promote less energy-intensive systems; tax incentives for income energy systems; removal of government subsidies to utilities (an estimated $10 billion per year in just the U.S.)[2]; strict enforcement of anti-trust laws; flat or inverted utility rate structures rather than discounts to large consumers; and power generation co-ops on the neighborhood level. Other strategies involve utilities which sell or rent wind and solar energy systems to their customers (instead of building more power plants); utilities that install insulation and other energy-conserving equipment; utilities that make loans for purchasing these facilities; and utilities that only exist as storage and distribution systems, not power-producing systems (power is purchased from decentralized, home, or neighborhood production units similar to the way the telephone communications utility does not generate the signals that go over the telephone lines but distributes the signals of the users).[15] Giving incentives to small private householders to install their own power generation equipment would greatly accelerate such a move. Such incentives as paying more for power purchased from the decentralized source would interest more people to invest their own capital in such facilities. Substituting income energy production units (hydroelectric, forests of windmills, solar electric, methane from urban wastes, etc.) for capital energy units, reducing and eliminating nuclear power, phasing out fossil fuels, diversifying energy sources, charging higher rates at peak-use times, matching electricity to what it is ideally suited for (and eliminating or reducing its inefficient use in such things as electrical resistance heating) thereby reducing electricity need to 5% (instead of 13%) of all U.S. energy end-use, meeting this 5% need with hydroelectric and cogeneration,[2] installing individualized electricity-use meters for every apartment in urban complexes so the apartment-dweller knows how much electricity is being used, and developing a global energy utility to help coordinate the global aspects of electricity production, research, development, standardization, safety, end-use, and income energy applications are also recommended.

Global Energy

In addition to the above areas of energy use and abuse, that is, in shelter, transportation, industry and commerce, and agriculture and utilities, there is another realm where appropriate strategies could increase efficiency of the global energy system: the international level. Strategies on this level include the already mentioned development of a global energy utility, plus the international development of

> *By mass-producing wind turbines and solar collectors, the price drops drastically. Utilities could place the type of order (large enough) for the cost for these type of energy systems to drop so far that they are cheaper than any other way of obtaining energy and everyone could afford them.*

Chart 1.

Energy Source	Before Globalization U.S. Per Capita Energy Supply in 10^6 kwh	After Globalization U.S. Per Capita Energy Supply in 10^6 kwh	World Per Capita Energy Supply in given units	World Per Capita Energy Supply in 10^6 kwh
Coal	51.3	15.2	1,900 mt	15.2
Petroleum	1.9	.8	500 bbl	.8
Natural Gas	1.6	.967	3.2×10^6 ft^3	.967
Uranium Ore	.4	.043	852 g	.043
Falling Water	.1	.025	.025	.025
Geothermal	.2	.125	.125	.125
Solar	1.47	17.7	17.7×10^6 kwh	17.7
Wind	.37	7.5		7.5
Tides	.0004	.009		.009
Waves	.003	.065		.065
Ocean Currents	.00002	.0004		.0004
Temp. Differential	.27	5.5		5.5
Total	57.613	47.93		47.93

energy resources; building an international fossil fuel reserve; instituting a global ban on nuclear power; creating world nuclear-waste storage facilities for the nuclear waste the world is already straddled with; establishing an international environmental court, an international energy welfare program (similar to the U.S. food stamp program), a world "energy-fund-for-food" that would allow energy-short and food-short regions to draw upon the energy stocks to help produce their food needs; geographical matching of energy production capabilities to energy produced (e.g., solar energy in a desert not nuclear); international prize(s) for most-energy-efficient designs and systems; the diversification of energy sources and carriers (such as hydrogen, methane, methanol, and compressed air); the horizontal divestiture by global energy corporations; strict enforcement of anti-trust laws on a world basis; curtailing the international war machine; quantifying military energy expenditure and then not allowing it (by SALT agreement) to exceed that level for five years, thereafter reducing it by 1% per year; and the globalization of energy resources. This last item being the literal division of the world's energy resources—coal, oil, gas, sun, wind, etc.—evenly among the four billion plus people on Earth. The average U.S. citizen's share before and after globalization is illustrated in Chart 1. As can be seen, after five years everyone is surprisingly better off when total figures are compared. We *are* on one planet—the sooner all of humanity realizes this fact, the quicker will come peace and constructive stability.

Critical Paths for the Global Energy Development Strategy

The critical path charts on the following pages outline what could and what needs to be done—"first things first"—to reach a preferred energy state of functioning for the world. Each chart deals with one aspect of the overall energy development strategy (e.g., wind, solar, tidal, etc.) over a ten-year period on five distinct levels of energy utilization and organization: Single Unit, for micro-energy use by single-family dwellings (home heating, cooling, lighting, food refrigeration and preparation, communication and recreation facilities); Community, where the energy needs are larger than the Single Unit (community maintenance, schools, community transportation facilities); Industry, where energy use is much larger (macro quantities of energy are used for materials extraction, refinement, processing, and manufacture into all industrial and commercial goods); Regional, where all the previous levels (Single Unit, Community, and Industry) are interconnected; and Global, where all energy use is part of the overall patterns of energy flow on the planet, and where all regional levels are interconnected. The Global and Regional levels are primarily levels of organization rather than use.

Starting in the present, each major "move" or "path" shows *current* technology or organizational know-how. Almost everything on the critical paths has already been built (e.g., 900,000 cubic-feet liquid hydrogen storage tanks for Saturn rocket fuel), is presently being built or prototyped, or is contracted to be built immediately (e.g., U.S. Department of Energy contract with Boeing for development of large 2.5 MW wind-power system). What has been changed is the level of commitment and the emphasis on each path. Wind power and all the other non-depletable energy sources are emphasized in a major way, rather than viewed as a faintly interesting sidelight. The intentional moves have altered the scope of their development; the change has been one of degree, not of kind. As was stated earlier, but bears reiteration, is the fact that what is postulated is only that which already exists or could easily be brought into demonstrable hard or software existence through already known, tried, and proven technological and organizational procedures. There is nothing in any of the moves that is not possible with today's technological and ecological know-how, know what, and know where. There are no fantasy fusion or anti-gravity energy sources proposed to fuel the world or miracle-assisted bureaucracies to service the needs of humanity. Every strategy is based on present-day know-how. It is all possible; we could do it, or something better.

Outline of Strategic Moves

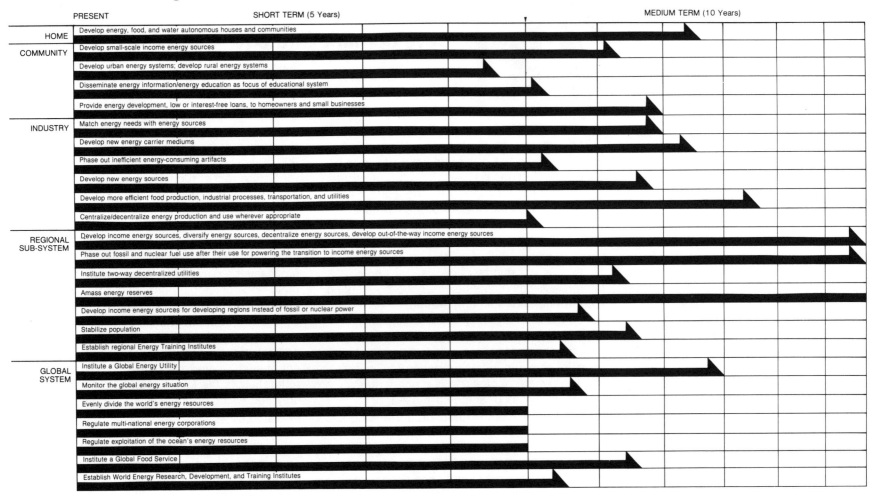

Wind

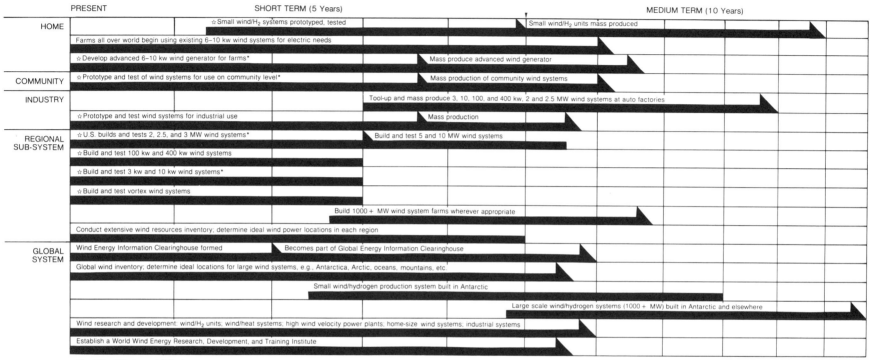

*Currently underway   ☆Artifact

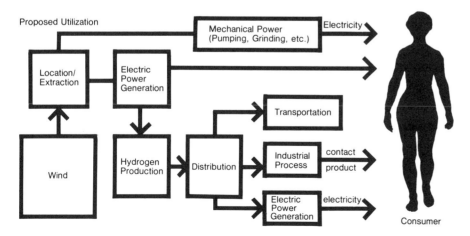

Proposed Utilization

Solar

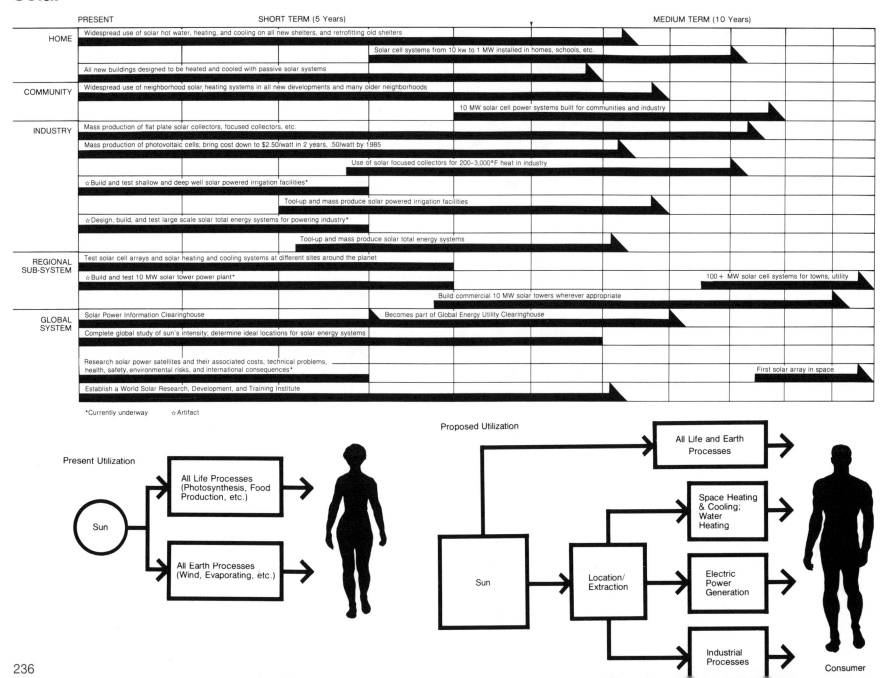

Geothermal

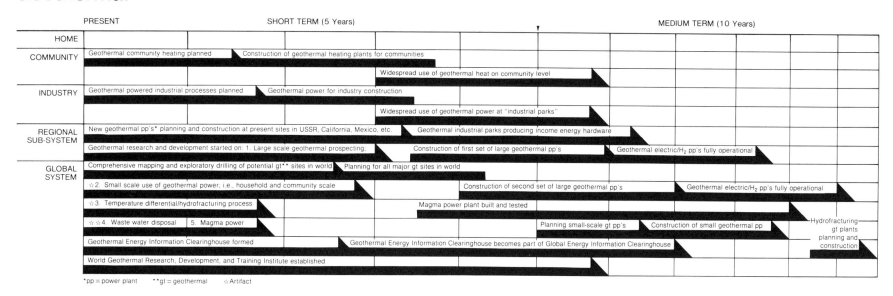

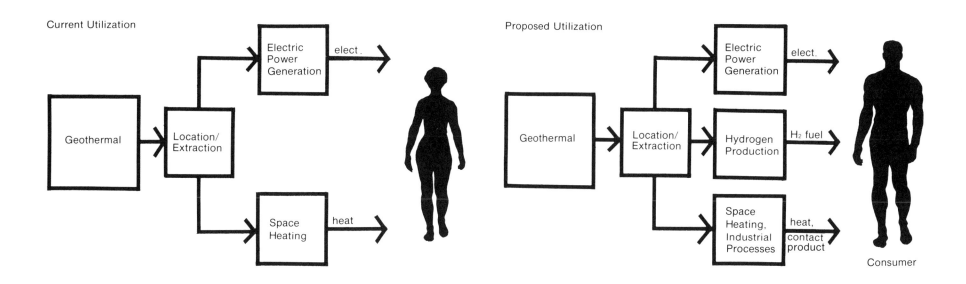

Tidal

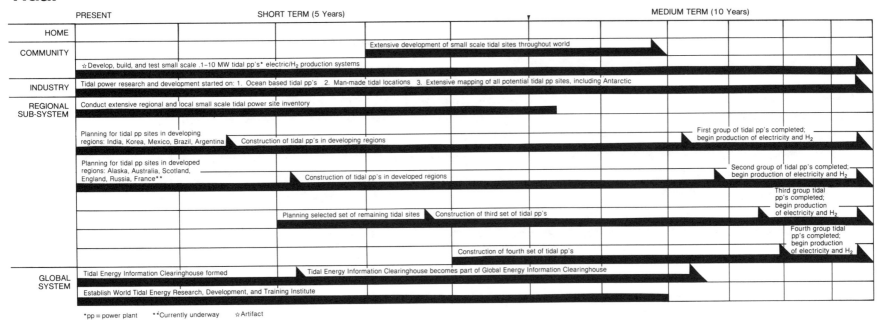

	PRESENT	SHORT TERM (5 Years)	MEDIUM TERM (10 Years)

HOME

COMMUNITY
- Extensive development of small scale tidal sites throughout world
- ☆ Develop, build, and test small scale .1–10 MW tidal pp's* electric/H_2 production systems

INDUSTRY
- Tidal power research and development started on: 1. Ocean based tidal pp's 2. Man-made tidal locations 3. Extensive mapping of all potential tidal pp sites, including Antarctic

REGIONAL SUB-SYSTEM
- Conduct extensive regional and local small scale tidal power site inventory
- Planning for tidal pp sites in developing regions: India, Korea, Mexico, Brazil, Argentina — Construction of tidal pp's in developing regions — First group of tidal pp's completed; begin production of electricity and H_2
- Planning for tidal pp sites in developed regions: Alaska, Australia, Scotland, England, Russia, France** — Construction of tidal pp's in developed regions — Second group of tidal pp's completed; begin production of electricity and H_2
- Planning selected set of remaining tidal sites — Construction of third set of tidal pp's — Third group tidal pp's completed; begin production of electricity and H_2
- Construction of fourth set of tidal pp's — Fourth group tidal pp's completed; begin production of electricity and H_2

GLOBAL SYSTEM
- Tidal Energy Information Clearinghouse formed — Tidal Energy Information Clearinghouse becomes part of Global Energy Information Clearinghouse
- Establish World Tidal Energy Research, Development, and Training Institute

*pp = power plant   **Currently underway   ☆ Artifact

Current Utilization

Proposed Utilization

Consumer

Consumer

Refuse Reduction

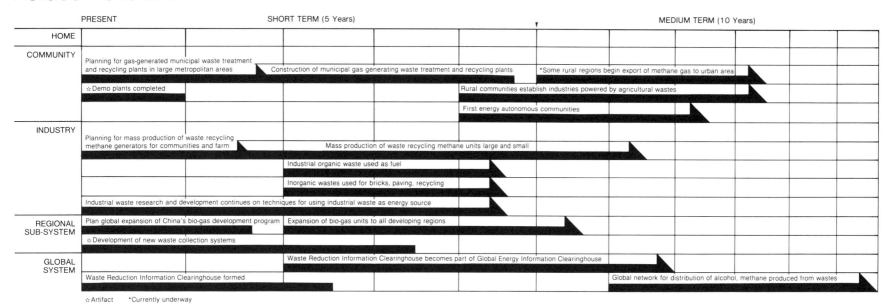

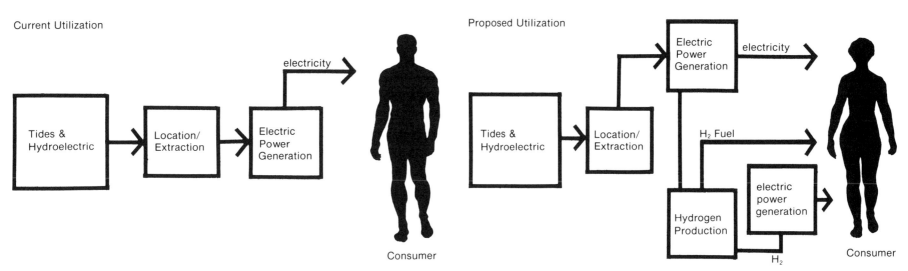

Temperature Differential

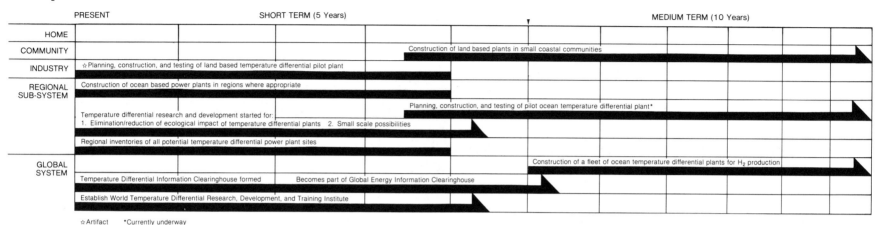

☆ Artifact *Currently underway

Bioconversion

☆ Artifact

Nuclear Power

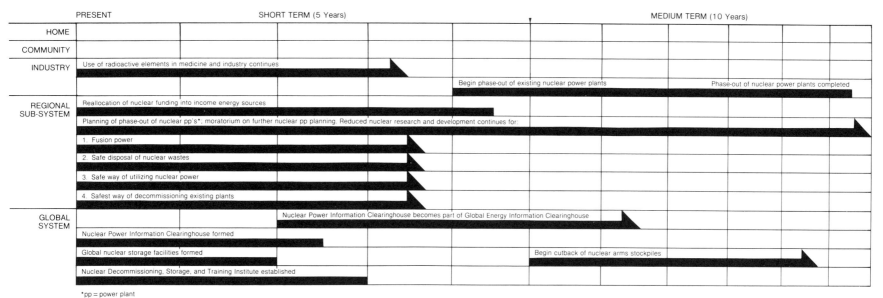

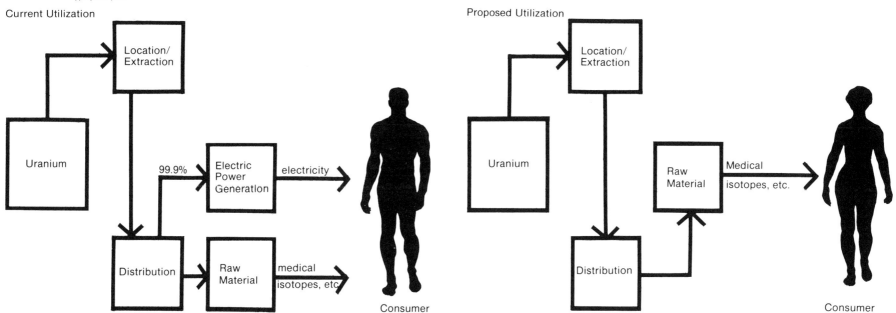

Hydroelectric

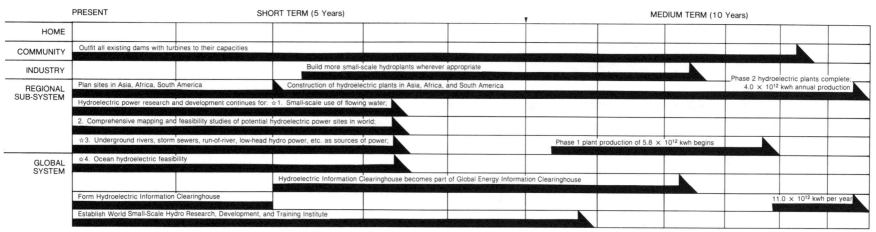

☆Artifact

Hydrogen

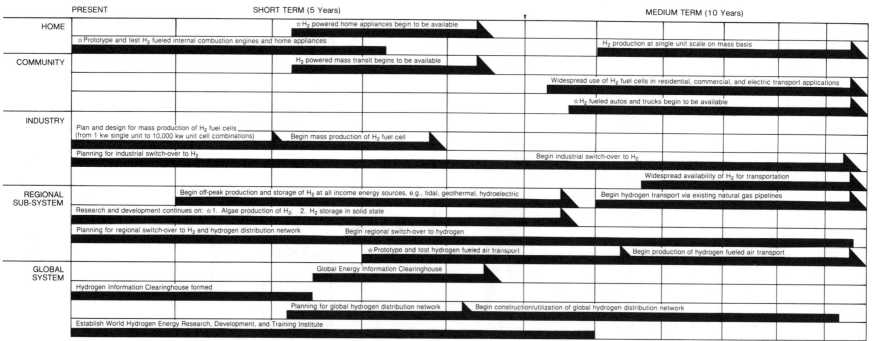

☆Artifact   *Currently underway

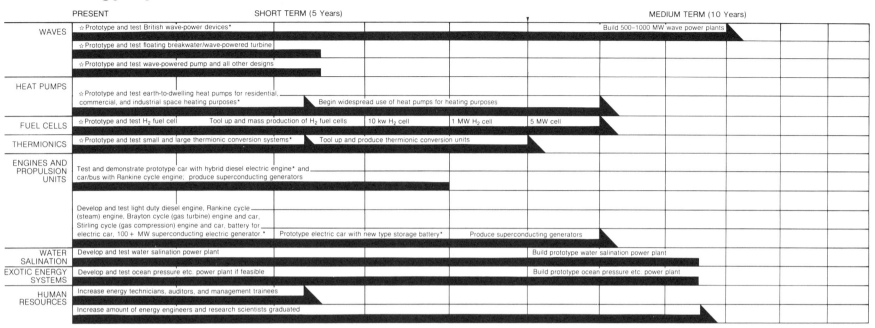
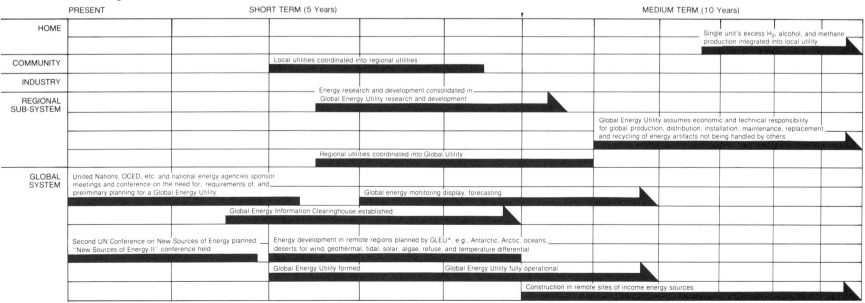

Fossil Fuels

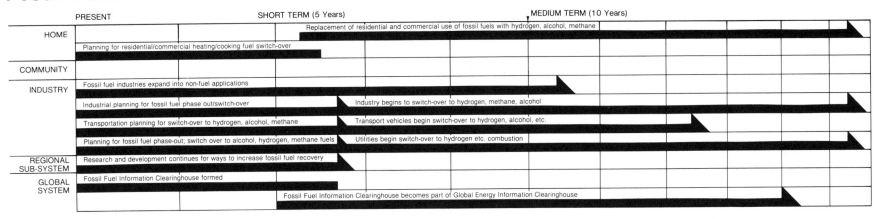

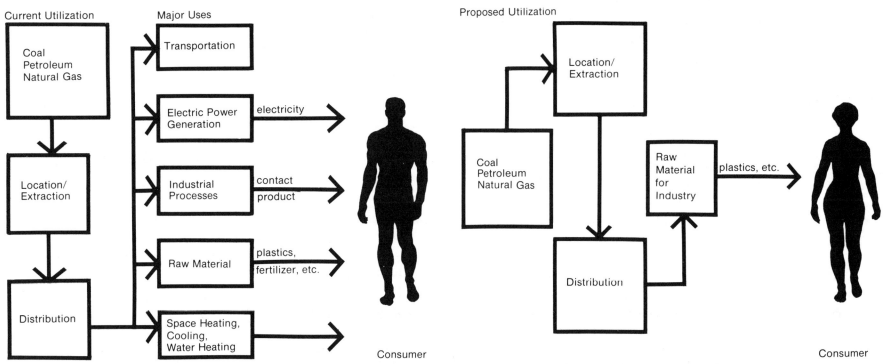

Conservation

	PRESENT	SHORT TERM (5 Years)		MEDIUM TERM (10 Years)
HOME	Implement substitutes for gas pilot lights	Single unit and larger housing begins government subsidized insulation additions	Energy performance criteria for new shelter increase energy conservation 50%	
COMMUNITY	Existing community mass transit expanded; new multi-modal mass transit planned	Mass transit constructed		
	Energy conservation education begins; evolution and institution of social customs and ethics which make energy conservation and efficiency a primary virtue			
	Energy consumption labeling on all appliances			
	Improve auto efficiency through use of low loss tires, improved engine tuning, car pools, decreased use in congested areas, and improved traffic flow, aerodynamic design, etc.			
		Single large heating and cooling plants to serve a number of single units, or district		
INDUSTRY	Industry energy performance criteria codes to go into effect			
	Industry begins switch-over to state-of-the-art efficiency processes in cement, glass, non-ferrous metals (ion exchange recovery and foam separation recovery), steel (continuous coking process and direct reduction process—taconite), aluminum (both processing and use of non-bauxite ores), petrochemical (heat reuse, efficient solvent recovery, waste as feedstock), paper, food, and agricultural processes, etc.			
	☆ Design and tooling up for more efficient appliances	Production of energy-efficient and durable appliances		
	Recycling and reuse of products and materials			
	Develop and test computer control of energy systems	Construction of large-scale demonstration plants for experimentation, tuning, and debugging of large-scale thermal processing equipment		
REGIONAL SUB-SYSTEM	Energy performance criteria for shelter, transportation, and appliances go into effect			
		Freight consolidation and containerization		
	Increased electricity rates for peak use periods go into effect			
	Regional mass transit expanded; new multimodal mass transit planned	Mass transit construction		
GLOBAL SYSTEM	Tax on sulfur content of fuel	Tax on extraction of capital energy sources from Earth (funds go into GLEU*)		
		Global Energy Information Clearinghouse		
	Energy Conservation Information Clearinghouse	Global energy performance criteria instituted		
	Establish World Conservation Research, Development, and Training Institute			

*GLEU = Global Energy Utility

Storage and Transport

	PRESENT	SHORT TERM (5 Years)		MEDIUM TERM (10 Years)
HOME		Small-scale flywheel unit used to store home generated energy		
COMMUNITY	Develop flywheel technology for peak storage	Develop prototype new flywheel system	Mass production of flywheel energy storage system	
		Develop and test superconducting magnets for peak storage		
		Develop and test Megawatt Li-Sulfur storage battery		
INDUSTRY	900,000 cubic foot hydrogen storage tanks constructed at numerous locations			
	Small-scale (single unit) H_2 storage plan and test for mass production	Small-scale hydrogen storage units produced		
	Construction of hydrogen tankers			
	Research and development continues on metal hydrides storage of hydrogen			
	Develop and test small-scale flywheel for single unit energy storage	Mass production of small-scale flywheel for single dwelling unit use		
REGIONAL SUB-SYSTEM	Prototype and test liquid H_2 pipeline/cryogenic electric transmission	Construction of liquid hydrogen pipeline/electric transmission		
	Develop and produce prototype liquid hydrogen tanker			
GLOBAL SYSTEM		Planning for hydrogen distribution network	Hydrogen distribution network construction	

☆ Artifact

5 Global Scenario

To speak of *global* energy strategies in a world of over 150 sovereign nation-states may be naive, but *not* to speak of global energy strategies on an interdependent spaceship is madness. To propose the extensive development of the Earth's income energy sources may be an expensive proposition, but it will be far cheaper than letting our existing fossil and nuclear systems exhaust the world's resources and then picking up the pieces after the crash. To seriously propose that the U.S. (and all other nation-states) internationalize their energy resources may strike some as altruistic or subversive, but not to do so is to assume a world of scarcity where it's "us or them." On a closed-system ship, the latter attitude is inevitably suicidal.

The following section is a synergetic and synoptic view of the variables that affect the world energy system. Its purpose is to weave together the various strategic and tactical moves of the previous sections and give a feel for the whole system of interacting policies, strategies, tactics, and their consequences.

The World Energy Data Chart at the end of the Energy Sources section helps to illustrate the fact that all the energy needs of the world could be met from non-depletable energy sources. Seventy-five trillion kilowatt hours worth of energy were consumed in the form of coal, oil, natural gas, and nuclear energy in 1976; more than three times this figure, about 284 trillion kwh, could be harnessed from income energy sources with current know-how in ten years, and theoretically more than 400–600 trillion could be harnessed in 25 to 50 years.[16]

Traditional methods for determining energy needs are usually based on the plotting of the historical trend of energy consumption in the past and extrapolating that into the future and then predicting energy demand. A somewhat different approach is taken here. There is no attempt to predict the future. As pointed out earlier, there are major difficulties with attempting to predict the future. In addition to what has already been mentioned, it can be said that attempting to predict the future is inherently impossible because of the "emergence of novelty by combination."[17] Perhaps the most important reason for not totally relying upon prediction of the future energy demand (by trend extrapolation or any other means) as a basis for energy planning is that the prediction of a future usually implies an acquiescence to that future, a subtly debilitating resignation to fate, rather than the challenge to bring about what is desired. Prediction implies and even brings about resignation; the future is the *last* place we should be resigned to (the past and present being perhaps more fitting contenders for the location of this vice). The future should be planned, not predicted. Planning implies and brings about responsibility, not resignation. When looking at the global energy situation and what we should do about it, we are in a fundamentally healthier position when we look at our values and ask ourselves what we want the future to look like rather than looking at our past and attempting to predict what we think the future will look like. It is for these reasons that this book takes the rather unorthodox perspective of not attempting to predict what the world's future energy demand will be, but what it *should* be, given present-day know-how. From the perspective used in this book, it is better to assume a goal, for instance a regenerative energy system in 25 years, and work from there to determine how to bring this state about than to accept the past and negative historical trends as fate.

Having said this about the limits of prediction and trend extrapolation, it should be pointed out that trends can be valuable tools in energy planning if used as aids in bringing about the desired future, not in determining that future. Instead of attempting to predict they can be used to anticipate the future bottlenecks and system overloads and breakdowns. Trends can be used to tell us what the future will or can *not* look like, rather than what it will look like.

To meet the whole world's needs for food, clothing, shelter, health care, education, etc. demands a lot of energy. To supply everyone with the amount an average North American consumes, the world would need 345.6×10^{12} kwh annually, more than five times the amount of energy now being used in the world.[18] Given that North America is a prodigious waster of energy, this is both not a reliable figure nor a socially, economically, or ecologically desirable one to use as a goal for any long-range planning. Both Sweden and New Zealand have managed to provide their people with the above mentioned life support amenities (in addition to many others) without using anywhere near the

---

The future is not to be predicted, it is to be planned.

amount of energy per capita that North America uses. To supply everyone in the world with the amount of energy the average Swedish person uses, the world would annually need 197.6 × 10^{12} kwh; using New Zealand as the reference figure the world would need 99.2 × 10^{12} kwh,[19] or roughly 45% more energy than is presently being used by the world. With adequate—that is, equal, ethical, and non-exploitative—distribution to the most energy-needing regions of the world and the *efficient* use of this energy, coupled with massive commitments to conservation and ephemerization in the energy-obese areas of the world, the New Zealand per capita energy figure (or one lower) would be able to provide everyone on Earth with all the energy needed to have all the previously mentioned necessities met.[20] It is therefore this goal, rather than energy demand prediction, that the various strategies in this book are designed to reach. As has been shown in the World Energy Data Chart, this figure is well below the amount that could be harnessed from our non-depletable sources of energy.

The overall plan calls for the existing hydrocarbon industrial systems to be used in a pump-priming role to power the production and distribution of most of the artifacts needed to switch over to power derived from income energy sources.

As pointed out earlier, there are two distinct energy conversion modes that match the two types of energy sources—those that are of large magnitude but are found relatively infrequently and those that are of low magnitude but are found very frequently. Both types would be developed and matched with the energy end-use that would be most appropriate. Large-scale energy-users and industrial high-temperature energy-users would be lumped together wherever possible to take advantage of cogeneration.

Because the income energy of decentralized and centralized tidal, solar, and wind sources is inherently intermittent and varies in intensity by location and time of season and day, energy carriers that store the income energy (much as fossil fuels store energy) and provide a transportable medium (much as the electrical transmission network functions) are highly desirable. There are a number of currently-available means of storing and transporting the various sources of income energy. Methane and methanol produced from biomass and urban wastes are two such energy carriers; hydrogen is another and a hydrogen/methane mixture is yet another. Hydrogen can be produced from

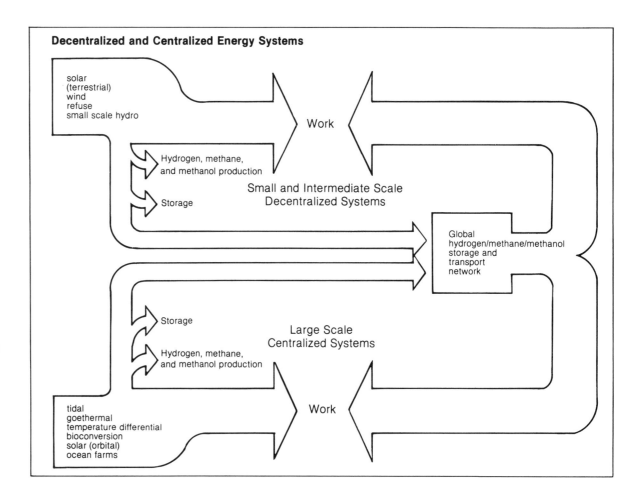

247

Strategy of Switchover to Income Energy

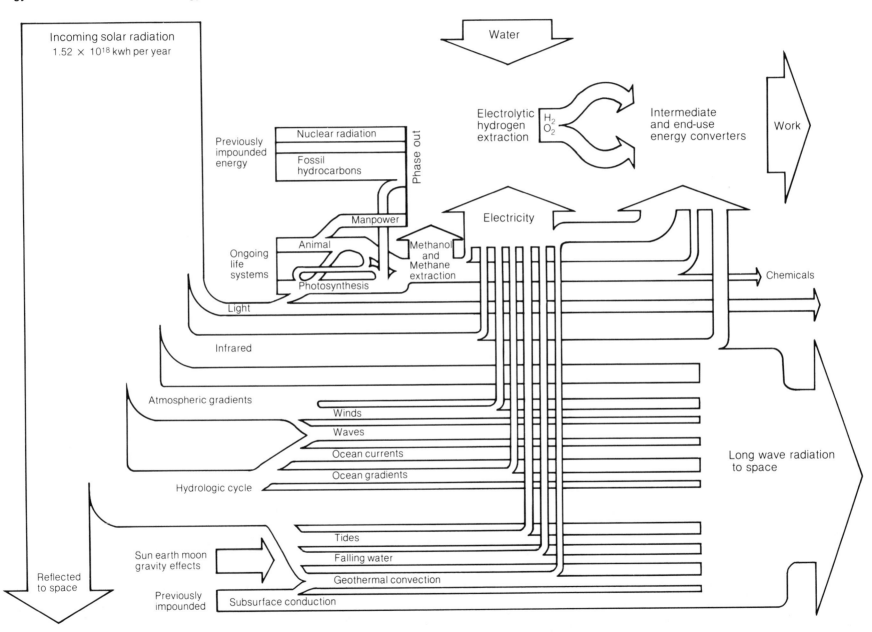

248

Proposed Energy System

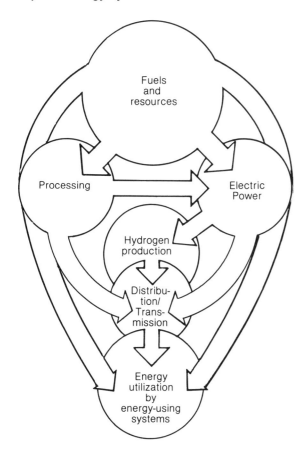

Hydrogen Economy Structure

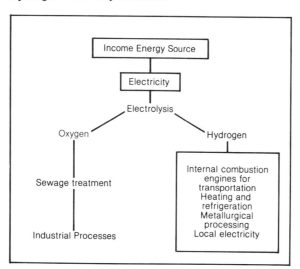

the scale of the backyard windmill-powered generator to a regional-scale solar-farm-powered hydrogen production plant. Off-peak hours at large-scale geothermal, tidal, and hydroelectric power plants could also be used for hydrogen production.[21] All three—methanol, methane, and hydrogen—can be used to fuel the processes that are currently run on natural gas, fuel oil, gasoline, kerosene, etc.

The overall energy strategy calls for the immediate and widespread development of a global hydrogen and methane storage and transport network, developed along the lines of the existing natural gas and electrical distribution network. Existing natural gas and oil pipelines can be used for hydrogen/methane distribution with little or no modification. Most, if not all, of our present-day natural gas and liquid fuel transport carriers, industrial processes, and residential appliances can be converted to use hydrogen.[22] In addition to the gases of methane and hydrogen, the liquid fuel alcohol can be developed as an energy carrier to supplement the gases, diversify the world's energy sources, and facilitate the transition to a post fossil-fueled world.

All the strategies are geared toward meeting the energy needs of 100% of humanity on a continually sustainable basis. The strategies are divided into four general stages: short term (first five years), medium term (ten years), long term (twenty-five years), and extended long term (fifty years). The emphasis in the short-term first five year development plan is on massive educational efforts and on getting basic life support energy to developing regions and energy conservation in developed regions. Energy conversion units for food production, preservation, and preparation; for health care stations; and for regional and local communication, education, and transportation facilities would all be emphasized in developing regions. New, efficient solar, wood, or biogas cooking stoves, along with solar food dryers, community windmills for grinding grain, pumping water, producing electricity for powering refrigeration units and communications facilities, community solar heating units for crop drying, neighborhood hot water and space heating and methane or alcohol production units to run tractors, buses, and local cottage industries that would produce the above items would be built during the short-range period. Preliminary local and regional energy plans would be developed by local expertise—drawn from local colleges, industry, concerned citizens, and governing officials—and printed in local newspapers and other appropriate public channels of communication after which they would be further refined by the local populace. All appropriate educational mediums would be used to educate people about energy. All new buildings would meet strict energy performance standards that require maximizing the use of passive solar systems and which use less than half the energy needed by the buildings built before 1980. All older buildings would begin to be retrofitted to reduce their energy use. Small-scale cogeneration units—producing heat and electricity—for single and multiple-family dwelling units will be in mass production and in widespread use. In the U.S. and other developed regions, these units would be powered by methane or hydrogen gas (produced from agricultural wastes, biomass crops, or electrolysis from income energy sources) that would be distributed through the existing natural gas pipeline network, as well as other fuels, such as alcohol, in areas not serviced by the gas pipeline.[6] Most electric utilities will decentralize, often synergetically coupling this decentralization with urban rehabilitation by providing small cogeneration units to dwellings that need new heating systems.[23,24] Large thermal storage of solar heat for communities will be in use, backed up with biomass and urban waste as a fuel source. Wind systems for heating and electricity production will be in widespread use. Cooling will be largely done through passive and active solar systems. Prototype energy-autonomous communities will be in operation; all institutions of higher learning around the world will use their investment resources— endowments, etc.—and their technical and educational expertise to publicly demonstrate the viability and economy of conservation and switching to income energy sources[25]; government contracts would be allocated to the lowest bidder who also could perform the job using the least amount of energy; excess electrical capacity will be used to electrify the railroads; large-scale solar collectors and concentrators for industrial use in both developing and developed regions will be in the prototype stage; there will be large-scale recycling of industrial wastes and large methane extractors in operation on the industrial, urban, and rural community levels; large three-megawatt wind-powered electric generators will be coming into use for hydrogen production; and off-peak extraction of hydrogen will be occurring at all but the smallest energy generation locations near a plentiful supply of water. A series of 100+ m.p.g., durable, safe mass-producible cars will be developed, prototyped, and tested, and the industrialized world's rail systems will be rehabilitated and electrified. During the same five-year period there will be intensive planning and construction of large-scale geothermal, tidal, hydroelectric, and biomass energy-producing facilities. Research and development will be under way for increasing the efficiency of conversion from all energy sources, the development of new energy sources, new energy storage and transmission mediums, new ways of conserving energy, and safe ways of storing nuclear wastes. No new nuclear energy installations will be started, unsafe existing installations will shut down and all nuclear plants will begin preparing for shut-down. Nuclear facilities in developing countries will be purchased and replaced with non-depletable energy sources, and the existing fossil hydrocarbon-fueled transportation, industrial, residential, agricultural, utility, and commercial systems will begin switching over to hydrogen, methane, and alcohol fuels.

Global Energy Utility

There are many things a local or regional utility or a multi-national fuel corporation cannot do because of the structure they are embedded within and the "rules of the game" they are playing. A corporation is bound to make a profit if it is to survive; the leaders of a nation-state are bound to protect their country's special interests if they are to stay in power. The problems facing the world are bigger than any nation-state or any corporation. To deal adequately with the world's energy problems a structure that can transcend cut-throat economic rivalry or narrow nation-state prejudice and suicidal competition is needed. The Global Energy Utility is intended to perform this function. It would be formed during the first five year period of the global energy development strategy and would coordinate the activities of the overall plan.

The major components or functions of the Global Energy Utility, outlined in the following diagram, would deal with the coordination of the world's currently existing regional and local electric utilities and the already global fuel corporations. It would coordinate the development and utilization of energy on a global basis. The purpose of the Utility would be to service the energy needs of 100% of humanity as quickly, efficiently, and with the least environmental and user costs as possible. The Utility would be a cooperative effort of the combined nations and energy corporations of the planet and would be non-political and non-profit; it would be designed from the start to only be in existence for a given, fixed amount of time, say ten or twenty-five years, and at that time to be phased out unless mandated at that time to continue. In addition, every five years there should be a review of the activities of the Global Utility to see if there are any activities that could be performed as well or better by private or nation-state enterprises. The Food and Agriculture Organization of the U.N. is one possible model for some aspects of such a Utility.

The Global Energy Utility would be divided into decentralized regional utilities, arranged according to special characteristics such as climate and energy resources and would straddle political boundaries. Each regional

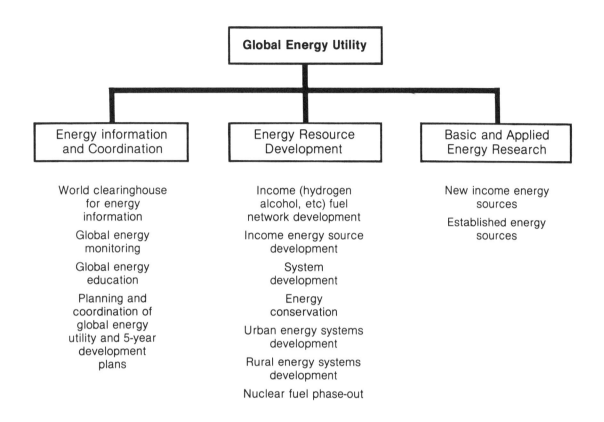

utility would be responsible for the special needs in each region. The "Global" part of the Energy Utility would coordinate these regional utilities as well as the development of "out-of-the-way" energy resources such as the wind energy resources of the Antarctic, Arctic, or the oceans; blending all the various sources of energy into one coordinated system. The "prime directive" of the Utility would be to coordinate the diversification and decentralization of the world's energy sources and to facilitate the transition to a post fossil-fueled world.

Energy Information and Coordination

In order to make any decision, information is necessary. The more accurate and comprehensive the information, the more likely that the decision will be "correct." The world energy situation is plagued by unreliable, conflicting, out-of-date, inaccurate, parochial, anachronistic, and misrepresentative masses of energy statistics that are mostly in different base units and only apply to one town or country. A global agency is needed to overcome these deficiencies.

The Energy Information and Coordination section of the Global Energy Utility will function in four areas. The first is as a world clearinghouse for energy information, collecting, correlating, and distributing data from all the world's energy research and economic organizations to the regional utilities, who will in turn pass the relevant information on to the people in their region.

The second function will be as a channel for feedback on the existing energy situation and for the actions taken by the Utility. It will monitor the key indicators (see Section One) of the energy environment for changes and watch to see if the desired changes resulting from a specific action are coming about. In this capacity, the energy information clearing-house will constantly inventory, monitor, display, and anticipate local, regional, and global energy resources, needs, and use.

The third mode of functioning will be in the area of energy education. All new energy systems and tactics need to be accompanied by broadly-based and extensive training programs. The regional level of the Utility will be responsible for developing appropriate educational programs in coordination with present educational systems so as to expose the consumer to the details of energy systems, tactics, and strategies. A primary emphasis will be on how energy autonomy can be achieved at the community and individual-dwelling-unit level.

The fourth function will be to coordinate all the functional divisions of the Utility and to compile all the national and regional energy plans from around the world. Such a compilation would be done immediately and would disclose any potential conflicts between national and regional plans. New local, national, and regional plans would be developed by the respective populations and coordinated into global energy development strategies.

Energy Resource Development

Energy Resource Development will be the agency for investigating the income energy resources of the planet. This agency will view the planet as one entity with certain power requirements and certain total energy resources for meeting the needs of the Earth's population. Energy Resource Development will be responsible for developing all the income energy sources of the Earth wherever they are found—the Arctic, Antarctica, oceans, mountains, shorelines, upper atmosphere, in orbit around the Earth, etc.—and for the means for getting this sometimes remotely located power to the places where it is needed.

The regional utilities will be responsible for the development of the income energy sources in their region and for the considerate use and orderly phasing out of all depletable energy sources.

Energy Source Development covers seven main categories: (1) income fuel development, (2) income energy source development, (3) system development, (4) energy conservation, (5) urban energy systems development, (6) rural energy systems development, and (7) nuclear phase out. Income fuel development is the development of large- and small-scale prototypes, and then full-scale mass-produced plants for the production and distribution of hydrogen, methane, and alcohol. Income energy source development will feature development of the Earth's non-depletable or not-soon-to-be-depleted energy sources, with special emphasis on those sources located in out-of-the-way parts of the planet. System development is the development of a global energy network, with electricity and hydrogen the two main energy carriers. Exisiting local and regional electric grids will be interconnected to take advantage of unique geographical features conducive to large-scale energy generation. Local energy supplies and needs will thus be interconnected with other local energy supplies and needs and with regional energy supplies and needs; regional supplies and needs will be interconnected with other regions and global supplies and needs. Energy system development will create a two-way energy system; that is, surplus energy on the local level will be available to the region, and regional energy supplies will be available to the local level. Surpluses at each scale of development will be available to the next higher level and vice versa.

Energy conservation will deal at first with reducing energy waste at the regional level, and then local Energy Extension Services offices will be set up throughout the world to act as catalysts and information services in bringing about conservation and non-depletable energy development measures. Urban energy systems development will develop those technologies that seem to be suited for centralized, high-density urban settings and their needs. Rural energy systems development will undertake similarly appropriate activities for the rural regions of the world. Nuclear fuel phase-out will coordinate the global phase-out of nuclear power—the decommissioning of power plants, storage of nuclear wastes, etc. All nuclear power plants in developing countries will be purchased and then replaced with income-energy-harnessing artifacts. One quick way of stopping the proliferation of nuclear power, and for instantly seeing who has nuclear power for power production and who has it for weapons production is to offer to purchase all nuclear power plants at profitable prices to the sellers and in addition offer to replace the nuclear power capacity with non-depletable energy sources of the same capacity at little or no costs. For countries that are thinking of purchasing nuclear power plants, other, subsidized, power plants could be sold at half the cost. In other words, undersell nuclear by so vast a margin no country could rationally justify purchase for power production. The threat of nuclear proliferation would be greatly reduced. Because of its large role in promoting nuclear power, the U.S. should bear a large part of the costs for such a program.

The regional utility or banks could provide loans to consumers wishing to install income-energy-harnessing or energy-conserving technology in their home or small business. Loans would be paid back with the savings that accrue due to the reduction or elimination of fuel and/or electricity bills.

Basic and Applied Energy Research

The Basic and Applied Energy Research Agency will allocate funds for basic and applied research into income energy sources and technologies. It will seek to eliminate economic and technological inefficiencies through the increase in know-how. It will begin large-scale funding for research into such proven income energy sources as solar, geothermal, tidal, wind, hydroelectric power, refuse reduction, bioconversion, hydrogen, waves, and temperature differential, and preliminary funding for such sources as electrostatics, ocean currents, gravity, nuclear fusion, and orbiting solar collectors.

The funds needed for basic and applied research (as well as for the entire Global Energy Utility) could be obtained from a capital energy depletion tax. That is, for the entropic degradation of a depletable energy source, a fee (determined by the energy content of the respective energy source) would be payable to the Global Energy Utility. Total degradation (e.g., from coal to ashes and smoke) would be the maximum fee; less of an increase in entropy (e.g., from coal to plastic) would be a lesser fee. A 1% tax on multi-national energy corporations' gross profits could also provide funds for the Utility, as could a 1% tax on nation-state military expenditures. Another source of funds could be the income from the out-of-the-way energy sources such as the Antarctic winds. An initial loan from the World Bank could start off the harnessing of diverse energy sources, and the income from the distribution of this energy could repay the loan and sustain the utility.[26]

By the end of the medium term (ten year) plan, almost all housing will be either solar heated and/or powered by small-scale cogeneration units. The units will be in all new housing being built at the end of the first five-year plan; with many of the older housing units retrofitted by the end of the medium term plan. All new appliances will use approximately half the amount of energy that pre-1980 appliances used. By the end of the short term (five-year) plan, large autonomous energy conversion units will be in mass production for communities and large buildings, and they will be in widespread use by the end of the medium term plan. Most nuclear power facilities will be decommissioned; those remaining will be used to produce isotopes for research and medicine. Also within ten years high-energy solar concentrators will be in industrial use; large-capacity wind-turbine complexes (100+ megawatts) will be in worldwide use; hydrogen/methane will become leading fuels and will be distributed throughout the world by pipeline and tanker. In the middle of the medium term plan all large-scale centralized energy production facilities that were begun in the early stages of the first five-year plan will come on-line. Much of the present-day energy-conversion equipment will be obsolete or quickly approaching obsolescence and will be phased out by higher efficiency plants. Fossil fuel combustion phase out will be started in a major way. Industries will be much more efficient, requiring on the average only about 80% as much energy per given output as in 1980. Energy use for transportation in urban areas has been reduced through mass transit and more efficient private vehicles, including widespread use of electric vehicles recharged with photovoltaic and wind-energy systems and conventional engines powered by methanol or hydrogen. Industrial processes needing temperatures of up to 350° F are solar powered. Chemical feedstock is beginning to be supplied by biomass. Agricultural energy needs are supplied almost totally by agricultural waste and biomass crops. New technologies, such as the space shuttle, will be available at acceptable costs for the first of new energy conversion prototypes such as orbital solar conversion satellites if these systems seem viable at that point in time.

By the end of the medium term (ten-year) plan, global energy development will be able to meet the basic life support needs of 100% of humanity. Starvation will be totally eliminated. Worldwide life support facilities will be at least equivalent to present-day New Zealand levels, Many of the environmental impacts associated with the 1960's and 1970's energy use are being eliminated.

The long range (twenty-five year) plan calls for continued phase out of fossil fuels, further refinements in energy conversion and utilization efficiencies, and continuing development of the total energy network. The more inefficient and environmentally impacting income energy-harnessing devices will be recycled, being replaced by less obtrusive and more efficient artifacts. Most of the present-day life support equipment will be phased out and recycled. Global living standards will continue to rise, population will stabilize, and energy consumption per capita will continue to decline. (There is a well-documented precedent for the increase of life-support despite declining resources per capita. Between 1900 and 1970 the world went from less than 1% "haves" to over 40% "haves" while the population was increasing from 1.9 billion to 3.6 billion, thus greatly lowering the per-capita resources available for such an increase in life support capabilities.) Developing countries will be powered almost entirely with non-depletable energy sources; malnourishment will be eliminated from the world; preventive health-care centers will be operating everywhere; educational, travel, and recreational facilities will be pervasive; nuclear weapons proliferation will be reversed; military budgets will be shrinking; national defense and sovereignty will be guaranteed by an independent international peace-keeping force; global communications will be to the 1970's what the 1970's were to the 1900's; the world's environment will be free of major environmental pollutants; and there will be a lasting, constructive, and evolving peace on Earth.

The extended long-range (fifty-year) scenario calls for the complete phase out of fossil fuels, the continuing progression of doing more with less, a stable population, the total elimination of malnourishment and substandard housing; the commensurate provision of ever better nutrition, preventive health care, housing, recreational facilities, unlimited educational opportunities for all the people of the world, communications facilities that allow for anyone anywhere finding out almost anything about anything, travel facilities that allow anyone to go almost anywhere on Earth at almost anytime; a global human culture that delights in and nurtures its diversity, a common and pervasive love and respect for the planet Earth and its sustaining life processes, a decentralized global democracy of sovereign individuals that are enriched, not separated, by their unique geographic origins, religions, cultures, and tastes, total elimination of human generated environmental overloads (pollution), elimination of nuclear weapons and the need for them, space colonization, all the treasures of all the museums of the world returned to the lands of their origins for display there (e.g., the British Museum's Egyptian collection back in Egypt); contact with non-human and non-Earth intelligences; and military forces (and their budgets) transformed into disaster relief armies and ecological armed forces who battle deserts, floods, earthquakes, tornadoes, hurricanes, environmental pollution accidents, soil erosion, etc. Or in the long-long-range visionary perspective and words of Buckminster Fuller:

There will be no thoughts whatsoever of *earning a living*. There will be no thoughts of, or even such words as *business competition, money,* or *lies* for such phenomena will be historically extinct. Such words as *politics, war, weapons, debt* will be only of historical significance.

Electronic means will have been highly developed for continual inventorying of all of humanity's thoughts, volitions and dispositions regarding all currently evolving problems. Humanity will know at all times what the unique majority volitions may be regarding each and every currently recognized and considered problem.

There will be one world management organization similar to but greatly improved over those of the 20th century U.S.A. "city manager" functions. The one world management will be taking its instructions directly from the computer read-out volitions of the majority. When the majority discovers a given decision is leading humanity into trouble, the popular realisation will be immediately computer manifest and the world management will alter the course accordingly. This feed-back, servo-mechanism is the same as that employed in "automatic" flight controls and in the steering of ships. The popular view will be immediately served by the management with no searching for scapegoats when erroneous decisions are discovered and corrected.

All human beings engaged in common wealth production or research and development will be doing so entirely on their own volition because that is what they will want to be doing. They will have to qualify for participation on production teams as in Olympic games. That which is plentiful will be socialized. That which is scarce must be used only for total advantage and must be used only in the research instruments and tools-that-make tools which produce the plentiful end-products for humanity.

All of humanity will be enjoying not only all of Earth but a great deal of local universe. "Where do you live?" "I live on the moon," or "I live on Mothership Earth," will be the kinds of answers.

Some large number of human beings will be engaged in archeological research as humanity will want to know a great deal more about the historical occupancy of our planet by humans. The important original buildings of antiquity will be rebuilt or restored as Babylon is now being rebuilt, and artifacts from museums around the world will be returned to original sites and reintroduced to function as of yore. Thus research teams can live experimentally at various historical control periods of history thus to elucidate much of the wisdom gained in the past.

While everybody will know much of what everybody is thinking, individuality will not cease but increase. What people are thinking spontaneously as a consequence of the interaction of the unique patterns of their inherited genes and their own experiences will make personalities even more interesting one to the other. Intuition will be fostered. Communication will probably be accomplished by thinking alone, ergo more swiftly and more realistically than by sound and words.

Omni-considerate, comprehensive, synergetic integrity will be the aesthetic criteria and its humanly evolved designs will come to do so much with so very little as to attain the ephemeral beauty heretofore manifest only by nature in her formulation of flowers, crystals, stars and the pure love of a child.[27]

As attractive as the above vision may be, it needs to be tempered by two considerations, one a caution, the other a celebration. The first is that the above deals with long-range possibilities, not the present-day situation wherein we teeter on the threshold of oblivion; where an accident, honest mistake, terrorist, malfunction, madman, or well-meaning president, prime minister, chairman, general secretary, general, admiral, commander, or enterprising individual could almost instantaneously create thousands of small stars around the planet that would light up the world and life and the memories of those who survived.

An explosion is just the *very* rapid release of energy. In a sense, an atomic holocaust is a fitting metaphor for the current energy-use patterns on our planet. Speeded up by the retrospection of future generations (or future civilizations and their archeologists) the rapid depletion of our fossil fuels is a veritable explosion. In addition to the ultimate threats posed by nuclear star-wars on our planet, we face numerous insidious threats to human well-being and evolution caused by starvation, malnourishment, illiteracy, lack of health care, and creative opportunities. Now is the time for humanity—you and me—to choose where we want to be and to go.

The other consideration dealing with the positive long-range vision described above is that the actual future, if we make it, will undoubtedly be very much different and better than anything we can imagine today.

References

Global Strategies for a Regenerative Energy System

1. Hayes, D., *Rays of Hope: The Transition to a Post-Petroleum World*, W. W. Norton, New York, 1977.

2. Lovins, A. B., *Soft Energy Paths: Towards a Durable Peace*, Ballinger, 1977.

3. Makhijani, A. B., and Lichtenberg, A. J., "Less Energy, Same Standard of Living," as reported in *Science News*, 3-11-72, p. 171.

4. A. D. Little, Inc., "An Impact Assessment of ASHRAE Standard 90-75," report to F.E.A., C-78309, December 1975.

5. U.S. Department of Energy, Weekly Announcements, Office of Public Affairs, Washington, D.C., 20461, Vol. 1, No. 9, 12-9-77, p. 6

6. Fiat Auto Group, "TOTEM: Total Energy Module," Sept. 1977, Fiat Motors of North America, Inc.

7. Metz, W. D., Hammond, A. L., *Solar Energy in America*, AAAS, 1978.

8. Shippner, L., Darmstadter, J., "The Logic of Energy Conservation," *Technology Review*, M.I.T., Cambridge, MA 02139, January 1978.

9. Pilati, D. A., "Airplane Energy Use and Conservation Strategies," Oak Ridge National Laboratory, Oak Ridge, TN 37830, 1974.

10. Fuller, R. B., "Geoview," *World Magazine*, 5-2-73, p. 27.

11. Brown, H., Cook, R., *Transition to the Regenerative Resource Economy*, Earth Metabolic Design, Box 2016, Yale Station, New Haven, CT 06520, 1975.

12. Henderson, C., *The Inevitability of Petroleum Rationing in the United States*, A Princeton Center for Alternative Futures Inc. Occasional Paper, Princeton, NJ, April 1978.

13. Bowman, R. L., Jack, J. R., "Energy Conservation Using Remote Thermal Scanning," NASA Tech Briefs, Summer 1978, p. 197.

14. See Gabel, M., *Ho-ping: Food for Everyone*, Anchor Press, Garden City, NY 1979.

15. Brown, H., "Real Utilities," unpublished paper, Earth Metabolic Design, Box 2016, Yale Station, New Haven, CT 06510.

16. These figures relate to just technological feasibility not economic or political probability; to make anything as extraordinary as this happen would require unprecedented and massive (except for war-time) economic and political commitments. Anyone looking at the automobile in the U.S. in 1910 could say something similar in terms of its technological capabilities of transporting people around in the 1970's but would be hard-pressed to justify the leap-of-faith required to say where the U.S. would get the billions of dollars for the Interstate Highway System or that such a fund would even exist. Similarly, today, the massive changeover to income energy sources postulated here and elsewhere is partly based on the assumption of popular demand, grassroots action and decentralized spontaneity coupled with over-riding necessity as seen "from the top" by regional, national, and international decision makers.

17. Georgescu-Roegen, N., *The Entropy Law and Economic Process*, Harvard University Press, Cambridge, MA, 1971.

18. *World Energy Supplies 1971–1975*, U.N. New York, 1977; each North American consumed 10,888 kg of coal equivalent in 1976; this comes to 86,400 kwh per person ($\times 4 \times 10^9 = 345.6 \times 10^{12}$ kwh).

19. *World Energy Supplies 1971–1975*, U.N. New York, 1977; Sweden consumed 6,178 kg of coal equivalent per capita in 1976; New Zealand 3,111 kg coal equivalent per capita.

20. Assming 1976 technology; any new innovations in doing-more-with-less energy would reduce this figure.

21. Meslan, F., Gordon, I. J., et al., "Geothermal Energy as a Primary Source in the Water Energy Economy," Bechtel Corporation, San Francisco, THEME, University of Miami, FL, April 1974.

22. Powell, J., Salzano, F., Sertan, W., "The Technology and Economics of Water Production from Fusion Reactors," Brookhaven National Laboratory, Upton, NY, THEME, University of Miami, FL, April 1974.

23. Commoner, B., "Reflections: The Solar Transition," *The New Yorker*, 4-30-79.

24. For a more thorough discussion of decentralized utilities, see Gabel, M., "The Decentralization of America: From Utility to Us-tility" in *Decentralizing Electric Utilities*, Yale University Press, New Haven, CT, 1980.

25. Craig, P., et al., editors, *Distributed Energy Systems in California's Future*, Interim Report, Prepared for U.S.D.O.E., Asst. Secy. for Environment, May 1978.

26. The concept of a global utility could and should be generalized to include other vital areas of human life support such as food, shelter, health and medical care, transportation, communication, and education.

27. Fuller, R. B., "2025, If . . . ," *The CoEvolution Quarterly*, Spring, 1975.

General References

Ackoff, R., *Redesigning the Future*, Wiley, New York, 1974.

Fuller, R. B., *Operating Manual for Spaceship Earth*, Simon and Schuster, New York, 1970.

_____, *Utopia or Oblivion: The Prospects for Humanity*, Bantam Books, New York, 1969.

_____, and McHale, J., "Inventory of World Resources." Document 1. 3500 Market St., Philadelphia, PA 19104, 1971.

Gabel, M., and Brown, H., *Design Science Information System, Introduction and Reference Guide*, New Haven, CT: Earth Metabolic Design, Box 2016, Yale Station, New Haven, CT 06520, 1973.

Georgescu-Roegen, N., *Entropy Law and the Economic Process*, Harvard University Press, Cambridge MA, 1973.

McHarg, I., *Design for Nature*, Doubleday, Garden City, NY, 1969.

Appendix A: Design Initiative

Soon-to-be-Needed and Necessary Artifacts for Global Energy Development Realization (or: What can be done by anyone willing to take the design initiative).

Energy conservation

1. Multi-mode mass transit.
   (a) Nonpetroleum multiple fuel individual and multi-passenger vehicles.
   (b) Hydrogen airship development.
2. Low-energy-using appliances.
3. Energy-efficient shelter and housing.
   (a) Energy-autonomous housing.
   (b) Energy-producing housing, i.e., housing as net energy source, not consumer.
   (c) Autonomous cluster housing.
4. Energy-efficient industrial processes.
5. Development of energy-efficient materials for manufactured goods.

Energy storage and transport

1. Cryogenic pipeline/electric transmission cable.
2. Mass-producible small-scale hydrogen gas, liquid, and solid (metal hydride) storage units.
3. Mass-producible small-scale hydrogen fuel cell.
4. Small-scale compressed-air storage units.
5. Compressed-air storage units with solar heaters to increase pressure.

Solar power

1. Low-energy materials, mass-producible solar space or hot water heaters for individual dwellings.
2. Special-use solar concentrators for industrial use in developing regions.
3. Large-scale solar thermal power plants.
4. Inexpensive mass-produced solar cells.
5. Solar-cell-powered autonomous energy-production units in modular form for single dwellings, apartment houses.
6. Solar-powered hydrogen production units.

Waves

Wave-power-harnessing devices; both large- and small-scale.

Wind

1. Low-energy-material mass-producible wind generators, blades, and towers for large-and small-scale use.
2. Wind turbine towers, hydrogen balloon-lifted wind turbines, and ocean-floating platforms and buoys for wind turbines.
3. Low rpm propeller/generator for power production.
4. Hydrogen production units for existing wind power units.
5. New wind power plants designed specially for hydrogen production.
6. 5 MW and larger-size wind power systems.
7. High rpm propeller/generator for power production in high-intensity areas.

Temperature differential

1. Ocean-based temperature differential power plant.
2. Land-based heat pumps for residential, commercial, and industrial space heating.

Hydrogen

1. Efficient, low-energy material, mass-producible small- and large-scale electrolysis units.
2. Internal combustion engines designed specifically for hydrogen.
3. Hydrogen switch-over units for existing cars, trucks, buses, trains, and planes.
4. Solid stored-hydrogen units for internal combustion engines and fuel cells.
5. Hydrogen-powered home appliances—refrigerators, stoves, air conditioners, furnaces.
6. Inexpensive, long-lasting hydrogen fuel cell.
7. Hydrogen-fueled industrial process units.
8. Hydrogen-fueled catalytic heaters.

Education

1. Education and training in ways of making dwellings and communities energy-autonomous.
2. Education about the total energy flows of the planet and the larger energy system of which the Earth is a part.
3. Posters, films, videotapes on wind, solar, tidal, geothermal, etc., energy sources for elementary schools, high schools, colleges, and the public.

Combinations

1. Mass-producible, integrated power unit for autonomous dwellings which is solar-heated, electrically powered from windmill, and capable of producing methane fuel from wastes.
2. Ocean-based energy/food farms utilizing wind, sun, and temperature differential.

Algae

1. Low-energy materials, mass-producible automated small-scale algae farms for hydrogen, methane, or alcohol production.
2. Large-scale algae farms.
3. More efficient hydrogen-producing algae.

Geothermal

1. Heat pump for heating individual dwelling, hospitals, schools, etc., from low-temperature Earth heat.
2. Industrial processes powered by geothermal sources.
3. Binary cycle power plants, low temperature utilization, downhole heat exchanger and pump, geopressure resource recovery.
4. Geothermal desalination.
5. Geothermal mineral resource recovery.

Bioconversion

1. Small- and large-scale low-energy materials, mass-producible automated waste powered gas-producing units.
2. Increased waste collection efficiency increases.
3. Small-scale wood-burning cogeneration systems appropriate for single-family dwellings and larger needs.

Falling water (hydroelectric)

1. Small-scale mass-producible hydroelectric units using low-energy materials, automated water-power units for individual dwellings and larger (1 kw–10 MW).
2. Underwater river power mechanisms.

Tidal

1. Small-scale tidal power units for communities and individual dwellings (1 kw–10 MW).

Research and Planning (or: the needed next steps)

Carry out similar comprehensive analysis for one or all *regions and localities* of the world, i.e., use different regions as basic unit of analysis rather than globe. Take this research to the next aggregate levels of analysis and planning detail.

(a) Determine energy resources for each separate global region and their exact locations, durations, intensities, and the environmental impacts of tapping them.
(b) Determine unique geographical characteristics of region re energy development.
(c) Determine unique economic, ecological, technological, cultural, and political restraints re energy of region.
(d) Identify and measure all sources of energy use and conservation within region (industry, transportation, commercial, residential).
(e) Identify and measure all local industries capable of producing income energy-harnessing artifacts (with local materials).
(f) Identify and measure all industrial processes capable of being tapped for energy: waste heat for use as space and water heating, processes that can be "stepped up" to produce useful power, etc.
(g) Identify and measure all present and projected energy needs of region; determine preferred levels of energy utilization/development.
(h) Obtain present energy, land use, development, etc. plans for locality/region.
(i) Formulate comprehensive energy development strategies for locality and region and integrate these with global energy development.
(j) Identify what specific changes need to be made in local/regional organization, planning, and regulation to bring strategies to fruition.
(k) Identify and contact all funding sources/decision-makers re energy development.
(l) Identify and contact all human resources re energy development.
(m) Make plan known to as large a group as possible in local region; involve as many people as possible in specific local planning process.

Appendix B

Glossary

Active Solar System An assembly of collectors, thermal storage devices and transfer fluid which converts solar energy into thermal energy, and in which energy in addition to solar is used to accomplish the transfer of thermal energy.

Aerobic Occurring in the presence of free oxygen.

Albedo Ratio of the radiation reflected from an object to the total radiation incident upon it, often expressed as a per cent.

Alternating Current (AC) An electric current whose direction is reversed at regular intervals. Electric power in the United States alternates with a frequency of 60 hertz, or cycles per second. Some European countries use 50 hertz.

Ampere A unit of measure for an electric current; the amount of current which flows in a circuit in which the electromotive force is one volt and the resistance is one ohm.

Anaerobic Occurring in the absence of air or free oxygen; sometimes called fermentation.

Animal Waste Conversion The process of obtaining oil from animal wastes. A Bureau of Mines experiment has obtained 80 gallons of oil per ton from cow manure. In comparison, average oil shale yields 25 gallons of oil per ton of ore.

Anthracite Coal A hard, black, lustrous coal that burns efficiently and is therefore valued for its heating quality. Coal of a calorific value of about 6,000 kcal/kg is composed of C: 67.9%; H; 0.8%; N: 1.5%; S: 0.5%; H_2O: 9%; ash: 15.5%.

Ash Noncombustible mineral matter contained in coal. There minerals are generally similar to ordinary sand, silt, and clay in chemical and physical properties.

Barrel (bbl) A liquid measure of oil, usually crude oil, equals to 42 American gallons or about 306 pounds. One barrel equals 5.6 cubic feet or 0.159 cubic meters. For crude oil 1 bbl is about 0.136 metric tons, 0.134 long tons, and 0.150 short tons. The energy values of petroleum products per barrel are: crude petroleum—5.6 million Btu/bbl; residual fuel oil—6.29; distillate fuel oil—5.83; gasoline—5.25; jet fuel (kerosene type)—5.67; jet fuel (naphtha type)—5.36; kerosene—5.67; petroleum coke—6.02; and asphalt—6.64.

Base Load The minimum load of a utility over a given period of time.

Bbls Barrels.

Biomass Total mass or weight of all organisms in a given area.

Bitumen Dense, sticky, semi-solid that is about 83% carbon.

Bituminous Coal Soft coal; coal that is high in carbonaceous and volatile matter. When volatile matter is removed from bituminous coal by heating in the absence of air, the coal becomes coke.

Bottoming Cycle A means to increase the thermal efficiency of a steam electric generating system by converting some waste heat from the condenser into electricity rather than discharging all of it to the environment.

Breakeven Costs The costs at which the price of a system's product is equal to the price of the equivalent energy product of the currently most economical energy-producing system.

Breeder Reactor A nuclear reactor so designed that it converts more uranium-238 or thorium into useful nuclear fuel than the uranium-235 or plutonium which it uses. The new fissionable materials are created by capture in the fertile materials of neutrons from the fission process. There are three types of breeder reactors: the liquid metal, fast breeder (LMFBR); the gas-cooled fast breeder (GCBR); and the molten salt breeder (MSBR).

British Thermal Unit (BTU) The quantity of heat necessary to raise the temperature of one pound of water one degree Fahrenheit. One BTU equals 252 calories, gram (mean), 778 foot-pounds, 1,055 joules, and 0.293 watt-hours.

Bunker "C" Fuel Oil A heavy residual fuel oil used by ships, industry, and for large-scale heating installations. In industry it is often referred to as No. 6 fuel.

Calorie A unit of heat energy equal to the amount of heat that will raise the temperature of one gram of water 1 degree centigrade. (cal.) The calorie is often used when temperature is measured on the Centigrade scale, while the British thermal unit is used when the measurement is on the Fahrenheit scale. One calorie equals 3.97×10^{-3} BTU, 4.18 joules, or 1.16×10^{-6} kilowatt hour. For energy issues, the usual term is the kilocalorie, or 1,000 calories.

Capital Energy Source Resources utilized by man which are exhaustible, depletable, and of a finite nature; there is no theoretical limit on the rate at which they can be used up, but there is a finite limit on the total quantity that can ever be used.

Centralized Energy Systems A few, large-scale production and conversion units; large capital investments, plants, and production and distribution capacities. Managerial/legal control of production and distribution in the control of relatively few people.

Coal Gas Manufactured gas made by distillation or carbonization of coal in a closed coal gas retort, coke oven, or other vessel.

Coal Gasification The conversion of coal (a solid) to a gas which is suitable for use as a fuel. The gas produced may be either a high-BTU or a low-BTU fuel. High-BTU gas is similar to natural gas and will range in energy content from 900 to 1,000 BTU per cubic foot. Low-BTU gas may range as low as 200 BTU per cubic foot.

Coal Liquefaction (Coal Hydrogenation) The conversion of coal into liquid hydrocarbons and related compounds by hydrogenation.

Coal Tar A gummy, black substance produced as a byproduct when bituminous coal is distilled.

Cogeneration The use of a single fuel source to generate power in one form, while the waste in the generating process is then recaptured and used to generate more power or to take the place of steam or heat that would otherwise have to be generated.

Coke A porous, solid residue resulting from the incomplete combustion of coal heated in a closed chamber, or oven, with a limited supply of air. Coke is largely carbon and is a desirable fuel in certain metallurgical industries.

Conduction The process by which energy is transferred directly from molecule to molecule; it is the way that electricity travels through a wire or heat moves from a warm body to a cool one when the two bodies are placed in contact.

Convection The transfer of heat by the circulation of a liquid or gas.

Coolant A substance circulated through a nuclear power plant to remove or transfer heat. Common coolants include water, air, carbon-dioxide, helium and liquid sodium.

Cooling Pond An artificial pond used to receive and dissipate waste heat, usually from a steam-electric power plant. Approximately an acre of pond surface is needed per megawatt of electric output for a modern steam-electric power plant.

Crude Oil A mixture of hydrocarbons that exist in natural underground reservoirs. It is liquid at atmospheric pressure after passing through surface separating processes and does not include natural gas products. It includes the initial liquid hydrocarbons produced from tar sands and oil shale.

Crude Oil Production The volume of crude oil flowing out of the ground. Domestic production is measured at the wellhead and includes lease condensate, which is a natural gas liquid recovered from lease separators or field facilities.

Crude Oil Imports The monthly volume of crude oil imported which is reported by receiving refineries, including crude oil entering the U.S. through pipelines from Canada.

Crude Oil Stocks Stocks held at refineries and at pipeline terminals. Does not include stocks held on leases (storage facilities adjacent to the wells), which historically total approximately 13 million barrels in U.S.

Cryogenics The study and production of very low temperatures and their associated phenomena.

Cubic Foot (cu. ft.) The most common unit of measurement of gas volume. It is the amount of gas required to fill a volume of 1 cubic foot under stated conditions of temperature, pressure, and water vapor. One cubic foot equals 28,317.01 cubic centimeters; 1,728 cubic inches; 7.48 gallons (U.S.); and 28.31 liters. One cubic foot/second equals 1.98 acre-feet/day; 448.8 gallons/minute; and 0.646 million gallons/day.

Decentralized Energy Systems Small-scale facilities located close to end-user, with minimal transportation/transmission costs but with storage costs. Managerial/legal control of production and distribution more in the hands of energy users.

Deep Mining The exploitation of coal or mineral deposits at depths exceeding about 1,000 feet. Coal is usually deep mined at not more than 1,500 feet. Mineral mines are deeper.

Depletion Allowance A tax allowance extended to the owner of exhaustible resources based on an estimate of the permanent reduction in value caused by the removal of the resource.

Direct Current (DC) Electricity that flows continuously in one direction, as contrasted with alternating current.

Direct Energy Conversion The generation of electricity from an energy source in a manner that does not involve transference of energy to a working fluid. Direct conversion methods have no moving parts and usually produce direct current. Some methods include photovoltaic conversion, thermionic conversion, and magnetohydrodynamic conversion.

District Heating The production of thermal energy by boilers or central station electricity-generating turbines which is then moved through a distribution system furnishing space heating, domestic hot water and/or steam for industrial processes.

Efficiency, Thermal Relating to heat, a percentage indicating the available BTU input that is converted to useful purposes. It is applied, generally, to combustion equipment. E = BTU output/BTU input.

Electricity A form of energy having chemical, radiant, and magnetic effects. It is a property of the basic particles of matter; a current of electricity is created by the flow of charged particles.

Electrostatic Precipitator Removes solids or liquids from air or gas by charging the particles and attracting them by the opposite charge to a boundary area where they can be collected.

Energy The capability of doing work. There are several forms of energy, including kinetic, potential, thermal, and electromagnetic. One form of energy may be changed to another such as burning coal to produce steam to drive a turbine which produces electricity.

Energy Conversion The transformation of energy from one form to another.

Energy Slave An inanimate energy source capable of producing the same amount of work as a man; 150,000 foot-pounds per 8-hour day, 250 days per year.

Energy Storage A process of turning energy into matter or valving energy into holding patterns so that it can be later reconverted to energy.

Enriched Uranium Uranium in which the amount of uranium-235 present has been artificially increased above the 0.71 percent found in nature. Uranium enriched between 3 and 6 percent is a common fuel for civil nuclear power stations. Uranium enriched to 90 percent or more is used for nuclear propulsion of warships and submarines, and in atomic bombs.

Enthalpy The heat content per unit mass, usually expressed in BTU per pound.

External Metabolics Technology, industrialization; the externalized life support of collective humanity, analogous function for function to the individual human's life-supporting internal metabolics.

Fast Breeder Reactor A nuclear reactor that operates with neutrons at the fast speed of their initial emission from the fission process, and that produces more fissionable material than it consumes.

Fission The splitting of a heavy nucleus into two approximately equal parts (which are radioactive nuclei of lighter elements), accompanied by the release of a relatively large amount of energy and generally one or more neutrons. Fission can occur spontaneously, but usually is caused by nuclear absorption of neutrons or other particles.

Fissionable Material Any material fissionable by slow neutrons. The three basic ones are uranium-235, plutonium-239 and uranium-233.

Flare Gas Unutilized natural gas burned in flares at an oil field; waste gas.

Flue Gas Gas from the combustion of fuel, the heating value of which has been substantially spent and which is, therefore, discarded to the flue or stack.

Fluidized Bed A fluidized bed results when a fluid, usually a gas, flows upward through a bed of suitably sized solid particles at a velocity high enough to buoy the particles, to overcome the influence of gravity, and to impart to them an appearance of great turbulence. Fluidized beds are used in the coal and petroleum industry. The Office of Coal Research is developing a coal-fired fluidized bed boiler which would permit use of Western U.S. low sulfur coals without slagging, and use of high sulfur coals without causing unacceptable environmental effects. Advantages include small size, efficient temperature control, little corrosion, high thermal efficiency (70%), considerable savings in investment costs (estimated at 21% for a 660 MW power plant), operating costs, and use of high sulfur coal.

Foot-pound The work required to lift 1 pound 1 vertical foot.

Fossil Fuels Decayed matter stored within the Earth; transformed over millions of years into coal, petroleum, natural gas, and peat.

Fuel Any substance that can be burned to produce heat. Sometimes includes materials that can be fissioned in a chain reaction to produce heat. The energy content of common fuels is as follows: 1 barrel (bbl.) of crude oil equals 5,600,000 BTU; 1 cubic foot of natural gas equals 1,031 BTU; 1 ton of coal equals 24,000,000 to 26,000,000 BTU.

Fuel Cell A device for combining fuel and oxygen in an electro-chemical reaction to generate electricity; chemical energy is converted directly into electrical energy without combustion.

Fuel Cycle The series of steps involved in supplying fuel for nuclear power reactors. It includes mining, refining of uranium, fabrication of fuel elements, their use in a nuclear reactor, chemical processing to recover remaining fissionable material, reenrichment of the fuel, refabrication into new fuel elements, and waste storage.

Fuel Oil Any liquid or liquefiable petroleum product burned for the generation of heat in a furnace or firebox, or for the generation of power in an engine; standard fuel oil is composed of C: 83.3%; H: 10.9%; O: 2.2%; S: 3.6% and produces about 38,000 BTU.

Fusion The formation of a heavier nucleus from two lighter ones, such as hydrogen isotopes, with the attendant release of energy.

Gas Turbine A prime mover in which gas, under pressure or formed by combustion, is directed against a series of turbine blades; the energy in the expanding gas is converted into mechanical energy supplying power at the shaft.

Gasoline A refined petroleum distillate, including naphtha, jet fuel or other petroleum oils (but not isoprene or cumene having a purity of 50 percent or more by weight, or benzene which meets the ASTM distillation standards for nitration grade) derived by refining or processing and having a boiling range at atmospheric pressure from 80 degrees to 400 degrees F.

Geothermal Energy Underground reservoirs of steam and scalding water found from a few hundred to 30,000 feet beneath the surface; the water is heated by the molten matter that oozes up from the interior of the Earth between the huge plates that make up the Earth's crust.

Geothermal Gradient The change in temperature of the earth with depth, expressed either in degrees per unit depth, or in units of depth per degree. The mean rate of increase in temperature with depth in areas that are not adjacent to volcanic regions is about 1.5 degrees F per 100 feet (2.75 degrees C per 100 meters), corresponding to about 80 degrees F per mile of depth.

Geothermal Steam Steam drawn from deep within the Earth. There are about 90 known places in the continental United States where geothermal steam could be harnessed for power. These are in California, Idaho, Nevada, and Oregon.

Greenhouse Effect The warming of the lower atmosphere and the surface of the Earth due to the absorption and reradiation of infrared radiation by the carbon dioxide and water vapor in the atmosphere.

Gross Energy The total amount of energy available.

Half-Life, Radioactive Time required for a radioactive substance to lose 50% of its activity by decay. Each radionuclide has a unique half-life.

Heat A form of kinetic energy, whose effects are produced by the vibration, rotation, and general motions of molecules.

Heat Engine An engine in which heat is transformed into mechanical energy.

Heat Pump A refrigeration machine that is used for heating rather than cooling. Expanding refrigeration fluid removes heat from a large heat source; the fluid is then compressed, and the heat resulting from compression is discharged to a heat exchanger next to the surroundings to be heated.

Horizontal Integration The control by one company or industry grouping of a number of related or substitutable products. In the energy industry this would be characterized by control of more than one fuel or energy source.

Horsepower (Hp.) A standard unit of power equal to 740 watts. One horsepower equals 2,545.08 BTU (mean)/hour, 550 foot-pounds/second.

Hydrocarbon A compound containing only carbon and hydrogen. The fossil fuels are predominantly hydrocarbons, with varying amounts of organic compounds of sulfur, nitrogen, oxygen, and some inorganic material.

Hydroelectric Powerplant An electric powerplant in which the turbine generator is driven by falling water.

In-Situ Recovery Refers to methods to extract the fuel component of a deposit without removing the deposit from its bed.

Joule A unit of energy or work which is equivalent to one watt per second or 0.737 foot-pounds.

Kerosene Any jet fuel, diesel fuel, fuel oil, or other petroleum oils derived by refining or processing crude oil or unfinished oils, in whatever type of plant such refining or processing may occur, which has a boiling range at atmospheric pressure from 400 degrees to 550 degrees F.

Kiloton (Kt) A measure of explosive force which represents the energy of 10^{12} calories, or 3.9×10^9 BTU or 4×10^{12} joules.

Kilowatt (Kw) 1,000 watts. A unit of power equal to 1,000 watts, or to the energy consumption at a rate of 1,000 joules per second. It is usually used for electrical power. An electric motor rated at one horsepower uses electrical energy at a rate of about 3/4 kilowatt.

Kilowatt-Hour (Kwh) A unit of work or energy equal to that expended by one kilowatt in one hour. It is equivalent to 3,415 BTU of heat energy.

Kinetic Energy The energy of motion; the ability of an object to do work because of its motion.

Langley Solar radiation is customarily measured in langleys per minute. 1 langley is equal to 1 calorie of radiation energy per square centimeter.

Latent Heat The amount of heat energy released when a gas is condensed to a liquid or a liquid is changed to a solid without change of temperature, expressed in calories per gram. Heat energy is stored as latent heat in the water vapor of the atmosphere.

Light-Water Reactor (LWR) Nuclear reactor in which water is the primary coolant/moderator with slightly enriched uranium fuel. There are two commercial light-water reactor types—the boiling water reactor (BWR) and the pressurized water reactor (PWR).

Lignite A low grade coal of a variety intermediate between peat and bituminous coal.

Liquefied Natural Gas (LPG) Natural gas that has been liquefied by cooling to about $-140°$ C. In this form, it occupies a relatively small volume and can be transported economically by ocean tanker.

Liquid Metal Fast Breeder (LMFBR) A nuclear breeder reactor cooled by molten sodium in which fission is caused by fast neutrons.

Liter The primary standard of capacity in the metric system, equal to the volume of one kilogram of pure water at maximum density, at approximately 4 degrees C, and under normal atmospheric pressure. One liter = 0.264 gallons (U.S.), 1.05 quarts (U.S.), or 2.11 pints (U.S.).

Magnetohydrodynamics (MHD) A branch of physics and power generation that deals with phenomena arising from the motion of electrically conducting fluids in the presence of electric and magnetic fields. In open-cycle MHD generators, the working fluid is exhausted to the atmosphere. In the closed-cycle MHD, the working fluid is continuously recirculated through a closed loop.

Megawatt (MW) 1,000 kilowatts, 1 million watts. Approximately 1,000 MW (the capacity of many modern nuclear power plants) is required to satisfy the electrical needs of an average city of 1 million people.

Megawatt-Day Per Ton (MWd/t) A unit that expresses the burnup of nuclear fuel in a reactor; specifically the number of megawatt-days of heat output per metric ton of fuel in the reactor.

Methane (CH_4) The lightest in the paraffin series of hydrocarbons. It is colorless, odorless, and flammable. It forms the major portion of marsh gas and natural gas.

Methyl Alcohol (CH_3OH) A poisonous liquid, also known as methanol, which is the lowest member of the alcohol series. Also known as wood alcohol, since its principal source is the destructive distillation of wood.

Metric Ton Coal Equivalent (mtce) The energy of 1 average metric ton of coal (28.8 million BTUs) which may be applied to the measure of energy transformations not necessarily having originated from coal.

Mine-Mouth Plant A steam-electric plant or coal gasification plant built close to a coal mine and usually associated with delivery of output via transmission lines or pipelines over long distances as contrasted with plants located nearer load centers and at some distance from sources of fuel supply.

Natural Gas Naturally occurring mixtures of hydrocarbon gases and vapors, the more important of which are methane, ethane, propane, butane, pentane, and hexane found in porous geologic formations beneath the earth's surface. The energy content of natural gas is usually taken as 1,032 BTU/cu. ft. (8,560 Kcal/m^3).

Net Energy Gross energy minus the energy cost of extracting and delivering the energy.

Nitrogen Oxides (NO_2) A product of combustion of fossil fuels whose production increases with the temperature of the process. It can become a serious air pollutant if concentrations are excessive.

Nuclear Power Plant Any device, machine, or assembly that converts nuclear energy into some form of useful power, such as mechanical or electrical power.

Nuclear Reactor A device in which a fission chain reaction can be initiated, maintained, and controlled. Its essential component is a core with fissionable fuel. It usually has a moderator, reflector, shielding, coolant, and control mechanisms. It is the basic machine of nuclear power.

OPEC Organization of Petroleum Exporting Countries. Founded in 1960 to unify and coordinate petroleum policies of the members. The members and the date of membership are: Abu Dhabi (1967); Algeria (1969); Indonesia (1962); Iran (1960); Iraq (1960); Kuwait (1960); Libya (1962); Nigeria (1971); Qatar (1961); Saudi Arabia (1960); and Venezuela (1960). OPEC headquarters is in Vienna, Austria.

Oil Shale A convenient expression used to cover a range of materials containing organic matter (kerogen) which can be converted into crude shale oil, gas, and carbonaceous residue by heating.

Outage The period in which a generating unit, transmission line, or other facility is out of service.

Outcrop (Cropline) A line on the Earth's surface where a coal bed or other rock strata comes to the surface. An outcrop does not necessarily require the visible presence of coal or other mineral at the surface but includes those places where they may be covered by a mantle of soil or other surface material.

Overburden Earth, rock, and other material overlying a seam of coal. The term is used where the coal is at sufficiently shallow depth to permit strip mining.

Particulate Matter Solid particles, such as ash, which are released from combustion process in exhaust gases at fossil-fuel plants.

Passive Solar System An assembly of natural and architectural components including collectors, thermal storage devices, and transfer fluid which converts solar energy into thermal energy in a controlled manner and in which no pumps are used to accomplish the transfer of thermal energy.

Peat A material formed in marshes and swamps from dead plants.

Petroleum An oily flammable bituminous liquid that may vary from almost colorless to black, occurs in many places in the upper strata of the Earth, is a complex mixture of hydrocarbons with small amounts of other substances, and is prepared for use as gasoline, naphtha, or other products by various refining processes.

Petroleum Refinery A plant that converts crude petroleum into the many petroleum fractions (asphalt, fuel oil, gasoline, etc.). Usually this conversion is accomplished by fractional distillation.

Photosynthesis A photochemical reaction by which carbon dioxide is reduced (fixed) to carbohydrate in the presence of chlorophyll. The process by which green plants utilize solar energy to convert carbon dioxide and water into new biomass.

Pilot Plant A small-scale industrial process unit operated to test the application of a manufacturing process or energy production facility under conditions that will yield information useful in the design and operation of full-scale equipment. The pilot unit serves to disclose the special problems to be solved in adapting a successful laboratory method to commercial-sized units.

Plutonium A fissionable element that does not occur in nature but is obtained by exposure of U^{238} to neutrons in a reactor.

Potential Energy Energy which a body possesses because of its position with respect to another body, or relative to other parts of itself. More generally, potential energy is energy in any form not associated with motion; thus the energy stored in chemical bonds is potential energy.

Power The rate at which work is performed or energy expended.

Primary Fuel Fuel consumed in original production of energy as contrasted to a conversion of energy from one form to another.

Primary Fuels Those fuels providing "original" energy, as contrasted to secondary energy sources and fuels such as electricity or hydrogen. At present primary fuels are all fossil fuels except for small amounts of uranium. For capture of continuous natural energy processes such as wind or falling water there is no primary fuel.

We CAN eliminate all fossil fuel and nuclear energy-use in ten years and harvest more energy per Earth individual than Americans currently misuse... Yay.

...IF we change all manner of habits (farewell, Big Car), speed up "progress" in some areas and stop it or run it backwards in other areas (such as soil-care), and *personally* do more with less... Boo.

"Boo" to contemplate. Living it may contain some deeper-throated Yays. Do you know anyone who didn't enjoy the "Energy Crisis" of '73? How about the thirties? I asked Buckminster Fuller that one. He remembers the Great Depression fondly as a period of trial and of abundant learning and creativity for him and everyone he knew.

The ability of people to learn is widely discredited. I've never seen it factored formally into forecasting schemes. The Future is usually treated as if it were a consumer product—the suckers will buy whatever we give them. When you're doing predicitons—rosy *or* dire—it's a seductive fallacy.

My medicine against that particular blindness is "co-evolution"—the understanding that everything alive learns from everything else constantly. Co-evolution—life—requires constant mistakes and constant correction. Forecast from the mistakes and you get deadly direness. Forecast from the corrections and you're a politician. Forecast from both and you get the cornucopia of complexity and uncertainty that is real life.

The central question is, "How do you convince people to change?"

You don't. What happens is that *after* a big change comes along, some people are found to have "pre-adapted" to it, and they flourish. Everyone else quickly learns from them. That's co-evolution—part of it.

Pre-adaption you *can* encourage, in a general way, by encouraging diversity, independence, inventiveness, difficulty (i.e., learning), and something that looks a lot like faith.

A pre-adaptive fanning out is also helped by doom-calling. "It's gonna be crowded and hungry and dark and cold and dangerous and uncertain!" Then come the "unlesses." This book is one—one of the best I think.

My expectation is that the sky *will* fall. My faith is that there's another sky behind it, made of accurate appraisal like this book, and accurate response such as only ever-dying life can make.

We'll see.

Stewart Brand
Sausalito, California

Primary Stocks of Refined Products Oil stocks held at refineries, bulk terminals, and pipelines.

Prime Movers Engines for converting fuel into mechanical energy.

Probable Reserves A realistic assessment of the reserves that will be recovered from known oil or gas fields based on the estimated ultimate size and reservoir characteristics of such fields. Probable reserves include those reserves shown in the proved category.

Proved Reserves The estimated quantity of crude oil, natural gas, natural gas liquids, etc. which analysis or geological and engineering data demonstrate with reasonable certainty to be recoverable from known oil or gas fields under existing economic and operating conditions.

Public Utility A business performing some public service and subject to special government regulation. Electric, water, and sewer service are examples.

Pumped Storage An arrangement whereby additional electric power may be generated during peak load periods by hydraulic means using water pumped into a storage reservoir during off-peak periods.

Quad A gross measure of energy standing for quadrillion BTU (10^{15}). One quadrillion is approximately equal to: 180 million barrels of petroleum, 42 million tons of bituminous coal, 980 billion ft^3 of natural gas, 300 billion kilowatt hours of electricity.

Recoverable Reserves Minerals expected to be recovered by present-day techniques and under present economic conditions.

Refinery A device (usually a tower) or process which heats crude oil so that it separates into chemical components, which are then distilled off as more usable substances. Simple structure components vaporize first. Typical crude fractions, from top to botton or simple to complex, are: methane and ethane (the gasolines); propane and butane; kerosene, fuel oil, and lubricants; jelly paraffin, asphalt, and tar.

Reprocessing Chemical recovery of unburned uranium and plutonium and certain fission products from spent fuel elements that have produced power in a nuclear reactor.

Scrubber Removes substances from exhaust gas from coal power plant by passing liquid (generally water) through it.

Solar Cell A device which converts solar radiation to a current of electricity.

Solar Energy The energy transmitted from the Sun, which is in the form of electromagnetic radiation. Although the Earth receives about one-half of one billionth of the total solar energy output, this amounts to about 1,500,000 trillion kilowatt-hours annually.

Solar Furnace An optical device with large mirrors that focuses the rays from the Sun upon a small focal point to produce very high temperatures.

Solar Power Useful power derived from solar energy.

Solar Total Energy Systems Solar energy systems that supply heat for industrial processes and electricity.

Strip Mining A type of surface mining whereby overburden and coal are removed from successive long parallel cuts, with overburden from the second and successive cuts being placed in the previously mined cut. This system is practicable only where the coal lies close to the surface. Strip mines operate where the depth of the minable coal beds increases gradually, allowing mining to proceed at distances of up to a mile or more into the outcropping seam. This method is common in the Midwest and West.

Super Tanker A very large oil tanker. The definition changes with advancing marine technology. In the late 1940's, 45,000 ton tankers were considered super tankers; in the 1950's, 100,000 ton was a super tanker; now common usage is 500,000 ton, and still larger ships are planned.

Tar Sands Hydrocarbon bearing deposits distinguished from more conventional oil and gas reservoirs by the high viscosity of the hydrocarbon, which is not recoverable in its natural state through a well by ordinary oil production methods; also known as oil sands and heavy oil, it is a mixture of 84–88% sand and mineral rich clays, 4% water and 8–12% bitumen.

Temperature The intensity of heat. The temperature of a body is proportional to the average kinetic energy of the molecules of which it is composed. Various properties of a body change when its temperature changes, and these changes are the bases for different temperature scales.

Tertiary Recovery Use of heat and other methods other than fluid injection to augment oil recovery (presumably occurring after secondary recovery).

Thermal Efficiency The ratio of the heat used to the total heat contained in the fuel consumed.

Thermal Pollution An increase in the temperature of water resulting from waste heat released by a thermal electric power plant.

Thermal Power Plant Any electric power plant which operates by generating heat and converting the heat to electricity.

Thermal Sea Power Heat energy from the Sun stored in the ocean and tapped via temperature differential power facilities.

Thermionics The direct conversion of heat into electricity by "boiling" electrons off from a hot metal surface and "condensing" them on a cooler surface.

Thermodynamics The science and study of the relationships between heat and mechanical work. First Law: Energy can neither be created nor destroyed. Second Law: Heat cannot pass from a colder to a warmer body without additional expenditure of energy.

Tidal Energy The kinetic energy of the motion of the Earth's large water bodies caused by the gravitational attraction of the Moon.

Ton A unit of weight equal to 2,000 pounds in the United States, Canada, and the Union of South Africa, and to 2,240 pounds in Great Britain. The American ton is often called the short ton, while the British ton is called the long ton. The metric ton, or 1,000 kilograms, equals 2,204.62 pounds.

Topping cycle A method of cogeneration in which fuel is burned to produce electricity and the exhaust heat from the generation system is used for process heat.

Total Gross Energy Consumption Total energy inputs into the economy, including coal, petroleum, natural gas, and the electricity generated by hydroelectric, nuclear, and geothermal power plants. Gross consumption includes conversion losses by the electric power sector.

Total Net Energy Consumption Inputs into the final consuming sectors, i.e., household and commercial, industrial, and transportation, and consisting of direct fuels and electricity distributed from the electric power sector. Conversion losses in the electric sector constitute the difference between net and gross energy.

Trillion 1 million million; 10^{12}.

Vertical integration The control by one company of the various stages of a commodity from primary production, through processing and marketing, to final consumer.

Volt A unit of electrical force equal to that amount of electromotive force that will cause a steady current of one ampere to flow through a resistance of one ohm.

Wastes, Radioactive Equipment and materials, from nuclear operations, which are radioactive and for which there is no further use. Wastes are generally classified as high-level (having radioactivity concentrations of hundreds to thousands of curies per gallon or cubic foot), low level (in the range of 1 microcurie per gallon or cubic foot), or intermediate.

Watt The rate of energy transfer equivalent to one ampere under an electrical pressure of one volt. One watt equals 1/746 horsepower, or one joule per second.

Watt-Hour The total amount of energy used in one hour by a device that uses one watt of power for continuous operation. Electrical energy is commonly sold by the kilowatt hour (1,000 watt-hours).

Wind Energy The kinetic energy of the motion of the Earth's enveloping air ocean caused by the Sun's heating of the atmosphere.

Wind Turbine Electrical generating windmill.

Appendix C

Conversion Table: Energy to Power

1 watt (w) = 1 joule (j) for 1 second
1 kilowatt (kw) = 1,000 watts = 1.3415
   hp = 738 ft. lb./sec. = .948 BTU/sec.
   = 3,415 BTU/hr.
1 megawatt (MW) = 1,000 kilowatt (10^6 watts)
1 gigawatt (GW) = 1,000 megawatts (10^9 watts)
1 terawatt (tw) = 1,000 gigawatts (10^{12} watts)
1 twh = 1 billion kwh
1 kw (capacity) = 8,760 kilowatt-hours (kwh; maximum annual production)
1 kwh = 1.34 horsepower-hours (hph)
1 kwh = 3,415 BTU
1 kwh = 3.6 × 10^6 watts per second
   = 3.6 × 10^6 joules (3.6 megajoules)
1 kwh = 2.66 × 10^6 ft. lbs.
1 kwh = 36 × 10^{12} ergs
1 hp = 745.7 watts = .7457 kw = 550 ft. lb./sec. = .707 BTU/sec. = 2,544 BTU/hr.
1 watt-hour = 3.6 × 10^3 joules
1 BTU = 1,055 joules = .252 kcal = 778.3 ft. lb. = .0002930 kwh
1 barrel = 42 gallons

Energy Content of Fuel

1 lb of TNT = 478 kcal = 1,890 BTU
1 lb of bread = 1,300 kcal = 5,100 BTU
1 lb of wood = 1,800 kcal = 7,100 BTU
1 lb of crude oil (.14 gal) = 4,800 kcal = 19,000 BTU
1 lb of natural gas (25 ft³) = 6,600 kcal = 26,000 BTU
1 lb of uranium-235 = 8.6 billion kcal = 34 × 10^9 BTU
anthracite coal = 26.0 million BTU/ton
bituminous coal = 24.8 million BTU/ton
sub-bituminous coal = 19.0 million BTU/ton
lignite = 13.4 million BTU/ton
peat = 7.6 × 10^6 BTU/ton to 13 × 10^6 BTU/ton
million metric tons of coal equivalent (mmtce)
   = 7 × 10^{12} kcal = 2.928 × 10^{16} j
crude petroleum = 5.60 million BTU/bbl (42 gal)
residual fuel oil = 6.29 million BTU/bbl
distillate fuel oil = 5.83 million BTU/bbl
gasoline (including aviation) = 5.25 million BTU/bbl
jet fuel (kerosene type) = 5.67 million BTU/bbl
jet fuel (naphtha type) = 5.36 million BTU/bbl
kerosene = 5.67 million BTU/bbl
asphalt and road oil = 6.64 BTU/bbl
petroleum coke = 6.02 million BTU/bbl
diesel = 5,825 million BTU/bbl
shale = 3.69 million BTU/ton
1 million metric tons of crude oil of petroleum products = 40 × 10^{12} BTU
natural gas (dry) = 1,031 BTU/cu ft at STP
natural gas (wet) = 1,103 BTU/cu ft at STP
natural gas (liquids—avg) = 4.1 million BTU/bbl
1,000 million m³ of natural gas = 36 × 10^{12} BTU
wood and drycrop waste = 14–18 × 10^6 BTU/ton

	kwhs	joules	calories	watt-years	BTUs
1 kwh =	1	3.6 × 10^6	.86 × 10^6	.114	3,410
1 joule =	.278 × 10^{-6}	1	.239	31.7 × 10^{-9}	.948 × 10^{-3}
1 calorie =	1.16 × 10^{-6}	4.18	1	13.3 × 10^{-6}	3.97 × 10^{-3}
1 watt-year =	8.77	31.6 × 10^6	7.54 × 10^6	1	24.4 × 10^3
1 BTU =	.293 × 10^{-3}	1,054	252	32.1 × 10^{-6}	1
1 metric ton (10^6 g) anthracite coal yields	7,630	27.5 × 10^9	6.56 × 10^9	879	26 × 10^6
1 barrel of crude petroleum yields	1,641	5.40 × 10^9	1.41 × 10^9	187	5.6 × 10^6
1 ft³ natural gas (dry) yields	.30	1.08 × 10^6	.259 × 10^6	.034	1,031
1 gram U-235 yields	23.0 × 10^3	82.8 × 10^9	19.8 × 10^9	2,620	78.4 × 10^6
1 gram deuterium yields	66.0 × 10^3	.238 × 10^{12}	56.8 × 10^9	753 × 10^3	.225 × 10^9
1 ton crude petroleum yields	11.0 × 10^3	.04 × 10^{12}	9.4 × 10^9	1.25 × 10^3	.375
1 ft³ hydrogen yields	.080	.0165 × 10^6	.069 × 10^6	.091	275

Energy Conversions

	Anthracite Coal (tons)	Oil (bbl)	Natural Gas (cu ft)	U-235 (g)	Deuterium (g)
1 kwh is derived from	.129 × 10^{-3}	.608 × 10^{-3}	3.51	43.5 × 10^{-6}	15.1 × 10^{-6}
1 joule is derived from	35.9 × 10^{-12}	.172 × 10^{-9}	1.03 × 10^{-6}	12 × 10^{-12}	4.21 × 10^{-12}
1 calorie is derived from	.15 × 10^{-9}	.707 × 10^{-9}	4.09 × 10^{-6}	58.2 × 10^{-12}	17.6 × 10^{-12}
1 watt-year is derived from	1.13 × 10^{-3}	5.33 × 10^{-3}	30.9	.38 × 10^{-3}	13.3 × 10^{-12}
1 BTU is derived from	38 × 10^{-9}	.178 × 10^{-6}	1.03 × 10^{-3}	12.8 × 10^{-9}	4.44 × 10^{-9}
1 metric ton of (10^6 BTU) anthracite coal =	1	5	25.4 × 10^3	.33	.12
1 barrel of oil (42 gal) =	.21	1	5.47 × 10^3	.071	.026
1 ft³ of natural gas (dry) =	39 × 10^{-6}	.18 × 10^{-3}	1	13.0 × 10^{-6}	4.54 × 10^{-6}
1 gram U-235 =	3.0	14	79.3 × 10^3	1	3.5
1 gram deuterium =	8.6	40	.227 × 10^6	2.9	1

Approximate Conversion Factors for Crude Oil

FROM	INTO						
	metric tons	long tons	short tons	barrels	kiloliters (m³)	1,000 gal (Imperial)	1,000 gal (U.S.)
	MULTIPLY BY						
metric tons	1	0.984	1.102	7.33	1.16	0.256	0.308
long tons	1.016	1	1.120	7.45	1.18	.261	.313
short tons	.907	.893	1	6.65	1.05	.233	.279
barrels	.136	.134	.150	1	.159	.035	.042
kiloliters (m³)	.863	.849	.951	6.29	1	.220	.264
1,000 gal (Imp.)	3.91	3.83	4.29	28.6	4.55	1	1.201
1,000 gal (U.S.)	3.25	3.19	3.58	23.8	3.79	.833	1

Calorif[ic]

gas
natural ga[s]
propane
refinery g[as]
coke-ove[n]
blast furn[ace]
producer
Hydrogen

The following are some frequently
equivalents in British Thermal Un[its]
actual natural gas equivalent has

1 cubic f[oot]
1 therm = 100 c[u ft]
1 Mcf = 1,000 cu [ft]
1 M Therm = 100,000
1 M Mcf = 1,000,000 c[u ft]
1,000 ft³ hy[drogen]

Primary Stocks of Refined Products Oil stocks held at refineries, bulk terminals, and pipelines.

Prime Movers Engines for converting fuel into mechanical energy.

Probable Reserves A realistic assessment of the reserves that will be recovered from known oil or gas fields based on the estimated ultimate size and reservoir characteristics of such fields. Probable reserves include those reserves shown in the proved category.

Proved Reserves The estimated quantity of crude oil, natural gas, natural gas liquids, etc. which analysis or geological and engineering data demonstrate with reasonable certainty to be recoverable from known oil or gas fields under existing economic and operating conditions.

Public Utility A business performing some public service and subject to special government regulation. Electric, water, and sewer service are examples.

Pumped Storage An arrangement whereby additional electric power may be generated during peak load periods by hydraulic means using water pumped into a storage reservoir during off-peak periods.

Quad A gross measure of energy standing for quadrillion BTU (10^{15}). One quadrillion is approximately equal to: 180 million barrels of petroleum, 42 million tons of bituminous coal, 980 billion ft^3 of natural gas, 300 billion kilowatt hours of electricity.

Recoverable Reserves Minerals expected to be recovered by present-day techniques and under present economic conditions.

Refinery A device (usually a tower) or process which heats crude oil so that it separates into chemical components, which are then distilled off as more usable substances. Simple structure components vaporize first. Typical crude fractions, from top to bottom or simple to complex, are: methane and ethane (the gasolines); propane and butane; kerosene, fuel oil, and lubricants; jelly paraffin, asphalt, and tar.

Reprocessing Chemical recovery of unburned uranium and plutonium and certain fission products from spent fuel elements that have produced power in a nuclear reactor.

Scrubber Removes substances from exhaust gas from coal power plant by passing liquid (generally water) through it.

Solar Cell A device which converts solar radiation to a current of electricity.

Solar Energy The energy transmitted from the Sun, which is in the form of electromagnetic radiation. Although the Earth receives about one-half of one billionth of the total solar energy output, this amounts to about 1,500,000 trillion kilowatt-hours annually.

Solar Furnace An optical device with large mirrors that focuses the rays from the Sun upon a small focal point to produce very high temperatures.

Solar Power Useful power derived from solar energy.

Solar Total Energy Systems Solar energy systems that supply heat for industrial processes and electricity.

Strip Mining A type of surface mining whereby overburden and coal are removed from successive long parallel cuts, with overburden from the second and successive cuts being placed in the previously mined cut. This system is practicable only where the coal lies close to the surface. Strip mines operate where the depth of the minable coal beds increases gradually, allowing mining to proceed at distances of up to a mile or more into the outcropping seam. This method is common in the Midwest and West.

Super Tanker A very large oil tanker. The definition changes with advancing marine technology. In the late 1940's, 45,000 ton tankers were considered super tankers; in the 1950's, 100,000 ton was a super tanker; now common usage is 500,000 ton, and still larger ships are planned.

Tar Sands Hydrocarbon bearing deposits distinguished from more conventional oil and gas reservoirs by the high viscosity of the hydrocarbon, which is not recoverable in its natural state through a well by ordinary oil production methods; also known as oil sands and heavy oil, it is a mixture of 84–88% sand and mineral rich clays, 4% water and 8–12% bitumen.

Temperature The intensity of heat. The temperature of a body is proportional to the average kinetic energy of the molecules of which it is composed. Various properties of a body change when its temperature changes, and these changes are the bases for different temperature scales.

Tertiary Recovery Use of heat and other methods other than fluid injection to augment oil recovery (presumably occurring after secondary recovery).

Thermal Efficiency The ratio of the heat used to the total heat contained in the fuel consumed.

Thermal Pollution An increase in the temperature of water resulting from waste heat released by a thermal electric power plant.

Thermal Power Plant Any electric power plant which operates by generating heat and converting the heat to electricity.

Thermal Sea Power Heat energy from the Sun stored in the ocean and tapped via temperature differential power facilities.

Thermionics The direct conversion of heat into electricity by "boiling" electrons off from a hot metal surface and "condensing" them on a cooler surface.

Thermodynamics The science and study of the relationships between heat and mechanical work. First Law: Energy can neither be created nor destroyed. Second Law: Heat cannot pass from a colder to a warmer body without additional expenditure of energy.

Tidal Energy The kinetic energy of the motion of the Earth's large water bodies caused by the gravitational attraction of the Moon.

Ton A unit of weight equal to 2,000 pounds in the United States, Canada, and the Union of South Africa, and to 2,240 pounds in Great Britain. The American ton is often called the short ton, while the British ton is called the long ton. The metric ton, or 1,000 kilograms, equals 2,204.62 pounds.

Topping cycle A method of cogeneration in which fuel is burned to produce electricity and the exhaust heat from the generation system is used for process heat.

Total Gross Energy Consumption Total energy inputs into the economy, including coal, petroleum, natural gas, and the electricity generated by hydroelectric, nuclear, and geothermal power plants. Gross consumption includes conversion losses by the electric power sector.

Total Net Energy Consumption Inputs into the final consuming sectors, i.e., household and commercial, industrial, and transportation, and consisting of direct fuels and electricity distributed from the electric power sector. Conversion losses in the electric sector constitute the difference between net and gross energy.

Trillion 1 million million; 10^{12}.

Vertical integration The control by one company of the various stages of a commodity from primary production, through processing and marketing, to final consumer.

Volt A unit of electrical force equal to that amount of electromotive force that will cause a steady current of one ampere to flow through a resistance of one ohm.

Wastes, Radioactive Equipment and materials, from nuclear operations, which are radioactive and for which there is no further use. Wastes are generally classified as high-level (having radioactivity concentrations of hundreds to thousands of curies per gallon or cubic foot), low level (in the range of 1 microcurie per gallon or cubic foot), or intermediate.

Watt The rate of energy transfer equivalent to one ampere under an electrical pressure of one volt. One watt equals 1/746 horsepower, or one joule per second.

Watt-Hour The total amount of energy used in one hour by a device that uses one watt of power for continuous operation. Electrical energy is commonly sold by the kilowatt hour (1,000 watt-hours).

Wind Energy The kinetic energy of the motion of the Earth's enveloping air ocean caused by the Sun's heating of the atmosphere.

Wind Turbine Electrical generating windmill.

Appendix C

Conversion Table: Energy to Power

1 watt (w) = 1 joule (j) for 1 second
1 kilowatt (kw) = 1,000 watts = 1.3415 hp = 738 ft. lb./sec. = .948 BTU/sec. = 3,415 BTU/hr.
1 megawatt (MW) = 1,000 kilowatt (10^6 watts)
1 gigawatt (GW) = 1,000 megawatts (10^9 watts)
1 terawatt (tw) = 1,000 gigawatts (10^{12} watts)
1 twh = 1 billion kwh
1 kw (capacity) = 8,760 kilowatt-hours (kwh; maximum annual production)
1 kwh = 1.34 horsepower-hours (hph)
1 kwh = 3,415 BTU
1 kwh = 3.6 × 10^6 watts per second = 3.6 × 10^6 joules (3.6 megajoules)
1 kwh = 2.66 × 10^6 ft. lbs.
1 kwh = 36 × 10^{12} ergs
1 hp = 745.7 watts = .7457 kw = 550 ft. lb./sec. = .707 BTU/sec. = 2,544 BTU/hr.
1 watt-hour = 3.6 × 10^3 joules
1 BTU = 1,055 joules = .252 kcal = 778.3 ft. lb. = .0002930 kwh
1 barrel = 42 gallons

Energy Content of Fuel

1 lb of TNT = 478 kcal = 1,890 BTU
1 lb of bread = 1,300 kcal = 5,100 BTU
1 lb of wood = 1,800 kcal = 7,100 BTU
1 lb of crude oil (.14 gal) = 4,800 kcal = 19,000 BTU
1 lb of natural gas (25 ft^3) = 6,600 kcal = 26,000 BTU
1 lb of uranium-235 = 8.6 billion kcal = 34 × 10^9 BTU
anthracite coal = 26.0 million BTU/ton
bituminous coal = 24.8 million BTU/ton
sub-bituminous coal = 19.0 million BTU/ton
lignite = 13.4 million BTU/ton
peat = 7.6 × 10^6 BTU/ton to 13 × 10^6 BTU/ton
million metric tons of coal equivalent (mmtce) = 7 × 10^{12} kcal = 2.928 × 10^{16} j
crude petroleum = 5.60 million BTU/bbl (42 gal)
residual fuel oil = 6.29 million BTU/bbl
distillate fuel oil = 5.83 million BTU/bbl
gasoline (including aviation) = 5.25 million BTU/bbl
jet fuel (kerosene type) = 5.67 million BTU/bbl
jet fuel (naphtha type) = 5.36 million BTU/bbl
kerosene = 5.67 million BTU/bbl
asphalt and road oil = 6.64 BTU/bbl
petroleum coke = 6.02 million BTU/bbl
diesel = 5,825 million BTU/bbl
shale = 3.69 million BTU/ton
1 million metric tons of crude oil of petroleum products = 40 × 10^{12} BTU
natural gas (dry) = 1,031 BTU/cu ft at STP
natural gas (wet) = 1,103 BTU/cu ft at STP
natural gas (liquids—avg) = 4.1 million BTU/bbl
1,000 million m^3 of natural gas = 36 × 10^{12} BTU
wood and drycrop waste = 14–18 × 10^6 BTU/ton

	kwhs	joules	calories	watt-years	BTUs
1 kwh =	1	3.6 × 10^6	.86 × 10^6	.114	3,410
1 joule =	.278 × 10^{-6}	1	.239	31.7 × 10^{-9}	.948 × 10^{-3}
1 calorie =	1.16 × 10^{-6}	4.18	1	13.3 × 10^{-6}	3.97 × 10^{-3}
1 watt-year =	8.77	31.6 × 10^6	7.54 × 10^6	1	24.4 × 10^3
1 BTU =	.293 × 10^{-3}	1,054	252	32.1 × 10^{-6}	1
1 metric ton (10^6 g) anthracite coal yields	7,630	27.5 × 10^9	6.56 × 10^9	879	26 × 10^6
1 barrel of crude petroleum yields	1,641	5.40 × 10^9	1.41 × 10^9	187	5.6 × 10^6
1 ft^3 natural gas (dry) yields	.30	1.08 × 10^6	.259 × 10^6	.034	1,031
1 gram U-235 yields	23.0 × 10^3	82.8 × 10^9	19.8 × 10^9	2,620	78.4 × 10^6
1 gram deuterium yields	66.0 × 10^3	.238 × 10^{12}	56.8 × 10^9	753 × 10^3	.225 × 10^9
1 ton crude petroleum yields	11.0 × 10^3	.04 × 10^{12}	9.4 × 10^9	1.25 × 10^3	.375
1 ft^3 hydrogen yields	.080	.0165 × 10^6	.069 × 10^6	.091	275

Energy Conversions

	Anthracite Coal (tons)	Oil (bbl)	Natural Gas (cu ft)	U-235 (g)	Deuterium (g)
1 kwh is derived from	$.129 \times 10^{-3}$	$.608 \times 10^{-3}$	3.51	43.5×10^{-6}	15.1×10^{-6}
1 joule is derived from	35.9×10^{-12}	$.172 \times 10^{-9}$	1.03×10^{-6}	12×10^{-12}	4.21×10^{-12}
1 calorie is derived from	$.15 \times 10^{-9}$	$.707 \times 10^{-9}$	4.09×10^{-6}	58.2×10^{-12}	17.6×10^{-12}
1 watt-year is derived from	1.13×10^{-3}	5.33×10^{-3}	30.9	$.38 \times 10^{-3}$	13.3×10^{-12}
1 BTU is derived from	38×10^{-9}	$.178 \times 10^{-6}$	1.03×10^{-3}	12.8×10^{-9}	4.44×10^{-9}
1 metric ton of (10^6 BTU) anthracite coal =	1	5	25.4×10^3	.33	.12
1 barrel of oil (42 gal) =	.21	1	5.47×10^3	.071	.026
1 ft³ of natural gas (dry) =	39×10^{-6}	$.18 \times 10^{-3}$	1	13.0×10^{-6}	4.54×10^{-6}
1 gram U-235 =	3.0	14	79.3×10^3	1	3.5
1 gram deuterium =	8.6	40	$.227 \times 10^6$	2.9	1

Calorific Values of Fuel Gases

gas	BTU/ft³	MJ/m³
natural gas	1,031	37.4
propane	2,450	91.6
refinery gases	1,600	59.8
coke-oven gas	500	18.7
blast furnace gas	90	3.4
producer gas	125	4.7
Hydrogen	275	10.8

Approximate Conversion Factors for Crude Oil

FROM	INTO metric tons	long tons	short tons	barrels	kiloliters (m³)	1,000 gal (Imperial)	1,000 gal (U.S.)
	MULTIPLY BY						
metric tons	1	0.984	1.102	7.33	1.16	0.256	0.308
long tons	1.016	1	1.120	7.45	1.18	.261	.313
short tons	.907	.893	1	6.65	1.05	.233	.279
barrels	.136	.134	.150	1	.159	.035	.042
kiloliters (m³)	.863	.849	.951	6.29	1	.220	.264
1,000 gal (Imp.)	3.91	3.83	4.29	28.6	4.55	1	1.201
1,000 gal (U.S.)	3.25	3.19	3.58	23.8	3.79	.833	1

The following are some frequently used units of measure and natural gas equivalents in British Thermal Units (BTU). For ease of computation, the actual natural gas equivalent has been rounded to the nearest hundred.

1 cubic foot = 1,031 BTU
1 therm = 100 cubic feet = 103,100 BTU
1 Mcf = 1,000 cubic feet = 1,031,000 BTU
1 M Therm = 100,000 cubic feet = 103,100,000 BTU
1 M Mcf = 1,000,000 cubic feet = 1,031,000,000 BTU
1,000 ft³ hydrogen = 80 kwh

Afterword

I've learned lately that news has a sort of a sex life. Hegel said the same.

If you take two items of antagonistic bad news and point them at each other, or *let* them at each other, you may get some surprising good news.

If, in Buckminster Fuller's favorite example, you take the poison sodium and the poison chlorine and combine them correctly, you get salt for your table (sodium chloride).

If, in my favorite example, you take the Energy Crisis and the Job Crisis and get adroitly out of their way, they may solve each other—with the bonus that what was done crudely by oil can be done exquisitely with hands-on human attention.

And, if you take the bad news of increasing centralized control over macro-energy systems and the bad news (based on catastrophic expectations) of individual homes and communities increasingly doing their own energy processing at the price of severe cutbacks . . . if you cheerfully combine those unforgiving antagonists in one report along with resourceful preliminary research and some bright ideas . . . then you get the surprisingly good news of this book.

However it is news which hasn't happened yet. It's equivalent to being informed, "If you stop smoking now you may live to be 75."

We CAN eliminate all fossil fuel and nuclear energy-use in ten years and harvest more energy per Earth individual than Americans currently misuse . . . Yay.

. . . IF we change all manner of habits (farewell, Big Car), speed up "progress" in some areas and stop it or run it backwards in other areas (such as soil-care), and *personally* do more with less . . . Boo.

"Boo" to contemplate. Living it may contain some deeper-throated Yays. Do you know anyone who didn't enjoy the "Energy Crisis" of '73? How about the thirties? I asked Buckminster Fuller that one. He remembers the Great Depression fondly as a period of trial and of abundant learning and creativity for him and everyone he knew.

The ability of people to learn is widely discredited. I've never seen it factored formally into forecasting schemes. The Future is usually treated as if it were a consumer product—the suckers will buy whatever we give them. When you're doing predicitons—rosy *or* dire—it's a seductive fallacy.

My medicine against that particular blindness is "co-evolution"—the understanding that everything alive learns from everything else constantly. Co-evolution—life—requires constant mistakes and constant correction. Forecast from the mistakes and you get deadly direness. Forecast from the corrections and you're a politician. Forecast from both and you get the cornucopia of complexity and uncertainty that is real life.

The central question is, "How do you convince people to change?"

You don't. What happens is that *after* a big change comes along, some people are found to have "pre-adapted" to it, and they flourish. Everyone else quickly learns from them. That's co-evolution—part of it.

Pre-adaption you *can* encourage, in a general way, by encouraging diversity, independence, inventiveness, difficulty (i.e., learning), and something that looks a lot like faith.

A pre-adaptive fanning out is also helped by doom-calling. "It's gonna be crowded and hungry and dark and cold and dangerous and uncertain!" Then come the "unlesses." This book is one—one of the best I think.

My expectation is that the sky *will* fall. My faith is that there's another sky behind it, made of accurate appraisal like this book, and accurate response such as only ever-dying life can make.

We'll see.

Stewart Brand
Sausalito, California